ORGANIC SYNTHESES

ORGANIC SYNTHESES

AN ANNUAL PUBLICATION OF SATISFACTORY
METHODS FOR THE PREPARATION
OF ORGANIC CHEMICALS

VOLUME 85
2008

SCOTT E. DENMARK
VOLUME EDITOR

A JOHN WILEY & SONS, INC., PUBLICATION

ORGANIC SYNTHESES

VOLUME	VOLUME EDITOR	PAGES
I*	†ROGER ADAMS	84
II*	†JAMES BRYANT CONANT	100
III*	†HANS THACHER CLARKE	105
IV*	†OLIVER KAMM	89
V*	†CARL SHIPP MARVEL	110
VI*	†HENRY GILMAN	120
VII*	†FRANK C. WHITMORE	105
VIII*	†ROGER ADAMS	139
IX*	†JAMES BRYANT CONANT	108
Collective Vol. I	A revised edition of Annual Volumes I–IX †HENRY GILMAN, *Editor-in-Chief* 2nd Edition revised by †A. H. BLATT	580
X*	†HANS THACHER CLARKE	119
XI*	†CARL SHIPP MARVEL	106
XII*	†FRANK C. WHITMORE	96
XIII*	†WALLACE H. CAROTHERS	119
XIV*	†WILLIAM W. HARTMAN	100
XV*	†CARL R. NOLLER	104
XVI*	†JOHN R. JOHNSON	104
XVII*	†L. F. FIESER	112
XVIII*	†REYNOLD C. FUSON	103
XIX*	†JOHN R. JOHNSON	105
Collective Vol. II	A revised edition of Annual Volumes X–XIX †A. H. BLATT, *Editor-in-Chief*	654
20*	†CHARLES F. H. ALLEN	113
21*	†NATHAN L. DRAKE	120
22*	†LEE IRVIN SMITH	114
23*	†LEE IRVIN SMITH	124
24*	†NATHAN L. DRAKE	119
25*	†WERNER E. BACHMANN	120
26*	†HOMER ADKINS	124
27*	†R. L. SHRINER	121
28*	†H. R. SNYDER	121
29*	†CLIFF S. HAMILTON	119
Collective Vol. III	A revised edition of Annual Volumes 20–29 †E. C. HORNING, *Editor-in-Chief*	890
30*	†ARTHUR C. COPE	115
31*	†R. S. SCHREIBER	122

*Out of print.
†Deceased.

Out of print.
†*Deceased.*

*Out of print.
†Deceased.

Collective Volumes, Collective Indices to Collective Volumes I–IX, Annual Volumes, 75–79 and Reaction Guide are available from John Wiley & Sons, Inc.

NOTICE

Beginning with Volume 84, the Editors of *Organic Syntheses* initiated a new publication protocol, which is intended to shorten the time between submission of a procedure and its appearance as a publication. Immediately upon completion of the successful checking process, procedures are assigned volume and page numbers and are then posted on the Organic Syntheses website (www.orgsyn.org). The accumulated procedures from a single volume are assembled once a year and submitted for publication in both hard cover and soft cover editions. The soft cover edition of this volume is produced by a rapid and inexpensive process, and is sent at no charge to members of the Organic Division of the American Chemical Society, Polskie Towarzystwo Chemiczne, Royal Society of Chemistry, and The Society of Synthetic Organic Chemistry, Japan. The soft cover edition is intended as the personal copy of the owner and is not for library use. The hard cover edition is published by John Wiley and Sons, Inc., in the traditional format, and it differs in content primarily by the inclusion of an index. The hard cover edition is intended primarily for library collections and is available for purchase through the publisher. Incorporation of graphical abstracts into the Table of Contents began with Volume 77. Annual volumes 70–74 and 75–79 have been incorporated into five-year versions of the collective volumes of *Organic Syntheses* that appeared as Collective Volume IX and X in the traditional hard cover format, available for purchase from the publishers. The Editors hope that the new Collective Volume series, appearing twice as frequently as the previous decennial volumes, will provide a permanent and timely edition of the procedures for personal and institutional libraries. The Editors welcome comments and suggestions from users concerning the new editions.

Organic Syntheses, Inc., joined the age of electronic publication in 2001 with the release of its free web site (www.orgsyn.org). Organic Syntheses, Inc., fully funded the creation of the free website at www.orgsyn.org in a partnership with CambridgeSoft Corporation and Data-Trace Publishing Company. The site is accessible to most internet browsers using Macintosh and Windows operating systems and may be used with or without a ChemDraw plugin. Because of continually evolving system requirements, users should review software compatibility at the website prior to use. John Wiley & Sons, Inc., and Accelrys, Inc., partnered with Organic Syntheses, Inc., to develop the new database (www.mrw.interscience.wiley.com/osdb) that is available for license with internet solutions from John Wiley & Sons, Inc. and intranet solutions from Accelrys, Inc.

Both the commercial database and the free website contain all annual and collective volumes and indices of *Organic Syntheses*. Chemists can draw structural queries and combine structural or reaction transformation queries with full-text and bibliographic search terms, such as chemical name, reagents, molecular formula, apparatus, or even hazard warnings or phrases. The preparations are categorized into reaction types, allowing search by category. The contents of individual or collective volumes can be browsed by lists of titles, submitters' names, and volume and page references, with or without reaction equations.

The commercial database at www.mrw.interscience.wiley.com/osdb also enables the user to choose his/her preferred chemical drawing package, or to utilize several freely available plug-ins for entering queries. The user is also able to cut and paste existing structures and reactions directly into the structure search query or their preferred chemistry editor, streamlining workflow. Additionally, this database contains links to the full text of primary literature references via CrossRef, ChemPort, Medline, and ISI Web of Science. Links to local holdings for institutions using open url technology can also be enabled. The database user can limit his/her search to, or ordered the search results by, such factors as reaction type, percentage yield, temperature, and publication date, and can create a customized table of reactions for comparison. Connections to other Wiley references are currently made via text search, with cross-product structure and reaction searching to be added in the coming year. Incorporations of new preparations will occur as new material becomes available.

INFORMATION FOR AUTHORS OF PROCEDURES

Organic Syntheses welcomes and encourages submissions of experimental procedures that lead to compounds of wide interest or that illustrate important new developments in methodology. Proposals for *Organic Syntheses* procedures will be considered by the Editorial Board upon receipt of an outline proposal as described below. A full procedure will then be invited for those proposals determined to be of sufficient interest. These full procedures will be evaluated by the Editorial Board, and if approved, assigned to a member of the Board for checking. In order for a procedure to be accepted for publication, each reaction must be successfully repeated in the laboratory of a member of the Editorial Board at least twice, with similar yields (generally ±5%) and selectivity to that reported by the submitters.

Organic Syntheses Proposals

A cover sheet should be included providing full contact information for the principal author and including a scheme outlining the proposed reactions (an *Organic Syntheses* Proposal Cover Sheet can be downloaded at orgsyn.org). Attach an outline proposal describing the utility of the methodology and/or the usefulness of the product. Identify and reference the best current alternatives. For each step, indicate the proposed scale, yield, method of isolation and purification, and how the purity of the product is determined. Describe any unusual apparatus or techniques required, and any special hazards associated with the procedure. Identify the source of starting materials. Enclose copies of relevant publications (attach pdf files if an electronic submission is used).

Submit proposals by mail or as e-mail attachments to:

Professor Charles K. Zercher
Associate Editor, Organic Syntheses
Department of Chemistry
University of New Hampshire
23 Academic Way, Parsons Hall
Durham, NH 03824

For electronic submissions: *org.syn@unh.edu*

Submission of Procedures

Authors invited by the Editorial Board to submit full procedures should prepare their manuscripts in accord with the Instructions to Authors which are described below or may be downloaded at orgsyn.org. Submitters are also encouraged to consult this volume of *Organic Syntheses* for models with regard to style, format, and the level of experimental detail expected in *Organic Syntheses* procedures. Manuscripts should be submitted to the Associate Editor. Electronic submissions are encouraged; procedures will be accepted as e-mail attachments in the form of Microsoft Word files with all schemes and graphics also sent separately as ChemDraw files.

Procedures that do not conform to the Instructions to Authors with regard to experimental style and detail will be returned to authors for correction. Authors will be notified when their manuscript is approved for checking by the Editorial Board, and it is the goal of the Board to complete the checking of procedures within a period of no more than six months.

Additions, corrections, and improvements to the preparations previously published are welcomed; these should be directed to the Associate Editor. However, checking of such improvements will only be undertaken when new methodology is involved.

NOMENCLATURE

Both common and systematic names of compounds are used throughout this volume, depending on which the Volume Editor felt was more appropriate. The Chemical Abstracts indexing name for each title compound, if it differs from the title name, is given as a subtitle. Systematic

Chemical Abstracts nomenclature, used in the Collective Indexes for the title compound and a selection of other compounds mentioned in the procedure, is provided in an appendix at the end of each preparation. Chemical Abstracts Registry numbers, which are useful in computer searching and identification, are also provided in these appendices. Whenever two names are concurrently in use and one name is the correct Chemical Abstracts name, that name is preferred.

ACKNOWLEDGMENT

Organic Syntheses wishes to acknowledge the contributions of Merck & Co. and Pfizer, Inc. to the success of this enterprise through their support, in the form of time and expenses, of members of the Board of Editors.

INSTRUCTIONS TO AUTHORS

All organic chemists have experienced frustration at one time or another when attempting to repeat reactions based on experimental procedures found in journal articles. To ensure reproducibility, *Organic Syntheses* requires experimental procedures written with considerably more detail as compared to the typical procedures found in other journals and in the "Supporting Information" sections of papers. In addition, each *Organic Syntheses* procedure is carefully "checked" for reproducibility in the laboratory of a member of the Board of Editors.

Even with these more detailed procedures, the experience of *Organic Syntheses* editors is that difficulties often arise in obtaining the results and yields reported by the submitters of procedures. To expedite the checking process and ensure success, we have prepared the following "Instructions for Authors" as well as a *Checklist for Authors* and *Characterization Checklist* to assist you in confirming that your procedure conforms to these requirements. These checklists, which are available at *www.orgsyn.org,* should be completed and submitted together with your procedure. Procedures submitted to *Organic Syntheses* will be carefully reviewed upon receipt and procedures lacking any of the required information will be returned to the submitters for revision.

Scale and Optimization

The appropriate scale for procedures will vary widely depending on the nature of the chemistry and the compounds synthesized in the procedure. However, some general guidelines are possible. For procedures in which the principal goal is to illustrate a synthetic method or strategy, it is expected, in general, that the procedure should result in at least 5 g and no more than 50 g of the final product. In cases where the point of the procedure is to provide an efficient method for the preparation of a useful reagent or synthetic building block, the appropriate scale may be larger, but in general should not exceed 100 g of final product. Exceptions to these guidelines may be granted in special circumstances. For example, procedures describing the preparation of reagents employed as catalysts will often be acceptable on a scale of less than 5 g.

In considering the scale for an *Organic Syntheses* procedure, authors should also take into account the cost of reagents and starting materials. In general, the Editors will not accept procedures for checking in which the cost of any one of the reactants exceeds $500 for a single full-scale run. Authors are requested to identify the most expensive reagent or starting material on the procedure submission checklist and to estimate its cost per run of the procedure.

It is expected that all aspects of the procedure will have been opti-mized by the authors prior to submission, and that each reaction will have been carried out at least twice on exactly the scale described in the procedure. It is appropriate to report the weight, yield, and purity of the product of each step in the procedure as a range. In any case where a reagent is employed in significant excess, a Note should be included explaining why an excess of that reagent is necessary. If possible, the Note should indicate the effect of using amounts of reagent less than that specified in the procedure.

Reaction Apparatus

Describe the size and type of flask (number of necks) and indicate how *every* neck is equipped.

"A 500-mL, three-necked, round-bottomed flask equipped with an overhead mechanical stirrer, 250-mL pressure-equalizing addition funnel fitted with an argon inlet, and a rubber septum is charged with. . . ."

Indicate how the reaction apparatus is dried and whether the reaction is conducted under an inert atmosphere. This can be incorporated in the text of the procedure or included in a Note.

"The apparatus is flame-dried and maintained under an atmosphere of argon during the course of the reaction."

In the case of procedures involving unusual glassware or especially complicated reaction setups, authors are encouraged to include a photo-graph or drawing of the apparatus in the text or in a Note (for examples, see *Org. Syn.*, Vol. 82, 99 and Coll. Vol. X, pp 2, 3, 136, 201, 208, and 669).

Reagents and Starting Materials

All chemicals employed in the procedure must be commercially available or described in an earlier *Organic Syntheses* or *Inorganic Syntheses* procedure. For other compounds, a procedure should be included either as one or more steps in the text or, in the case of relatively straightforward preparations of reagents, as a Note. In the latter case, all requirements with regard to characterization, style, and detail also apply.

In one or more Notes, indicate the purity or grade of each reagent, solvent, etc. It is desirable to also indicate the source (company the chemical was purchased from), particularly in the case of chemicals where it is suspected that the composition (trace impurities, etc.) may vary from one supplier to another. In cases where reagents are purified, dried, "activated" (e.g., Zn dust), etc., a detailed description of the procedure used should be included in a Note. In other cases, indicate that the chemical was "used as received".

"Diisopropylamine (99.5%) was obtained from Aldrich Chemical Co., Inc. and distilled under argon from calcium hydride before use. THF (99+%) was obtained from Mallinckrodt, Inc. and distilled from sodium benzophenone ketyl. Diethyl ether (99.9%) was purchased from Aldrich Chemical Co., Inc. and purified by pressure filtration under argon through activated alumina. Methyl iodide (99%) was obtained from Aldrich Chemical Co., Inc. and used as received."

The amount of each reactant should be provided in parentheses in the order mL, g, mmol, and equivalents with careful consideration to the correct number of significant figures.

The concentration of solutions should be expressed in terms of molarity or normality, and not percent (e.g., 1 N HCl, 6 M NaOH, not "10% HCl").

Reaction Procedure

Describe every aspect of the procedure clearly and explicitly. Indicate the order of addition and time for addition of all reagents and how each is added (via syringe, addition funnel, etc.).

Indicate the temperature of the reaction mixture (preferably internal temperature). Describe the type of cooling (e.g., "dry ice-acetone bath") and heating (e.g., oil bath, heating mantle) methods employed. Be careful to describe clearly all cooling and warming cycles, including initial and final temperatures and the time interval involved.

Describe the appearance of the reaction mixture (color, homogeneous or not, etc.) and describe all significant changes in appearance during the course of the reaction (color changes, gas evolution, appearance of solids, exotherms, etc.).

Indicate how the reaction can be monitored to determine the extent of conversion of reactants to products. In the case of reactions monitored by TLC, provide details in a Note, including eluent, R_f values, and method of visualization. For reactions followed by GC, HPLC, or NMR analysis, provide details on analysis conditions and relevant diagnostic peaks.

"The progress of the reaction was followed by TLC analysis on silica gel with 20% EtOAc-hexane as eluent and visualization with *p*-anisaldehyde. The ketone starting material has $R_f = 0.40$ (green) and the alcohol product has $R_f = 0.25$ (blue)."

Reaction Workup

Details should be provided for reactions in which a "quenching" process is involved. Describe the composition and volume of quenching agent, and time and temperature for addition. In cases where reaction mixtures are added to a quenching solution, be sure to also describe the setup employed.

"The resulting mixture was stirred at room temperature for 15 h, and then carefully poured over 10 min into a rapidly stirred, ice-cold aqueous solution of 1 N HCl in a 500-mL Erlenmeyer flask equipped with a magnetic stirbar."

For extractions, the number of washes and the volume of each should be indicated.

For concentration of solutions after workup, indicate the method and pressure and temperature used.

"The reaction mixture is diluted with 200 mL of water and transferred to a 500-mL separatory funnel, and the aqueous phase is separated and extracted with three 100-mL portions of ether. The combined organic layers are washed with 75 mL of water and 75 mL of saturated NaCl solution, dried over $MgSO_4$, filtered, and concentrated by rotary evaporation (25°C, 20 mmHg) to afford 3.25 g of a yellow oil."

"The solution is transferred to a 250-mL, round-bottomed flask equipped with a magnetic stirbar and a 15-cm Vigreux column fitted with a short path distillation head, and then concentrated by careful distillation at 50 mmHg (bath temperature gradually increased from 25 to 75°C)."

In cases where solid products are filtered, describe the type of filter funnel used and the amount and composition of solvents used for washes.

"... and the resulting pale yellow solid is collected by filtration on a Büchner funnel and washed with 100 mL of cold (0°C) hexane."

When solid or liquid compounds are dried under vacuum, indicate the pressure employed (rather than stating "reduced pressure" or "dried *in vacuo*").

"... and concentrated at room temperature by rotary evaporation (20 mmHg) and then at 0.01 mmHg to provide. ..."

"The resulting colorless crystals are transferred to a 50-mL, round-bottomed flask and dried overnight in a 100°C oil bath at 0.01 mmHg."

Purification: Distillation

Describe distillation apparatus including the size and type of distillation column. Indicate temperature (and pressure) at which all significant fractions are collected.

"... and transferred to a 100-mL, round-bottomed flask equipped with a magnetic stirbar. The product is distilled under vacuum through a 12-cm, vacuum-jacketed column of glass helices (Note 16) topped with a Perkin triangle. A forerun (ca. 2 mL) is collected and discarded, and the desired product is then obtained, distilling at 50–55°C (0.04–0.07 mmHg). ..."

Purification: Column Chromatography

Provide information on TLC analysis in a Note, including eluent, R_f values, and method of visualization.

Provide dimensions of column and amount of silica gel used; in a Note indicate source and type of silica gel.

Provide details on eluents used, and number and size of fractions.

"The product is charged on a column (5 × 10 cm) of 200 g of silica gel (Note 15) and eluted with 250 mL of hexane. At that point, fraction collection (25-mL fractions) is begun, and elution is continued with 300 mL of 2% EtOAc-hexane (49:1 hexanes:EtOAc) and then 500 mL of 5% EtOAc-hexane (19:1 hexanes:EtOAc). The desired product is obtained in fractions 24–30, which are concentrated by rotary evaporation (25°C, 15 mmHg). ..."

Purification: Recrystallization

Describe procedure in detail. Indicate solvents used (and ratio of mixed solvent systems), amount of recrystallization solvents, and temperature protocol. Describe how crystals are isolated and what they are washed with.

"The solid is dissolved in 100 mL of hot diethyl ether (30°C) and filtered through a Büchner funnel. The filtrate is allowed to cool to room temperature, and 20 mL of hexane are added. The solution is cooled at −20°C overnight and the resulting crystals are collected by suction filtration on a Buchner funnel, washed with 50 mL of ice-cold hexane, and then transferred to a 50-mL, round-bottomed flask and dried overnight at 0.01 mmHg to provide. ..."

Characterization

Physical properties of the product such as color, appearance, crystal forms, melting point, etc. should be included in the text of the procedure. Comments on the stability of the product to storage, etc. should be provided in a Note.

In a Note, provide data establishing the identity of the product. This will generally include IR, MS, ^{1}H-NMR, and ^{13}C-NMR data, and in some cases UV data. Copies of the proton NMR spectra for the products of each step in the procedure should be submitted showing integration for all resonances. Submission of copies of the carbon NMR spectra and other nuclei are encouraged as appropriate.

In the same Note, provide data establishing the purity of the product. This should generally include elemental analysis and/or GC and/or HPLC data. Provide details on equipment and conditions for GC and HPLC analyses.

In procedures involving non-racemic, enantiomerically enriched products, optical rotations should generally be provided, but enantiomeric purity must be determined by another method such as chiral HPLC or GC analysis.

In cases where the product of one step is used without purification in the next step, a Note should be included describing how a sample of the product can be purified and providing characterization data for the pure material. Copies of the proton NMR spectra of both the product both *before* and *after* purification should be submitted.

Hazard Warnings

Any significant hazards should be indicated in a statement at the beginning of the procedure in italicized type. Efforts should be made to avoid the use of toxic and hazardous solvents and reagents when less hazardous alternatives are available.

Discussion Section

The style and content of the discussion section will depend on the nature of the procedure.

For procedures that provide an improved method for the preparation of an important reagent or synthetic building block, the discussion should focus on the advantages of the new approach and should describe and reference all of the earlier methods used to prepare the title compound.

In the case of procedures that illustrate an important synthetic method or strategy, the discussion section should provide a mini-review on the new methodology. The scope and limitations of the method should be discussed, and it is generally desirable to include a table of examples. Competing methods for accomplishing the same overall transformation should be described and referenced. A brief discussion of mechanism may be included if this is useful for understanding the scope and limitations of the method.

Style and Format

Articles should follow the style guidelines used for organic chemistry articles published in the ACS journals such as *J. Am. Chem. Soc., J. Org. Chem., Org. Letters*, etc. as described in the the ACS Style Guide (3rd Ed.). The text of the procedure should be constructed using a standard word processing program, like MS Word, with 14-point Times New Roman font. Chemical structures and schemes should be drawn using the standard ACS drawing parameters (in ChemDraw, the parameters are found in the "ACS Document 1996" option) with a maximum width of 6 inches. The graphics files should be inserted into the document at the correct location and the graphics files should also be submitted separately. All Tables that include structures should be entirely prepared in the graphics (ChemDraw) program and inserted into the word processing file at the appropriate location. Tables that include multiple, separate graphics files prepared in the word processing program will require modification.

Biographies and Photographs of Authors

Photographs and 100-word biographies of all authors should be submitted as separate files at the time of the submission of the procedure.

The format of the biographies should be similar to those in the Volume 84 procedures found at the orgsyn.org website. Photographs can be accepted in a number of electronic formats, including tiff and jpeg formats.

HANDLING HAZARDOUS CHEMICALS

A Brief Introduction

General Reference: *Prudent Practices in the Laboratory*; National Academy Press; Washington, DC, 1995.

Physical Hazards

Fire. Avoid open flames by use of electric heaters. Limit the quantity of flammable liquids stored in the laboratory. Motors should be of the nonsparking induction type.

Explosion. Use shielding when working with explosive classes such as acetylides, azides, ozonides, and peroxides. Peroxidizable substances such as ethers and alkenes, when stored for a long time, should be tested for peroxides before use. Only sparkless "flammable storage" refrigerators should be used in laboratories.

Electric Shock. Use 3-prong grounded electrical equipment if possible.

Chemical Hazards

Because all chemicals are toxic under some conditions, and relatively few have been thoroughly tested, it is good strategy to minimize exposure to all chemicals. In practice this means having a good, properly installed hood; checking its performance periodically; using it properly; carrying out all operations in the hood; protecting the eyes; and, since many chemicals can penetrate the skin, avoiding skin contact by use of gloves and other protective clothing at all times.

a. Acute Effects. These effects occur soon after exposure. The effects include burn, inflammation, allergic responses, damage to the eyes, lungs, or nervous system (e.g., dizziness), and unconsciousness or death (as from overexposure to HCN). The effect and its cause are usually obvious and so are the methods to prevent it. They generally arise from inhalation or skin contact, so should not be a problem if one follows the admonition "work in a hood and keep chemicals off your hands". Ingestion is a rare route, being generally the result of eating in the laboratory or not washing hands before eating.

b. Chronic Effects. These effects occur after a long period of exposure or after a long latency period and may show up in any of numerous organs. Of the chronic effects of chemicals, cancer has received the most attention lately. Several dozen chemicals have been demonstrated to be carcinogenic in man and hundreds to be carcinogenic to animals. Although there is no simple correlation between carcinogenicity in animals and in man, there is little doubt that a significant proportion of the chemicals used in laboratories have some potential for carcinogenicity in man. For this and other reasons, chemists should employ good practices at all times.

The key to safe handling of chemicals is a good, properly installed hood, and the referenced book devotes many pages to hoods and ventilation. It recommends that in a laboratory where people spend much of their time working with chemicals there should be a hood for each two people, and each should have at least 2.5 linear feet (0.75 meter) of working space at it. Hoods are more than just devices to keep undesirable vapors from the laboratory atmosphere. When closed they provide a protective barrier between chemists and chemical operations, and they are a good containment device for spills. Portable shields can be a useful supplement to hoods, or can be an alternative for hazards of limited severity, e.g., for small-scale operations with oxidizing or explosive chemicals.

Specialized equipment can minimize exposure to the hazards of laboratory operations. Impact resistant safety glasses are basic equipment and should be worn at all times. They may be supplemented by face shields or goggles for particular operations, such as pouring corrosive liquids. Because skin contact with chemicals can lead to skin irritation or sensitization or, through absorption, to effects on internal organs, protective gloves should be worn at all times.

Laboratories should have fire extinguishers and safety showers. Respirators should be available for emergencies. Emergency equipment should be kept in a central location and must be inspected periodically.

MSDS (Materials Safety Data Sheets) sheets are available from the suppliers of commercially available reagents, solvents, and other chemical materials; anyone performing an experiment should check these data sheets before initiating an experiment to learn of any specific hazards associated with the chemicals being used in that experiment.

DISPOSAL OF CHEMICAL WASTE

General Reference: *Prudent Practices in the Laboratory* National Academy Press, Washington, D.C. 1996

Effluents from synthetic organic chemistry fall into the following categories:

1. **Gases**

   1a. Gaseous materials either used or generated in an organic reaction.
   1b. Solvent vapors generated in reactions swept with an inert gas and during solvent stripping operations.
   1c. Vapors from volatile reagents, intermediates and products.

2. **Liquids**

   2a. Waste solvents and solvent solutions of organic solids (see item 3b).
   2b. Aqueous layers from reaction work-up containing volatile organic solvents.
   2c. Aqueous waste containing non-volatile organic materials.
   2d. Aqueous waste containing inorganic materials.

3. **Solids**

   3a. Metal salts and other inorganic materials.
   3b. Organic residues (tars) and other unwanted organic materials.
   3c. Used silica gel, charcoal, filter aids, spent catalysts and the like.

The operation of industrial scale synthetic organic chemistry in an environmentally acceptable manner* requires that all these effluent categories be dealt with properly. In small scale operations in a research or

---

*An environmentally acceptable manner may be defined as being both in compliance with all relevant state and federal environmental regulations *and* in accord with the common sense and good judgment of an environmentally aware professional.

academic setting, provision should be made for dealing with the more environmentally offensive categories.

1a. Gaseous materials that are toxic or noxious, e.g., halogens, hydrogen halides, hydrogen sulfide, ammonia, hydrogen cyanide, phosphine, nitrogen oxides, metal carbonyls, and the like.
1c. Vapors from noxious volatile organic compounds, e.g., mercaptans, sulfides, volatile amines, acrolein, acrylates, and the like.
2a. All waste solvents and solvent solutions of organic waste.
2c. Aqueous waste containing dissolved organic material known to be toxic.
2d. Aqueous waste containing dissolved inorganic material known to be toxic, particularly compounds of metals such as arsenic, beryllium, chromium, lead, manganese, mercury, nickel, and selenium.
3. All types of solid chemical waste.

Statutory procedures for waste and effluent management take precedence over any other methods. However, for operations in which compliance with statutory regulations is exempt or inapplicable because of scale or other circumstances, the following suggestions may be helpful.

Gases

Noxious gases and vapors from volatile compounds are best dealt with at the point of generation by "scrubbing" the effluent gas. The gas being swept from a reaction set-up is led through tubing to a large trap to prevent suck-back and into a sintered glass gas dispersion tube immersed in the scrubbing fluid. A bleach container can be conveniently used as a vessel for the scrubbing fluid. The nature of the effluent determines which of four common fluids should be used: dilute sulfuric acid, dilute alkali or sodium carbonate solution, laundry bleach when an oxidizing scrubber is needed, and sodium thiosulfate solution or diluted alkaline sodium borohydride when a reducing scrubber is needed. Ice should be added if an exotherm is anticipated.

Larger scale operations may require the use of a pH meter or starch/iodide test paper to ensure that the scrubbing capacity is not being exceeded.

When the operation is complete, the contents of the scrubber can be poured down the laboratory sink with a large excess (10–100 volumes) of water. If the solution is a large volume of dilute acid or base, it should be neutralized before being poured down the sink.

Liquids

Every laboratory should be equipped with a waste solvent container in which *all* waste organic solvents and solutions are collected. The contents of these containers should be periodically transferred to properly labeled waste solvent drums and arrangements made for contracted disposal in a regulated and licensed incineration facility.**

Aqueous waste containing dissolved toxic organic material should be decomposed *in situ*, when feasible, by adding acid, base, oxidant, or reductant. Otherwise, the material should be concentrated to a minimum volume and added to the contents of a waste solvent drum.

Aqueous waste containing dissolved toxic inorganic material should be evaporated to dryness and the residue handled as a solid chemical waste.

Solids

Soluble organic solid waste can usually be transferred into a waste solvent drum, provided near-term incineration of the contents is assured.

Inorganic solid wastes, particularly those containing toxic metals and toxic metal compounds, used Raney nickel, manganese dioxide, etc. should be placed in glass bottles or lined fiber drums, sealed, properly labeled, and arrangements made for disposal in a secure landfill.** Used mercury is particularly pernicious and small amounts should first be amalgamated with zinc or combined with excess sulfur to solidify the material.

Other types of solid laboratory waste including used silica gel and charcoal should also be packed, labeled, and sent for disposal in a secure landfill.

Special Note

Since local ordinances may vary widely from one locale to another, one should always check with appropriate authorities. Also, professional disposal services differ in their requirements for segregating and packaging waste.

**If arrangements for incineration of waste solvent and disposal of solid chemical waste by licensed contract disposal services are not in place, a list of providers of such services should be available from a state or local office of environmental protection.

Albert I. Meyers
1932–2007

Albert I. Meyers, John K. Stille Professor Emeritus, Colorado State University Distinguished Professor Emeritus, died from a long-standing heart condition on October 23, 2007, at the age of 74. Al was born in New York City on November 22, 1932, and spent his childhood and young adulthood in New York, earning a bachelor's degree in 1954, from New York University, and his Ph.D. from the same institution in 1957, under the direction of Professor J. J. Ritter.

He began his academic career in 1958 as an assistant professor at Louisiana State University in New Orleans. In 1965 he was a National Institutes of Health Postdoctoral Fellow for one year at Harvard University with Professor E. J. Corey, after which he returned to LSU. In 1970 he moved to Wayne State University and, after a brief two years, moved to Colorado State University where he remained for the rest of his career and life.

With his outgoing personality and ability to talk to anybody and everybody, Al was instrumental in the dramatic development and improvement of the Chemistry Department at Colorado State University, during the difficult 1970's. He was a staunch supporter of the Chemistry Department both internally and externally.

His major research interests were in the application of heterocycles to organic synthesis, asymmetric synthesis and the total synthesis of biologically important compounds utilizing the methodology he developed. During his long career (he reluctantly retired in 2002, due to failing health) he mentored over 80 Ph. D. students and 200 postdoctoral co-workers in his research program, and taught organic chemistry to thousands of undergraduates. In addition to his close relationship with his own research group, he was available and helpful to any student who would walk into his office. His enthusiasm for organic chemistry was boundless.

Al served as a consultant for a number of pharmaceutical companies including Bristol-Myers-Squibb, GlaxoSmithKline and Hoffmann-La Roche, and was on the advisory board of several biotech companies. He received more than 75 national and international awards, including the ACS Award for Creative Work in Synthetic Organic Chemistry in 1985, an ACS Arthur C. Cope Scholar Award in 1987, the Yamada Prize from the Chemical Society of Japan in 1996, and the International Award in Heterocyclic Chemistry in 1997. He was elected to the National Academy of Sciences in 1994.

Al was an associate editor for the *Journal of the American Chemical Society* from 1980–1984, and was the editor of Volume 70 of *Organic Syntheses*.

Beyond all of this, he was a warm and outgoing human being, exuberant and extravagant at times, and always entertaining. He is sorely missed.

Al's wife of 50 years, Joan, a talented artist in her own right, died of cancer shortly after his death. They are survived by a son, Harold, two daughters, Jill Bombel and Lisa Thompson, and seven grandchildren.

LOUIS S. HEGEDUS
Fort Collins, CO

Nelson J. Leonard
1916–2006

The passing of Nelson Jordan Leonard on October 9, 2006 at age 90 deprived us of a chemist who produced highly significant research and possessed those human qualities that led to a wide personal popularity. Leonard was a primary contributor to fundamental knowledge of the chemistry of nitrogen-containing organic molecules. The sweep of his research covered alkaloids, nitrogen heterocycles, small- and medium-ring compounds, transannular interactions and reactions, natural and synthetic cytokinins that facilitate cell growth, cell division and cell differentiation, chemical, spatial, fluorescent and dimensional probes of enzyme-coenzyme interactions and of nucleic acid structure and function and fluorescent covalently-linked DNA/RNA cross sections of normal, narrow and wide dimensions. He was a master in the application of organic synthesis to the solution of important problems in biochemistry and plant physiology.

Nelson Leonard joined Organic Syntheses Inc. as a member of the Board of Editors in 1951 and was the Editor of Vol. 36 published in 1957. He served on the Board of Directors (1969–2001), as Vice President (1978–1980) and President (1980–1988). In gratitude for his service, Organic Syntheses, Inc. provided funds to assist in the endowment of the Nelson J. Leonard Distinguished Lectureship at the University of Illinois and sponsors the Nelson J. Leonard Graduate Fellowship administered by the ACS Division of Organic Chemistry.

Shortly before his death, Nelson completed and published his auto-biography entitled *More Than a Memoir* (ISBN 1-59926-791-8). In addition to this book, Nelson left an extensive document recalling in his own view the most significant personal and professional events, influences and achievements of his life's journey. This article is largely based on that document.

Nelson was born on September 1, 1916, in Newark, New Jersey, to Harvey Nelson Leonard and Olga Pauline Jordan. His father's ances-tors had come from England in the first half of the seventeenth century while his mother's forebears were Huguenots who left France for Ger-many in the sixteenth century and sailed to the United State in the mid-nineteenth century. Both parents' families followed a puritan work ethic that was relieved by music (piano, organ, mandolin and singing) and an appreciation of nature (hiking, swimming, etc.). Nelson's father was a salesman of men's clothing and he developed a large and loyal following of customers who transferred allegiance from store to store in New York City whenever he shifted employers.

Nelson attended public schools in Mount Vernon, New York. In high school Nelson was active in student politics, in the glee club, in the pre-sentation of operettas and in solo recitals. His graduation year 1933 was in the trough of the depression. Political storm clouds were gathering over Europe with the accession of Hitler to power and, in the United States, the banks were closed. When the Mount Vernon bank in which Nelson had deposited his hard-earned savings was allowed to reopen, he had half enough money for a first year in college. His matriculation to Lehigh University in Bethlehem, Pennsylvania, was made possible by a scholarship. He played varsity soccer, was class president in his junior year, was active in theater and operetta productions, in glee club concerts and in radio appearances. An intended chemical engineer through his junior year, he shifted to a B.S. in chemistry curriculum. As a senior in 1937 at Lehigh University and in anticipation of his study as a Rhodes Scholar at Oxford, Nelson had prepared a summary of the scientific papers of Neville Sidgwick and had been fascinated by Sidgwick's *Organic Chemistry of Nitrogen*. This interest proved to be quite anticipatory of Nelson's later research on nitrogen compounds. While at Oxford, Nelson's continuing interest in music was evident in his participation in the Oxford Bach Choir, the Opera Club, and the Lincoln College Choir. His sport shifted to rowing.

The beginning of World War II in September 1939, forced Nelson's return to the United States and the termination of his research with Leslie Sutton on the use of fluoro compounds in the determination of valency angles by electric dipole moment measurements. Chemistry was not the only part of Nelson's life that was interrupted by the war. Through family connections, he had met and fallen in love with Louise Cornelie Vermey of the Netherlands. After a year and a half, they became engaged but were not to see each other again until the end of the war, (1945 in Holland) and were not able to arrange for her journey to the U.S. and marriage until 1947.

Nelson was able to continue his graduate education in chemistry, concluding with a Ph.D. (1942) at Columbia University, New York. His research, which consisted of structure establishment and partial synthesis of alstonine, a naturally occurring antimalarial, was done under the direction of Robert C. Elderfield.

A postdoctoral research assistantship brought Nelson to the University of Illinois, Urbana-Champaign, where he worked with Roger Adams on *Senecio* alkaloids. Teaching duties were added in 1943, which grew to include U.S. Navy and U. S. Army units passing through the University of Illinois. He joined the team of Charles C. Price, III, and Harold R. Snyder on antimalarial research to help with the synthesis and production of chloroquine in time for its use in the Pacific. At the end of the war, during 1945–1946, Nelson served as a Scientific Consultant and Special Investigator to the Field Intelligence Agency Technical, U.S. Army and U.S. Department of Commerce, European Theater. He returned to the University of Illinois and remained on the teaching staff until his retirement in 1986, by which time he was Reynold C. Fuson Professor of Chemistry, Professor of Biochemistry and Member of the Center for Advanced Study.

Very early in his research career, he adopted as a guiding concept, *organic synthesis with a purpose*. To this end, he developed a catalytic reductive cyclization of intermediates that led directly to *Senecio* and *Lupin* alkaloid components and establishment of their relative configurations. Synthesis of selected 1,2-diketones established the dependence of their spectroscopic properties upon the dihedral angle between the carbonyl groups. Fundamental study of the electrolytic reduction of aminoketones provided a new route to medium-ring compounds containing nitrogen. New functionalities were invented based upon transannular reactions across medium rings, which included defining the ring-size and the electronic limitation of transannular interactions. He became well known for his recognition of the iminium group (the product of enamine protonation) as a fast-acting ionic carbonyl equivalent and for assembly of new families of reactions based upon this functional group. By reaction of iminium salts with diazomethane Nelson and his students made stable aziridinium salts available. These three-membered ring compounds, postulated previously to be unstable intermediates, were investigated systematically.

From 1943 until 1955, Nelson's musical career had flourished in the Midwest along with his academic career in chemistry. Solo appearances as a bass-baritone in choral works with the Chicago, Cleveland and St. Louis Symphony Orchestras, in Bach festivals, at other universities and with many different choruses were interspersed with recitals at the University of Illinois, Washington University, St. Louis, and Illinois Wesleyan, Springfield. When, in 1955 at age thirty-eight, Nelson was elected to membership in the National Academy of Sciences, he felt that, if his peers had chosen to recognize him as a chemist, then he had better do something about it. He realized that there was more scope for originality in full-time devotion to chemistry and a more lasting contribution through the literature of science. The heavy professional demands of chemistry meant that there were no more singing performances. He and Louise (affectionately known as Nell) had four children by that time: Kenneth Jan, Marcia Louise, James Nelson, and David Anthony. Another decision point was reached in 1960, when Nelson was on sabbatical leave, aided by a Guggenheim Fellowship, in Basel, Switzerland. He had concluded that organic chemistry, *per se*, was not enough, a dictum that he passed along to his students thereafter, and he devoted much more of his time to reading the current literature of biochemistry. The broadening of his interests soon appeared in research publications emanating from the Illinois laboratory and as well as those of the scientists with whom he collaborated.

By synthesis of the cytokinin, $\underline{N}^6$-isopentenyladenine, and collaboration with Folke Skoog, plant physiologist at the University of Wisconsin, it was found that this compound occurred naturally in the plant pathogen *Corynebacterium fascians* as a major component responsible for its biological activity. The combined Illinois-Wisconsin search for other natural cell-growth, cell-differentiation factors uncovered eight additional highly active substances from plant, animal, bacterial, and fungal sources. Stereoselective syntheses to produce these compounds, and their structure/activity investigations led to agents more active than the naturally occurring cytokinins. In very low concentrations, the cytokinins initiate plant, flower, and tree growth from tissue culture that is basic to horticultural and agricultural developments. This collaboration lasted eighteen years and resulted in some fifty publications.

Leonard's laboratory provided fundamental findings on the reaction of diethyl pyrocarbonate (DEP) with adenosine and adenosine-containing nucleotides and dinucleoside phosphates, culminating in 1973 with the proposition that DEP, as a chemical probe, could serve the purpose of detecting adenosine or deeoxyadenosine modification at exposed sites in RNA or DNA.

Concomitant with the research on triacanthine (3-isopentenyladenine) and the related cytokinins, research on spatial probes of enzyme-coenzyme interactions was initiated with the synthesis of 3-isoadenosine and its phosphates. Leonard and coworkers, with the collaboration of other laboratories, showed that the range of similar biological activities for the 3-isoadenylates with the adenylates (9-substituted on purine), while initially surprising, turned out to be readily understandable in spatial terms. Thus, the superposition of the purine ring of a 3-isoadenosine derivative over that of adenosine illustrates the close spatial relationship that exists between the two, especially the proximate location of the individual nitrogens in each. A definitive study at Illinois (1996) confirmed the hydrogen-bonding pattern that had been postulated thirty years earlier and helped in the understanding of the parameters limiting early nucleic acid development in nature. Nelson deduced that nature might have discarded the N3 (vs. N9) attachment site for purines because of chemical instability, but not of structural incompatibility.

Leonard's derivatization of nucleosides, nucleotides, and coenzymes by fluorescent probes, placed him among the most often quoted scientists. He was successful in providing reagents for 4-thiouridine, cytidine, adenosine, and guanosine. In a fruitful collaboration with Illinois biochemist Gregorio Weber, incorporation of fluorescence moieties into the related coenzymes provided details as to both size and locus of enzyme-coenzyme binding sites. The idea involved testing of the dimensional restrictions of enzyme-active sites by using synthesis to stretch the cofactor by known magnitudes.

His final forays into the synthesis of nitrogen heterocycles included the synthesis, chemical behavior, and valence orbital structure of tri-s-triazine and 1,2,4,6-tetraazapentalene. The first of these is a fundamental nitrogen aromatic ring system consisting of a coplanar arrangement of three fused s-triazine rings, with a 2π-electron periphery. This long-sought nucleus, first conceived in correct formulation by Pauling and

Sturdevant in 1937, finally became available in a remarkably abbreviated synthesis.

All tolled Nelson published 438 papers and his roster of coworkers included 120 Ph.D. students and 90 postdoctorates.

Following his retirement from the University of Illinois, in 1987, Nell passed away and before long Nelson underwent serious cancer surgery himself. After recovery, he became a Fogarty Scholar-in-Residence at the National Institutes of Health, Bethesda, Maryland (1989–1990) in association with Arnold Brossi. This was followed by an appointment as Distinguished Visiting Professor at the University of California, San Diego, under the auspices of D. R. Kearns and M. Goodman and, in 1991, as a Sherman Fairchild Distinguished Scholar in the Division of Chemistry and Chemical Engineering at the California Institute of Technology, Pasadena, where he was retained as a Faculty Associate starting in 1992. He had earlier connections with Caltech through collaborations with J. D. Roberts, which originated on mutual ski vacations and resulted in four joint publications. A particularly meaningful collaboration with Jack Roberts and his wife, Edith, however, had its roots in their introducing him to Margaret Taylor Phelps, which resulted in Nelson and Peggy's marriage in 1992. Peggy Phelps introduced him, in his California years, to the world of contemporary art and to travels centered on art and architecture and shared with Nelson an avid interest in skiing. Nelson remained involved in the musical world serving on the board of the Pasadena Symphony whose piano chair was endowed in his honor by family and friends on the occasion of his 85th birthday.

During the course of Nelson's career, he served at a consultant for Phillips Petroleum Company, Monsanto Chemical Company, Eli Lilly and Company, in that order. He lectured widely, nationally and internationally. In addition to his membership in the National Academy of Science (1955), he became a fellow of the American Academy of Arts and Sciences in 1961. He was elected a Member of the American Philosophical Society in 1996. He received the ACS Award for Creative Work in Synthetic Organic Chemistry in 1963, the Medal of the Synthetic Organic Chemical Manufacturers Association in 1970 and the ACS Roger Adams Award in Organic Chemistry in 1981. Awards continued after Nelson's retirement, including the George W. Wheland Award of the University of Chicago (1991), the (first) Creativity Award of the University of Oregon (1994) and an ACS Arthur C. Cope Scholar Award (1995). When Nelson received the (first) Paul G. Gassman Distinguished Service Award of the ACS Division of Organic Chemistry in 1994, he was being recognized for the years of his time that he had contributed, *inter alia*, to editorial work *(Journal of Organic Chemistry, Organic Syntheses, Journal of the American Chemical Society, Biochemistry, Chemistry International*, and *Pure and Applied Chemistry)* and to committee work on foundations (National Science Foundation, National Research Council, Alfred P. Sloan Foundation, National Institutes of Health, John Simon Guggenheim Memorial Foundation, Searle Scholars Program in the Chicago Community Trust). He was active in the Division of Organic Chemistry of the American Chemical Society and the International Union of Pure and Applied Chemistry. Along with his earned degrees (B.S., 1937, Lehigh University; B.Sc., 1940 and D.Sc., 1983, University of Oxford; Ph.D., Columbia University, 1942), Nelson's honorary degrees included Sc.D. (1963) Lehigh University, Doctor Hon. Causa (1980) Adam Mickiewicz University, Poznan, Poland and D.Sc. (Hon) (1988) University of Illinois, Urbana-Champaign.

Over a period of five decades, Nelson Leonard's research showed him to be a leader rather than a follower, a successful explorer in areas other than organic chemistry but with the advantage of a firm base in synthetic and structural organic chemistry. He was a major force in organic and bioorganic chemistry. Nelson was an awe inspiring personality – an uncommonly handsome man and a genuine gentleman. He was my Ph.D. mentor, my tireless supporter, my role model, my colleague on the Board of Directors of Organic Syntheses, Inc. and my dear friend.

CARL R. JOHNSON
Hartfield, Virginia

PREFACE

It is the weight, not numbers of experiments that is to be regarded.
                                                              Sir Isaac Newton

These words capture the essence of what the series *Organic Syntheses* has represented in the 87 years since Roger Adams' compilation of 16 tested preparations appeared in 1921. Adams and the founding fathers of *Organic Syntheses* recognized that the foundation of modern organic chemistry and its ability to transform the natural world rested squarely on the shoulders of sound, reproducible and robust experimentation. Whether to provide access to a valuable building block or to enable a key transformation in a synthetic endeavor, the success of chemical synthesis requires the availability of well-documented and vetted procedures. Their vision has created the lasting legacy of *Organic Syntheses* as the "gold standard" of experimentation in organic chemistry.

Newton's insight that it is the "weight (i.e. quality), not numbers of experiments" that matters is relevant today in a way that he could never have anticipated. The exponential growth of the chemical literature, that has continued unabated since the latter part of the 20th century, has led to an undreamed of edifice of experimental descriptions. The practitioner in search of a reliable protocol to accomplish a specific chemical manipulation, faces the daunting task of culling through the mountain of information and selecting that recipe which appears to be most relevant, best described and hopefully, most transportable. Now more than ever, the procedures found in the pages of *Organic Syntheses* serve as an oasis in the sea of irreproducibility.

Volume 85 contains thirty detailed procedures that reflect a broad swath of synthetic organic chemistry. The procedures compiled herein are presented chronologically in the order that checking and final editing was completed. This process allows the procedures to be posted immediately on the *Organic Syntheses* website (www.orgsyn.org.) and thus available to the community prior to the compilation of the annual volume. Therefore, my description of the contents of this volume does not follow their sequential appearance, but rather attempts to identify common themes and targets among the collected preparations.

The procedures in *Organic Syntheses* can be roughly divided into two broad families. The first (and historically the most prevalent) are those that produce useful compounds as building blocks, reagents, auxiliaries and catalysts. The second (that have taken on greater significance with the rise of modern synthetic methodology) are those that illustrate newly developed synthetic transformations of both a specific and general nature.

In the family of useful compounds, some of the more interesting building blocks include Ragan and co-workers' preparation of 1,3-cycloheptanedione (p. 138), Lautens and co-workers' synthesis of a methylenecyclopropane carboxylic ester (p. 172), the silyl glyoxylate conjunctive agent described by Johnson and co-workers (p. 278) and the enantiomerically pure pipecolic acid derivative prepared from aspartate by Guichard and co-workers (p. 147). Two other enantiopure building blocks from the family of β^2-amino acid derivatives are described by Whiting and co-workers (β-amino nitrile, p. 231) and Seebach and co-workers (β^2-phenylalanine, p. 295). The Seebach procedure nicely highlights the use of their DIOZ auxiliary previously described in *Organic Syntheses* (**2003**, *80*, 57) along with a useful aminomethylating agent for Mannich reactions described in this volume (p. 287). Finally the procedures from Moore (cyclam, p. 1) and Pagenkopf (2,5-dibromosilole, p. 53) provide access to building blocks useful for applications in macrocycle synthesis (bifunctional chelators) and organic light emitting diodes (OLEDS), respectively.

The category of reagents features a safer and milder preparation of diphenyldiazomethane by Brewer (p. 189). Procedures representing three major classes of auxiliaries are included. Glorius and co-workers describe a simple process for the preparation of chiral oxazolines from common aldehydes (p. 267). Enantiomerically pure 2,3-dimethyl-1,4-butanediol (the related dimethylsuccinic acid) is prepared in an interesting oxidative homo-coupling process of titanium enolates as described by Zakarian and co-workers (p. 158). Classical resolution still provides preparatively useful access to enantiomerically pure compounds as illustrated by the preparation both enantiomers of *trans* 2-amino-1-cyclohexanol by Bolm and co-workers (p. 106).

Chiral ligands and catalysts and their applications are increasingly represented in *Organic Syntheses*. This volume features the preparation of the tetrazole analog of L-proline by Ley and co-workers

(p. 72) that has found considerable utility in enantioselective transformations. Another extremely useful ligand for many transition metal transformations is the BINOL derived phosphoramidite (Feringa ligand), which is described by RajanBabu and co-workers (p. 238) and is also employed by them in a companion procedure (p. 248). The family of *N*-heterocyclic carbenes (NHCs) has risen to prominence both as ligands for transition metals and as Lewis basic catalysts in their own right. Fürstner and co-workers provide a new procedure for the synthesis of less accessible imidazolium ions (NHC precursors) that bear both aryl and alkyl substituents (p. 34). Finally, the importance of copper catalysis in synthetic chemistry has not abated and the need for discrete, soluble, copper catalysts for organic transformations is addressed by van Koten and co-workers' preparation of a copper thiolate (p. 209).

Procedures describing new synthetic methods are also featured in this volume. The synthesis of heterocyclic compounds is the subject of procedures by Miller and McNaughton (Friedlander quinoline synthesis, p. 27), Movassaghi (precursors for pyridine synthesis, p. 88), Mani and Deng (novel pyrazole synthesis, p. 179) and Yu (pyrroline carbamate, p. 64). The synthesis and manipulation of alkynes is described in the preparation of alkynyl ethers by Kocienski and Snaddon (p. 45), the enantioselective synthesis of propargyl alcohols by addition of alkynylindium reagents to aldehydes from Shibasaki and co-workers (p. 118), the formation of carboxylic amides from terminal alkynes and tosyl azide by Chang and co-workers (p. 131), and the diastereoselective bromination of tetrolic acid by Thibonnet and co-workers (p. 231).

Two novel olefination methods are described in this volume as well. The first from Hodgson and co-workers employs an unusual combination of organolithium agents with lithiated epoxides (p. 1) and the second from Taylor and co-workers (p. 15) illustrates the ability to telescope the homologation of an ester to the next higher vinylog through a reduction-oxidation-Wittig olefination sequence.

Finally, transition metal-catalyzed transformations are increasingly apparent in the practice of organic synthesis and three procedures that employ Group 10 metals are illustrated here. The first is a method for the formation of allylphosphinous acids by palladium catalyzed C-P bond formation from Montchamp and co-workers (p. 96). The second is a novel decarboxylative biaryl synthesis that employs carboxylic acids as donors in palladium-catalyzed, cross-coupling reactions with aryl bromides by Gooßen and co-workers (p. 196). Finally, RajanBabu and co-workers describe a remarkable nickel-catalyzed, enantioselective hydrovinylation reaction that sets quaternary stereogenic centers (p. 248).

Although the advent of electronic media has greatly reduced the impact of the printed publications, it is hoped that the familiar maroon volumes of *Organic Syntheses* still find their way (free of charge to members of the ACS Division of Organic Chemistry) onto the desks and shelves of practicing organic chemists. The contents of these volumes provides not only the critical "know how" for a specific reaction, but also serve as a model for young chemists to emulate in the execution, documentation and publication of their own experimental procedures.

The contents of this volume represent the combined efforts of the submitters and the Board of Editors and their co-workers (for checking), all under the steady guidance of Rick Danheiser (Editor in Chief) and Chuck Zercher (Associate Editor). I want to especially acknowledge my colleagues on the Board of Editors for their assistance in soliciting and selecting many of the procedures included in this volume. Most importantly, I thank the junior checkers whose painstaking experimentation and reporting constitute the backbone of the *Organic Syntheses* enterprise. All of these various activities are masterfully orchestrated and directed by Rick and Chuck under whose reign, the vitality of *Organic Syntheses* has flourished and considerably improved over the eight years of my tenure on the Board.

Finally I want to thank the past Board of Editors who gave me the opportunity to be a part of one of Roger Adams' most lasting legacies. The spirit of service to the organic chemistry community that characterized "The Chief" runs deep at the University of Illinois. It has been a pleasure and a privilege to add my name to the list of those who honor that tradition.

SCOTT E. DENMARK
Urbana, Illinois

CONTENTS

FACILE SYNTHESIS OF 2-ETHYL-3-QUINOLINECARBOXYLIC ACID HYDROCHLORIDE

Brian R. McNaughton and Benjamin L. Miller

PREPARATION OF A NON-SYMMETRICAL IMIDAZOLIUM SALT: 1-ADAMANTYL-3-MESITYL-4,5-DIMETHYLIMIDAZOLIUM TETRAFLUOROBORATE

Alois Fürstner, Manuel Alcarazo, Vincent César, and Helga Krause

PREPARATION OF A 1-ALKOXY-1-ALKYNE FROM REACTION OF A 2,2,2-TRIFLUOROMETHYL ETHER WITH AN ALKYLLITHIUM REAGENT: 1-BENZYLOXYMETHOXY-1-HEXYNE

Philip J. Kocienski and Thomas N. Snaddon

DIRECT SYNTHESIS OF 2,5-DIHALOSILOLES 53
Nicholas A. Morra and Brian L. Pagenkopf

ONE-POT CONVERSION OF LACTAM CARBAMATES 64
TO CYCLIC ENECARBAMATES: PREPARATION OF
1-*TERT*-BUTOXYCARBONYL-2,3-DIHYDROPYRROLE
Jurong Yu, Vu Truc, Peter Riebel, Elizabeth Hierl, and
Boguslaw Mudryk

(*S*)-5-PYRROLIDIN-2-YL-1*H*-TETRAZOLE 72
Valentina Aureggi, Vilius Franckevičius, Matthew O. Kitching,
Steven V. Ley, Deborah A. Longbottom, Alexander J. Oelke and
Gottfried Sedelmeier

CATALYTIC ENANTIOSELECTIVE ADDITION OF TERMINAL ALKYNES TO ALDEHYDES: PREPARATION OF (S)-(-)-1,3-DIPHENYL-2-PROPYN-1-OL AND (S)-(-)-4-METHYL-1-PHENYL-2-PENTYN-1,4-DIOL

Ryo Takita, Shinji Harada, Takashi Ohshima, Shigeki Matsunaga, and Masakatsu Shibasaki

COPPER-CATALYZED THREE-COMPONENT REACTION OF 1-ALKYNES, SULFONYL AZIDES, AND WATER: N-(4-ACETAMIDOPHENYLSULFONYL)-2-PHENYLACETAMIDE

Seung Hwan Cho, Seung Jun Hwang, and Sukbok Chang

PREPARATION OF CYCLOHEPTANE-1,3-DIONE VIA REDUCTIVE RING EXPANSION OF 1-TRIMETHYLSILYLOXY-7,7-DICHLOROBICYCLO [3.2.0]HEPTAN-6-ONE

Nga Do, Ruth E. McDermott, and John A. Ragan

2-METHYLENECYCLOPROPANECARBOXYLIC ACID   172
ETHYL ESTER
Mark E. Scott, Nai-Wen Tseng, and Mark Lautens

1.5-2.0:1 mixture of
trans and cis

REGIOSELECTIVE SYNTHESIS OF 1,3,5-TRISUBSTITUTED   179
PYRAZOLES BY THE REACTION OF N-
MONOSUBSTITUTED HYDRAZONES WITH
NITROOLEFINS
Xiaohu Deng and Neelakandha S. Mani

DIPHENYLDIAZOMETHANE   189
Muhammad I. Javed and Matthias Brewer

Lukas J. Gooßen, Nuria Rodríguez, Christophe Linder, Bettina Zimmermann, and Thomas Knauber

Method A

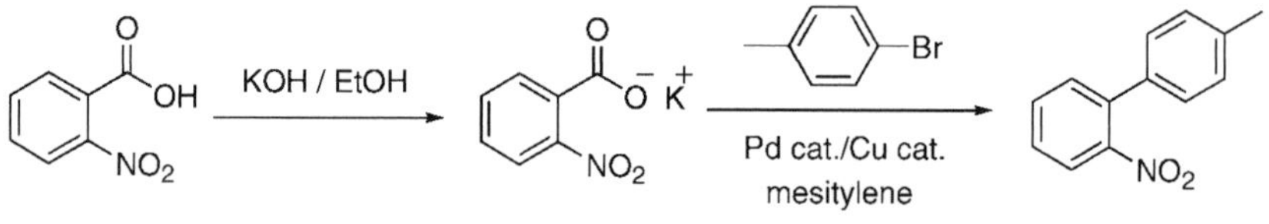

Method B

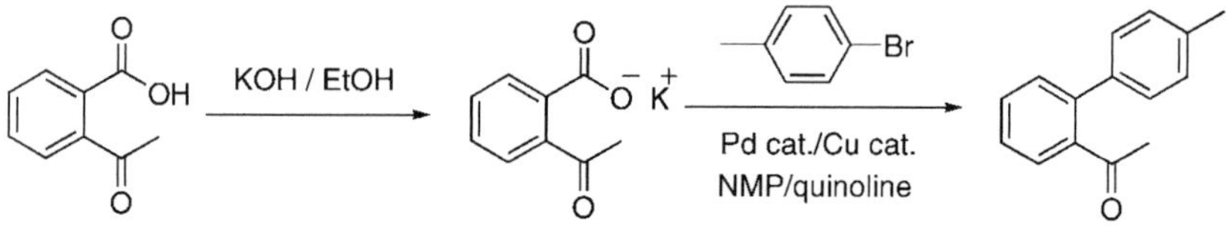

SYNTHESIS OF AMINOARENETHIOLATO-COPPER(I)   **209**
COMPLEXES

Elena Sperotto, Gerard P.M. van Klink and Gerard van Koten

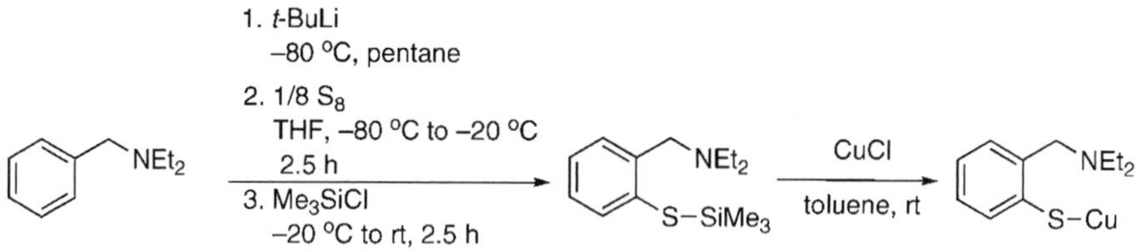

1

(R)-2,2'-BINAPHTHOYL-(S,S)-DI(1-PHENYLETHYL) AMINOPHOSPHINE. SCALABLE PROTOCOLS FOR THE SYNTHESES OF PHOSPHORAMIDITE (FERINGA) LIGANDS

Craig R. Smith, Daniel J. Mans and T. V. RajanBabu

(R)-3-METHYL-3-PHENYL-1-PENTENE *VIA* CATALYTIC ASYMMETRIC HYDROVINYLATION

Craig R. Smith, Aibin Zhang, Daniel J. Mans and T. V. RajanBabu

EFFICIENT OXIDATIVE SYNTHESIS OF (-)-2-*TERT*-BUTYL-(4*S*)-BENZYL-(1,3)-OXAZOLINE

Björn T. Hahn, Kirsten Schwekendiek and Frank Glorius

TERT-BUTYL *TERT*- BUTYLDIMETHYLSILYL-GLYOXYLATE: A USEFUL CONJUNCTIVE REAGENT

David A. Nicewicz, Guillaume Brétéché, and Jeffrey S. Johnson

BENZYL ISOPROPOXYMETHYL CARBAMATE – AN AMINOMETHYLATING REAGENT FOR MANNICH REACTIONS OF TITANIUM ENOLATES

Hartmut Meyer, Albert K. Beck, Radovan Sebesta, and Dieter Seebach

ORGANOLITHIUMS AND LITHIUM 2,2,6,6-TETRAMETHYLPIPERIDIDE IN REDUCTIVE ALKYLATION OF EPOXIDES: SYNTHESIS OF (E)-ALKENES [(E)-2-METHYLTETRADECA-1,3-DIENE]

Submitted by David M. Hodgson,[1] Philip G. Humphreys and Matthew J. Fleming.
Checked by Hiroyuki Morimoto, Hisashi Mihara, and Masakatsu Shibasaki.

1. Procedure

(E)-2-Methyltetradeca-1,3-diene.   Caution!   tert-Butyllithium   is extremely pyrophoric and care must be taken to avoid exposure to air at all times. A 2-L, single-necked, round-bottomed flask (Note 1) is equipped with a magnetic stirrer bar and a rubber septum pierced by an argon inlet needle and an outlet needle.   The flask, in which a positive flow of argon is maintained throughout the entire procedure, is flame-dried under argon, allowed to cool, and then charged by syringe with dry tetrahydrofuran (75 mL) (Note 2) and 2-bromopropene (6.24 mL, 70.2 mmol, 1.30 equiv) (Note 3). The solution is cooled to –78 °C (bath temperature) in a dry ice/acetone bath and then a 1.58 M solution of *tert*-butyllithium in pentane (89.2 mL, 140 mmol, 2.6 equiv) (Note 4) is added dropwise by syringe over 20 min in ten roughly equal portions (Note 5). The resulting bright-yellow suspension is stirred for 10 min at –78 °C, then an additional portion of 2-bromopropene (0.65 mL, 7.3 mmol, 0.14 equiv) is added dropwise over 7 min. The resulting pale-yellow suspension is stirred for an additional 30 min at –78 °C (during which time the preparation of the lithium 2,2,6,6-tetramethylpiperidide (LTMP) solution is initiated) and is warmed to 0 °C by transfer into an ice bath for 10 min. The resulting pale-yellow solution is re-cooled to –78 °C (bath temperature) (Note 6).

A 1-L, single-necked, round-bottomed flask is equipped with a magnetic stirrer bar and a rubber septum pierced by an argon inlet needle and an outlet needle. The flask, in which a positive flow of argon is maintained throughout the entire procedure, is flame-dried under argon, then is charged with dry pentanes (800 mL) (Note 7) and 2,2,6,6-tetramethylpiperidine (18.2 mL, 108 mmol, 2.0 equiv) by syringe (Note 8). The colorless solution is cooled to 0 °C in an ice bath and then a 1.57 M solution of *n*-butyllithium in hexanes (68.8 mL, 108 mmol, 2.0 equiv) (Note 4) is added dropwise by syringe over 15 min in seven roughly equal portions (Note 9). The resulting yellow LTMP solution is stirred at 0 °C for 30 min further and then is added *via* cannula (Note 10) over 15 min to the stirring pale-yellow suspension of 2-propenyllithium at –78 °C (Note 11). The resulting yellow suspension is transferred to an ice bath, is stirred for 30 min and then 1,2-epoxydodecane (11.8 mL, 54.0 mmol) (Note 8) is added dropwise by syringe in three roughly equal portions over 5 min. The ice bath is removed and the resulting yellow suspension is stirred for 2 h (Note 12).

The bright-yellow solution is re-cooled to 0 °C and 50 mL of a 1 M aqueous hydrochloric acid solution is added by syringe over 5 min (Note 13). The reaction mixture is poured into a 3-L separatory funnel containing 1 M aqueous hydrochloric acid solution (200 mL). The reaction flask is rinsed with diethyl ether (2 x 100 mL), which is added to the separatory funnel. After thorough mixing, the aqueous layer is separated and extracted with diethyl ether (3 x 200 mL). The combined organic extracts are washed with brine (200 mL), dried over magnesium sulfate, filtered and evaporated under gradually reduced pressure (40 °C, 760 mmHg to 15 mmHg) using a rotary evaporator to leave 18.6 g of a pale-yellow residue (Note 14).

The yellow residue is loaded neat onto a pad of silica gel (15 cm x 6 cm) (Note 15) that has already been wetted with *n*-pentane, and the flask is rinsed with pentane (2 x 10 mL) (Note 16). The pad is rinsed with 1000 mL of *n*-pentane and the eluent is collected directly as a single fraction. Rotary evaporation under reduced pressure (40 °C, 760 mmHg to 15 mmHg) for 2 h (Note 17) gives 6.47 g (58%) (Note 18) of (*E*)-2-methyltetradeca-1,3-diene as a colorless oil (Note 19). The *E*/*Z* ratio of this product is determined to be 97:3 (Note 20).

2. Notes

1.   All apparatus was oven-dried overnight.

2.   The submitters purchased tetrahydrofuran (99+%) from Rathburn Chemicals Ltd. and the solvent was freshly distilled from sodium and benzophenone under an atmosphere of nitrogen. The checkers purchased tetrahydrofuran (anhydrous, 99.9%, inhibitor free) from Aldrich Chemical Co. and freshly distilled the solvent from sodium and benzophenone under an atmosphere of argon. The checkers determined the amount of water in the distilled tetrahydrofuran to be 4.0 ppm by Karl-Fischer titration.

3.   2-Bromopropene (99%) was purchased from Aldrich Chemical Co. The submitters used it as received. The checkers distilled it (bp 50–51 °C) from $CaCl_2$ under an atmosphere of argon before use.

4.   The submitters purchased *tert*-butyllithium (1.5 M in pentane) and *n*-butyllithium (1.6 M in hexanes) from Fisher Scientific Worldwide Co. Checkers purchased *tert*-butyllithium (1.57 M in *n*-pentane) from Kanto Chemical Co. and *n*-butyllithium (1.6 M in hexanes) from Aldrich Chemical Co., respectively. The organolithium reagents were used as received. Fresh 100-mL bottles of the organolithiums were used for each reaction run. The solutions were titrated prior to use using *s*-BuOH and 1,10-phenanthroline according to an established procedure.[2]

5.   *tert*-Butyllithium   formed   a   bright   yellow   complex   with tetrahydrofuran at –78 °C, whereas 2-propenyllithium was colorless. If a bright (rather than pale) yellow color persisted following the addition of *tert*-butyllithium, the excess *tert*-butyllithium was consumed by slow addition of additional 2-bromopropene (0.65 mL by the checker, a few drops by the submitter) until a pale yellow/colorless solution remained.[3] When the bright yellow solution was used directly, the product was obtained in lower purity with a detectable amount of by-products (presumably derived from the excess *tert*-butyllithium). Addition rate of *tert*-butyllithium (within 20 min in a full scale and within 10 min in a half scale) is also important to obtain reproducible results. When the checkers added *tert*-butyllithium slowly over 50 min, the product was obtained in lower yield and purity.

6.   At this concentration 2-propenyllithium was a solution in THF at 0 °C and a suspension at –78 °C.

7.   To prevent precipitation of LTMP, 800 mL of pentane was required. Pentane was purchased by the submitters from Rathburn Chemicals Ltd. and was degassed with nitrogen and passed through alumina under an atmosphere of nitrogen directly into the pre-calibrated flask.[4] The checkers purchased pentane (anhydrous, 99+%) from Aldrich Chemical Co.

and used it as received. The checkers determined the amount of water in the pentane to be 6.0 ppm by Karl-Fischer titration.

8. 2,2,6,6-Tetramethylpiperidine (99%) was purchased from Aldrich Chemical Co. 1,2-Epoxydodecane (95%) was purchased from Aldrich Chemical Co. 2,2,6,6-Tetramethylpiperidine was distilled from CaH_2 under an atmosphere of argon (bp 150–152 °C) and 1,2-epoxydodecane was distilled from CaH_2 under reduced pressure (bp 110–112 °C, 7.5 mmHg).

9. Use of 2.0 equiv of LTMP was necessary to minimize by-product formation. Carrying out the reaction with 1.3 equivalents of LTMP gave a 37% yield of (E)-2-methyltetradeca-1,3-diene, a 5% yield of dodecanal[5] and a 41% yield of 2-methyltetradec-1-en-4-ol. If LTMP was omitted from the reaction, 2-methyltetradec-1-en-4-ol was isolated in 96% yield.

10. An oven-dried, 14 gauge cannula was used.

11. A bleed needle was used in the flask containing the 2-propenyllithium suspension to allow transfer to occur rapidly.

12. The reaction was monitored by TLC analysis on aluminum-backed plates pre-coated with silica containing a fluorescent indicator (0.2 mm, Merck 60 F_{254}, purchased from Merck KGAA) using n-pentane/diethyl ether (20:1) as the eluent. The plates were visualized following staining with basic potassium permanganate solution and gentle heating. The product diene had $R_f = 0.96$ (weakly UV active; bright yellow after staining); the epoxide starting material had $R_f = 0.38$ (UV inactive; white to yellow); dodecanal had $R_f = 0.35$ (UV inactive; white to yellow) and 2-methyltetradec-1-en-4-ol had $R_f = 0.12$ (UV inactive; yellow).

13. Although the submitters observed evolution of gas, the checkers did not observe significant evolution of gas.

14. It was important to remove solvent well at this stage to avoid separation problems during silica gel column chromatography.

15. Silica gel (Kieselgel 60) was purchased from Aldrich Chemical Co.

16. n-Pentane (>98% purity) was purchased from Wako Chemical Co. and used as received. The submitters used light petroleum ether (30–40 °C, 99+%) from Rathburn Chemicals Ltd. The checkers found that the use of normal petroleum ether (b.p. 30–60 °C) caused problems due to incomplete removal of petroleum ether residue at the last stage of evaporation.

17. The control of pressure during the evaporation (20 kPa) was important to remove n-pentane: when evaporation was performed at 27 kPa

4

for 2 h, the residual *n*-pentane was observed in ^{1}H and ^{13}C NMR spectra of the product.

18. Submitters obtained the product in 65% yield. Checkers obtained the product in 58% yield on a half-scale reaction. Purity of the product was >95% by gas chromatographic analysis using a 30 m, 0.25 mm internal diameter, 95% dimethyl-5% diphenylpolysiloxane capillary column (t_R = 10.5 min, initial temp. 200 °C, column temp. 150 °C, detection temp. 250 °C). A minor component (presumably the *Z*-isomer) was detected at t_R = 8.6 min (1.4 %). If necessary, the product can be further purified by distillation under reduced pressure (9 mmHg, bp 120-121 °C, >95% purity by GC analysis).

19. The diene had the following characteristics: bp 120–121 °C, 9 mmHg; IR (neat) 3082 (m), 3015 (m), 2925 (s), 2853 (s,) 1767 (w), 1647 (w), 1609 (m), 1456 (s), 1377 (m), 1313 (w), 1085 (w), 964 (s), 881 (s), 721 (m) cm^{-1}; ^{1}H NMR (500 MHz, CDCl$_3$) δ: 0.89 (t, *J* = 6.7 Hz, 3 H), 1.21–1.35 (m, 14 H), 1.35–1.45 (m, 2 H), 1.84 (brs, 3 H), 2.10 (brtd, *J* = 7.0, 7.0 Hz, 2 H), 4.86 (brs, 2 H), 5.67 (td, *J* = 7.0, 15.6 Hz, 1 H), 6.14 (brd, *J* = 15.6 Hz, 1 H); ^{13}C NMR (125 MHz, CDCl$_3$) δ: 14.1, 18.7, 22.7, 29.3, 29.4, 29.5, 29.5, 29.6, 29.6, 31.9, 32.8, 114.0, 131.1, 132.7, 142.2; LRMS (EI) *m/z* 208 (M$^+$); HRMS (EI) *m/z* calcd for C$_{15}$H$_{28}^+$ (M$^+$), 208.2191, found 208.2190. Anal. Calcd for C$_{15}$H$_{28}$: C, 86.46; H, 13.54. Found: C, 86.06; H, 13.75.

20. The *E/Z* ratio determined by NMR analysis [^{1}H NMR (500 MHz, CDCl$_3$); minor *Z* isomer: δ: 2.20–2.31 (m, 2H, 5-C$\underline{H}_2$), major *E* isomer: δ: 2.10 (brtd, *J* = 7.0, 7.0 Hz, 2 H, 5-C$\underline{H}_2$).

3. Discussion

The ability to prepare carbon-carbon double bonds in a regio- and stereocontrolled manner is of central importance in organic synthesis. One such method originally reported by Crandall and Lin,[6] and subsequently studied in more detail by Mioskowski and co-workers,[7] involves addition of an organolithium to an epoxide to give the corresponding alkene. However, the reaction in its original manifestation suffers from three significant limitations: (1) only simple alkyllithiums are effective partners; (2) high *E*-selectivity is observed only with secondary and tertiary alkyllithiums; and (3) at least 2 equiv of the organolithium are required.

The LTMP-modified reductive alkylation of epoxides reported herein provides a solution to the limitations mentioned above allowing rapid, convergent, and position-specific access to alkenes in high E/Z purity from readily available starting materials (Table 1).[8]

Table 1. Alkenes from epoxides using LTMP and organolithiums.[a]

Entry	LiR²	Alkene	Yield (%)[b]	Ratio[c]
1	LiPh		93	98:2
2	LiC₆H₄-p-OCH₃		70	98:2
3			73	98:2
4			85	E-only
5			82	98:2
6			84	99:1
7			72	98:2
8			70	90:10
9			80	91:0
10			85	91:9
11[d,e]			65	97:3
12[e]			74	E-only

[a] Reactions performed on 0.43-1.25 mmol scale. [b] Isolated yield. [c] Determined by GCMS. Ratios refer to $E:Z$ (entries 1-4, 11, 12), $E,E:Z,E$ (entries 5-7, 10), $Z,E:E,E$ (entry 8) and Z,E:other isomers (entry 9). [d] Et₂O as solvent (in hexane 84% yield, $E:Z$, 65:35). [e] 1.3 equiv of LTMP used.

The use of alkenyllithiums gives the corresponding diene in high isomeric purity, while employing aryllithiums gives arylated alkenes. *E*-Allylsilanes, which are valuable intermediates in synthesis,[9] are also conveniently prepared by addition of a terminal epoxide to a mixture of LTMP and an α-silyl-organolithium.

We propose that LTMP functions as a base to form *trans*-α-lithiated epoxide 1, then the organolithium can act as a nucleophile on this transient carbenoid (Scheme 1). Subsequent elimination of Li_2O gives the observed alkenes in generally high isomeric purity.

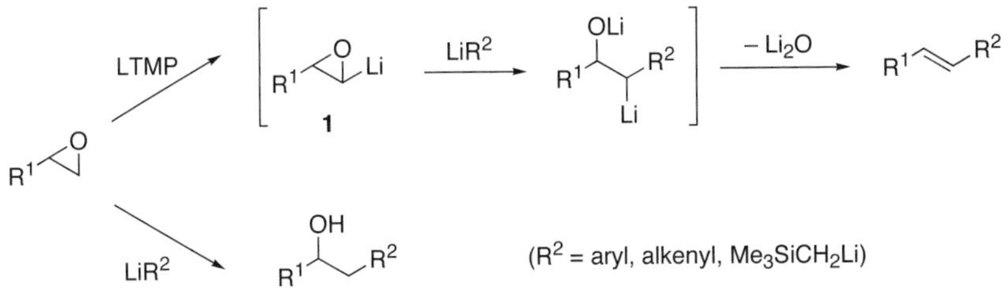

Scheme 1

Addition of the organolithium reagent to the terminal epoxide in the absence of LTMP gives the corresponding secondary alcohol from direct ring-opening of the epoxide as the major product.

1. Department of Chemistry, University of Oxford, Chemistry Research Laboratory, Mansfield Road, Oxford OX1 3TA, UK. E-mail: david.hodgson@chem.ox.ac.uk.
2. (a) Watson, S. C.; Eastham, J. F. *J. Organomet. Chem.* **1967**, *9*, 165–168. (b) Schlosser, M. *Organometallics in Synthesis*, 2nd ed. Wiley, New York, 2002, pp. 671–672.
3. Morwick, T.; Paquette, L. A. *Org. Synth., Coll. Vol. IX* **1998**, 670–675.
4. Pangborn, A. B.; Giardello, M. A.; Grubbs, R. H.; Rosen, R. K.; Timmers, F. J. *Organometallics* **1996**, *15*, 1518–1520.
5. Hodgson, D. M.; Bray, C. D.; Kindon, N. D. *J. Am. Chem. Soc.* **2004**, *126*, 6870–6871.
6. Crandall J. K.; Lin, L. -H. C. *J. Am. Chem. Soc.* **1967**, *89*, 4527–4528.
7. Doris, E.; Dechoux, L.; Mioskowski, C. *Tetrahedron Lett.* **1994**, *35*, 7943–7946.

8. (a) Hodgson, D. M.; Fleming, M. J.; Stanway, S. J. *J. Am. Chem. Soc.* **2004**, *126*, 12250–12251. (b) Hodgson, D. M.; Fleming, M. J.; Stanway, S. J. *J. Org. Chem.* **2007**, *72*, 4763-4773.
9. (a) Fleming, I.; Dunoguès, J.; R. Smithers, R. *Org. React.* **1989**, *37*, 57–575. (b) Sarkar, T. K. In *Science in Synthesis*; Fleming, I., Ed.; Thieme; Stuttgart, 2001; Vol. 4, pp. 837–925.

Appendix
Chemical Abstracts Nomenclature; (Registry Number)

2-Bromopropene; (557-93-7)

tert-Butyllithium: Lithium, (1,1-dimethylethyl)-; (5944-19-4)

n-Butyllithium: Butyllithium; (109-72-8)

2,2,6,6-Tetramethylpiperidine: (768-66-1)

1,2-Epoxydodecane: Oxirane, decyl-; (2855-19-8)

David Hodgson obtained his first degree in Chemistry at Bath University. After a PhD at Southampton University in the field of natural product synthesis (with P. J. Parsons) and a research position at Schering, he was appointed in 1990 to a lectureship at Reading University. In 1995 he moved to the Chemistry Department at Oxford University, where he is now a Professor of Chemistry. His research interests are broadly in the development and application of synthetic methods, particularly (asymmetric) generation and transformations of carbenoids.

Philip Humphreys was born in 1981 in Worcester, UK. After obtaining his MSci in Biochemistry and Biological Chemistry (first class honors) in 2003 from the University of Nottingham, he moved to the University of Oxford (Oriel College) for his PhD studies under the supervision of David Hodgson. His work was sponsored by GlaxoSmithKline and concerned the reactions of alpha-lithiated aziridines. Phil is currently a postdoctoral researcher with Larry Overman at UC Irvine, supported by a Merck Sharp & Dohme Postdoctoral Research Fellowship.

Matthew Fleming was born in 1978 in London, UK. After obtaining his MSci in Chemistry (1st class honors) in 2002 from King's College London, he moved to the University of Oxford (Hertford College) and in 2005 completed his PhD under the supervision of David M. Hodgson. His research was sponsored by GlaxoSmithKline and was concerned with the development of new reaction methodology involving metalated heterocycles and sulfur ylides. He is currently a postdoctoral research assistant with Prof. Mark Lautens at the University of Toronto, working in the area of natural product synthesis.

Hiroyuki Morimoto was born in 1981 in Hiroshima, Japan. He graduated in 2004 and received his M. S. degree in 2006 from the University of Tokyo under the direction of Professor Masakatsu Shibasaki. The same year he started his Ph. D. study under the supervision of Professor Shibasaki. His current interest is development of novel direct asymmetric catalytic systems using ester equivalents as donor substrates.

Hisashi Mihara was born in 1981 in Shiga, Japan. He received his M. S. degree in 2006 from the University of Tokyo. He is pursuing Ph.D. degree at the Graduate School of Pharmaceutical Sciences, The University of Tokyo, under the guidance of Professor Masakatsu Shibasaki. His current interest is catalytic enantioselective total synthesis of biologically active compounds.

SELECTIVE TRIALKYLATION OF CYCLEN WITH
tert-BUTYL BROMOACETATE
[1,4,7,10-Tetraazacyclododecane-1,4,7-triacetic
acid, Tri-*tert*-butyl Ester Hydrobromide]

Submitted by Dennis A. Moore.[1]
Checked by Leslie Patterson and Marvin J. Miller.

1. Procedure

Warning: tert-Butyl bromoacetate is a lachrymator. The reagent, reaction and its work-up should be handled in an adequately ventilated fume hood while wearing gloves, safety glasses and laboratory coat.

A 250-mL, four-necked, round-bottomed flask, fitted with a mechanical stirrer (Teflon paddle, 6 x 1.8 cm), a condenser fitted with a nitrogen inlet on top, thermometer and pressure-equalizing addition funnel is charged with cyclen (5.0 g, 0.029 mol), sodium acetate trihydrate (13.0 g, 0.096 mol) and dimethylacetamide (40 mL) (Note 1). The heterogeneous mixture is stirred for 30 min, allowing the initial endotherm to subside. To this mixture is added, by means of the addition funnel, a solution of *tert*-butyl bromoacetate (18.7 g, 14.1 mL, 0.096 mol), in dimethylacetamide (20 mL), dropwise, over 30 min. The rate of the addition is adjusted so as to keep the temperature of the reaction mixture 20–25 °C (Notes 2 and 3). After complete addition, the mixture is allowed to stir for 60 h (Note 4). The mixture is then diluted with diethyl ether (20 mL), cooled to –10 to –15 °C (bath temperature), by means of an ice-methanol bath, and is stirred for 2 h. The resulting precipitated solid is collected by filtration on a coarse glass frit, then is washed with cold (–10 to –15 °C), fresh dimethylacetamide (10 mL), suctioned dry, and then is washed with cold diethyl ether (2 x 25 mL) and suctioned dry again. The crude, white solid (19.7 g) is dissolved in

10

chloroform (100 mL) and the solution washed with water (2 x 15 mL) and sat. aq. NaBr solution (1 x 15 mL) (Note 5). The organic phase is collected, then is dried over magnesium sulfate and filtered with suction through Whatman #4 paper on a porcelain Buchner funnel. The filtrate is transferred to a 250-mL single necked round-bottomed flask and is concentrated by rotary evaporation (25–35 °C water bath, 3–15 mmHg) to a thin, colorless oil (approximately 40 g) (Note 6). A magnetic stir bar is added to the evaporation flask and the oil is diluted with hexanes, (80 mL) with stirring. Crystallization begins after a few minutes of stirring. The mixture is stirred at room temperature for 3 h, then is cooled to –10 to –15 °C (bath temperature), by means of an ice-methanol bath, and is stirred for an additional 2 h. The resulting white solid is collected by filtration, then is washed with cold, hexanes/chloroform (4/1, 25 mL), suctioned dry and is dried in vacuo (15 mmHg) overnight at room temperature to afford 11.2 g (65%) of the product as a white, amorphous solid (Notes 7, 8 and 9).

2. Notes

1. Cyclen, 1,4,7,10-tetraazacyclododecane (min. 98%), was purchased from Strem. *tert*-Butyl bromoacetate (98%) dimethylacetamide (99.9%) and anhydrous magnesium sulfate (Analytical Reagent) were obtained from the Aldrich Chemical Co. and were used as received. Chloroform (100%, with 0.75% ethanol added) (Acros), hexanes (95% n-hexane) (Fisher), sodium acetate trihydrate (99.28%) (Acros) were all used as received.

2. Short temperature excursions above 25 °C were not overly harmful to the yield, but heating above 35 °C for an extended period of time, e.g. overnight, reduced both the yield and purity of the product. The temperature was kept at 25–26 °C. A few pieces of ice were needed to allow the temperature to remain under 26 °C and for the addition to be complete in 30 min.

3. The initial third of the addition was quite exothermic, so a water cooling bath was needed intermittently. During the course of the addition, most of the solids dissolved. Near the end of the addition, the crude product began to precipitate from the pale, amber mixture as a white solid.

4. Shorter reaction time, e.g. 24–48 hours, decreased the yield by 10–15%. Longer reaction time did not improve the yield.

5.   Using saturated sodium chloride in the final washing allowed for halide exchange, resulting in the isolation of a mixed Br⁻, Cl⁻ salt according to the submitter. The material prepared by the checkers analyzed correctly for the bromide salt (Note 8).

6.   On this scale of reaction the target mass of evaporated oil, consisting of product plus chloroform, should be 36–44 g. If too much chloroform was removed by evaporation, the difference was made up by adding enough chloroform to achieve approximately 40 g of solution, at which time it was possible to proceed with the crystallization as normal.

7.   The submitters performed the same procedure with 25 g of cyclen and when everything else was scaled appropriately, they obtained 69.5 g (80% yield).

8. ^{1}H NMR (500 MHz, CDCl$_3$) δ: 1.42 (s, 9 H), 1.43 (s, 18 H), 2.81–2.95 (br, m, 1 2H), 3.07 (br, m, 4 H), 3.26 (s, 2 H), 3.35 (s, 4), 9.99 (br, s, 2 H). ^{13}C NMR (125.70 MHz, CDCl$_3$) δ: 28.07, 28.11, 47.4, 48.6, 49.0, 51.1, 51.2, 58.1, 81.6, 81.7, 169.5, 170.4.   mp: 179–181 °C (d).   Anal. Calcd. For C$_{26}$H$_{51}$BrN$_4$O$_6$: C, 52.43; H, 8.63; N, 9.41. Found: C, 52.63; H, 8.43; N, 9.36.

9.   The product appeared to be stable at room temperature, with no extraordinary measures taken, for at least several months.

Waste Disposal Information

All hazardous materials should be handled and disposed of in accordance with "Prudent Practices in the Laboratory"; National Academy Press; Washington, DC, 1995.

3. Discussion

DO3A tris(t-Bu ester) continues to be a compound of interest as a useful scaffold for the synthesis of bifunctional chelates.[2] Our procedure is an adaptation of the earliest, most detailed, reported preparations. However, these procedures call for much longer reaction times or more complicated isolation procedures.[3] For instance, Berg, et al. reported a procedure that required a reaction time of six days and chromatographic purification prior to isolation. Whereas, Himmelsbach, et al. allowed the reaction to proceed for nineteen days and needed to isolate a second crystallization-crop of product to achieve a reasonable yield. More recently, Cong, et al,[4] reported performing the reaction with chloroform and triethylamine, curiously giving the HCl salt after chromatographic purification. Very recently, Mishra[5] reported that the reaction could be carried out in chloroform without the need for a base, like sodium acetate. Though the yield appeared to be good, the procedure was not very detailed.

The origin of the observed selectivity appears to be the precipitation of the trialkylated HBr salt from the reaction mixture, thereby preventing tetraalkylation. Our technique is a routine, scalable procedure in our laboratory, affording material in good yield and purity.

1. Mallinckrodt, Tyco Healthcare, P.O. Box 5840, 675 McDonnell Boulevard, St. Louis, Missouri, 63134. dennis.moore@tycohealthcare.com.

2. Some recent citations include, but are not limited to: Quici, S.; Cavazzini, M.; Raffo, M. C.; Armelao, L.; Bottaro, G.; Accorsi, G.; Sabatini, C.; Barigelletti, F. *J. Mat. Chem.* **2006**, *16*, 741-747. Aarons, R. J.; Notta, J. K.; Meloni, M. M.; Feng, J.; Vidyasagar, R.; Narvainen, J.; Allan, S.; Spencer, N.; Kauppinen, R. A.; Snaith, J. S.; Faulkner, S. *Chem. Commun.* **2006**, 909-911. Duimstra, J. A.; Femia, F. J.; Meade, T. J. *J. Am. Chem. Soc.* **2005**, *127*, 12847-12855. Gunnlaugsson, T.; Leonard, J. P. *Dalton Trans.* **2005**, 3204-3212. Aime, S.; Botta, M.; Cravotto, G.; Frullano, L.; Giovenzana, G. B.; Crich, S. G.; Palmisano, G.; Sisti, M. *Helv. Chim. Acta* **2005**, *88*, 588-603.

3. Berg, A.; Almen, T.; Klaveness, Jo; Rongved, P.; Thomassen, T. US Patent 5,198,208, March 30, 1993. Himmelsbach, R. J.; Rongved, P.; Klaveness, J.; Strande, P.; Dugstad, H. World Patent Application, PCT WO 93/02045, July 17, 1991. Berg, A.; Almen, T.; Thomassen, T.; Klaveness, J.; Rongved, P. European Patent Application, EP 299795 A2, July 15, 1988.

4. Li, C.; Wong, W.-T. *Tetrahedron* **2004**, *60*, 5595-5601.

5. Mishra, A. K. European Patent Application, EP 1637524 A1 September 20, 2004.

Appendix
Chemical Abstracts Nomenclature; (Registry Number)

Cyclen: 1,4,7,10-Tetraazacyclododecane; 294-90-6

Sodium acetate trihydrate; (6131-90-4)

tert-Butyl bromoacetate; (5292-43-3)

1,4,7,10-Tetraazacyclododecane-1,4,7-triacetic acid, Tri-*tert*-butyl Ester
   Hydrobromide: 1,4,7,10-Tetraazacyclododecane-1,4,7-tricarboxylic
   acid, 1,4,7-tris(1,1-dimethylethyl) ester; (175854-39-4)

Dennis Moore received his B.S. degree in Chemistry from Southeast Missouri State University in 1982, his A.M., 1984, and his Ph.D., 1987, in Chemistry from Washington University in St. Louis with John Bleeke. Following a short postdoctoral stint at Washington University School of Medicine with Michael Welch, he joined Mallinckrodt in St. Louis performing research for the search for new magnetic resonance contrast agents. From 1999 to 2003, he held positions with AnorMed, British Columbia, and Cambridge-Major Laboratories, Wisconsin. In 2004 he returned to Mallinckrodt, where he has since been a Principal Scientist in contrast media research. His current research is focused on the synthesis of macrocyclic metal complexes for use in diagnostic radiology.

Leslie Patterson is a 4th year graduate student at the University of Notre Dame. She is co-advised by Marvin J. Miller and Paul Helquist. She received her B.S. in Biology-Chemistry from Manchester College in North Manchester, Indiana. She is currently working to prepare synthetic siderophore conjugates for therapeutic and diagnostic applications.

(±) *trans*-3,3'-(1,2-CYCLOPROPANEDIYL)BIS-2-(*E*)-PROPENOIC ACID, DIETHYL ESTER : TANDEM OXIDATION PROCEDURE (TOP) USING MnO₂ OXIDATION-STABILIZED PHOSPHORANE TRAPPING

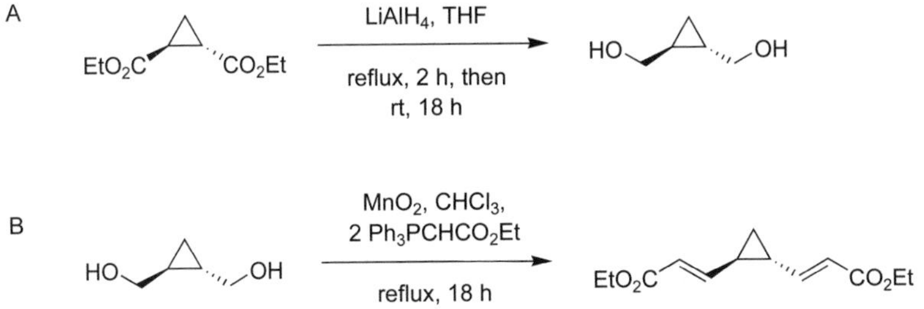

A

EtO₂C ◁ CO₂Et →[LiAlH₄, THF / reflux, 2 h, then / rt, 18 h] HO ◁ OH

B

HO ◁ OH →[MnO₂, CHCl₃, / 2 Ph₃PCHCO₂Et / reflux, 18 h] EtO₂C ◁ CO₂Et

Submitted by Richard J. K. Taylor, Leonie Campbell, and Graeme D. McAllister.[1]
Checked by Robert Webster and Mark Lautens.

1. Procedure

A. *trans 1,2-Bis-(hydroxymethyl)cyclopropane.* (Note 1) A 500-mL, three-necked flask equipped with a Teflon-coated magnetic stirrer, a reflux condenser, a pressure-equalizing dropping funnel, and a gas inlet adaptor attached to a nitrogen line is charged with tetrahydrofuran (100 mL) (Note 2) and lithium aluminium hydride (4.17 g, 109.7 mmol) (Note 3), and the mixture is cooled to 0 °C (bath temperature) in an ice-water bath. A solution of diethyl *trans*-1,2-cyclopropanedicarboxylate (13.59 g, 73.1 mmol) (Note 4) in THF (25 mL) is added over 1 h via the dropping funnel. After complete addition, the mixture is allowed to warm to room temperature and then is heated to reflux in an oil bath under nitrogen for 2 h. After allowing to cool, it is stirred at room temperature for 18 h. After being cooled in an ice-water bath, the mixture is treated cautiously with sat. aq. NH₄Cl solution (30 mL), whereupon the reaction mixture precipitates granular aluminium salts. Ethyl acetate (30 mL) is added to the mixture and a glass stir rod is used to break up the solid salts and facilitate stirring. The mixture is stirred for 5 h and then is filtered (Note 5). The insoluble salts are re-suspended in ethyl acetate (50 mL) and stirred for 2 h, then filtered again followed by further washing of the salts with ethyl acetate (2 x 50 mL). The combined filtrates are then dried over Na_2SO_4 (15 g), then are filtered and transferred to a 1-L round-bottomed flask, are evaporated (<40 °C, 8 mmHg) to give 7.45 g of a pale-yellow oil containing the title compound and its mono-acetate derivative in an approximate 3:1 ratio. The crude product is dissolved in methanol (100 mL) and sodium methoxide is added (100 mg) (Note 6). The solvolysis of the acetate can be monitored by TLC and is complete after 3 h at room temperature (Note 7). Neutralization of the mixture using IR-120 acidic ion exchange resin (2 g) (Note 8) followed by filtration and evaporation of the solvent gives 6.81 g (91%) of the crude diol (Note 9), which can be used without purification in the next step.

B. *(±)-Diethyl* trans-*(E,E)-cyclopropane-1,2-acrylate*. A flame-dried, 2-L, one-necked flask equipped with a Teflon-coated magnetic stirrer and a reflux condenser is charged with *trans-1,2-bis*(hydroxymethyl)cyclopropane (5.10 g, 50.0 mmol), manganese(IV) oxide (86.94 g, 1.00 mol) (Note 10), (carbethoxymethylene)triphenylphosphorane (41.81 g, 120.0 mmol) (Note 11), and chloroform (300 mL) (Note 12). The resulting suspension is stirred vigorously with heating in an oil bath to maintain gentle solvent reflux for 18 h. After allowing to cool to room temperature, the suspension is filtered through Celite (Note 13), and is washed with additional chloroform (4 x 100 mL). The filtrate is concentrated under reduced pressure (<40 °C, 8 mmHg), and then diethyl ether (200 mL) is added to the yellow, solid residue. This suspension is stirred at room temperature for 30 min, then the white precipitate (triphenylphosphine oxide) is removed by filtration and is washed with pentane (200 mL) (Note 14). Silica gel (40 g) (Note 15) is added to the combined organic filtrates, and the suspension is concentrated under reduced pressure (<40 °C, 8 mmHg) until the solid is free flowing. This mixture is loaded onto a silica gel column (Note 16) and is eluted with pentane/Et_2O, 3/1 (Note 17). The fractions containing the product (R_f = 0.41; pentane/Et_2O, 2/1) are combined and evaporated under reduced pressure to afford a pale-yellow solid. The solid is dissolved in hot pentane (50 mL) and the solution is allowed to cool to room temperature. The precipitated crystals are collected by suction filtration on a Büchner funnel, then are washed with ice-cold pentane. The crystals are transferred to a 50-mL, round-bottomed flask and are dried at room temperature under high vacuum to provide 7.24 g (61%) of pure diester (Notes 18, 19) as colorless needles. The mother liquors are concentrated under reduced pressure (<40 °C, 8 mmHg), and the yellow residue is purified by chromatography on silica gel

16

(Note 20), eluting with pentane/diethyl ether, 3/1. The fractions containing the product (R_f = 0.41; pentane/Et$_2$O, 2:1) are combined and evaporated under reduced pressure to afford 1.12 g (9%), total yield, 70%) of the diester.

2. Notes

1.   The preparation of *trans*-1,2-*bis*(hydroxymethyl)cyclopropane was based upon the method of Ashton, *et al.*[2]

2.   Tetrahydrofuran was purchased from Fisher Scientific UK, Ltd, and distilled from sodium/benzophenone prior to use.

3.   Lithium aluminium hydride was purchased from Aldrich Chemical Company, Inc. (reagent grade, 95%, powder, cat. no. 199877) and was used as received.

4.   Diethyl *trans*-1,2-cyclopropanedicarboxylate was purchased from Aldrich Chemical Company, Inc. (97%, cat. no. 157295) and was used as received.

5.   Filtration was performed through use of a 7-cm diameter Büchner funnel (fitted with a 6.5 cm-diameter Fisherbrand QL-100 filter paper) attached to a 500-mL Büchner flask.

6.   Sodium methoxide was purchased from Aldrich Chemical Company, Inc. (powder, 95% cat. no. 164992) and used as received.

7.   (2-(Hydroxymethyl)cyclopropyl)methyl acetate was present as a side product in amounts ranging from 20–30 mol% when using ethyl acetate as an extraction solvent. Extraction yields were improved using this protocol, presumably because of the increased lipophilicity of the side product. The yield decreased when additional water was used during the work-up due to the hydrophilicity of the desired product. It was possible to separate the two compounds using flash chromatography on silica gel with ethyl acetate as an eluent. The methanolysis of the side product[3] proceeded in quantitative yield using the conditions stated in the procedure. The physical properties of the side product were as follows: clear colorless oil; ^{1}H NMR (CDCl$_3$, 400 MHz) δ: 0.55 (ddd, 1 H, J = 5.6, 7.2 Hz), 1.07 (m, 2 H), 1.54 (s, 1 H, broad), 2.06 (s, 3 H), 3.47 (dddd, 2 H, J = 11.2, 17.6 Hz), 3.93 (ddd, 2 H, J = 7.2, 11.6, 17.8 Hz); ^{13}C NMR (CDCl$_3$, 100 MHz) δ: 8.6, 15.9, 20.1, 21.2, 66.2, 68.0, 171.4; IR (film): 3406, 1735 cm^{-1}; LRMS (EI) *m/z* (relative intensity): 145 (MH$^+$) (27%), 127 (100%), 113 (28%), 103

(57%), 84 (36%), 67 (40%), 55 (69%); HRMS [M + H$^+$] calcd for C$_7$H$_{13}$O$_3$: 145.0865. Found: 145.0866

8.   Amberlite® IR-120 (plus) ion-exchange resin was purchased from Aldrich Chemical Company, Inc. (cat. no. 216534) and used as received.

9.   The crude reduction product was of sufficient purity for direct use in the next step. It can also be purified by bulb-to-bulb distillation in a Kugelrohr apparatus under vacuum (5 mmHg), or alternatively using flash chromatography on silica gel using ethyl acetate as an eluent. The physical properties were as follows: bp 125–128 °C / 5 mmHg (Lit.[4] 90 °C / 1 mmHg); ^{1}H NMR (CDCl$_3$, 400 MHz) δ: 0.44 (t, 2 H, J = 6.4 Hz), 0.96–1.07 (m, 2 H), 3.07 (dd, 2 H, J = 8.4, 11.2 Hz), 3.70 (s, 2 H, broad), 3.82 (dd, 2 H, J = 4.8, 11.2 Hz); ^{13}C NMR (CDCl$_3$, 100 MHz) δ: 7.5, 20.2, 66.4; IR (film) cm^{-1}: 3330, 3002, 2873, 1427, 1065, 1024 cm^{-1}; LRMS (EI) m/z (relative intensity): 103 (MH$^+$) (10%), 88 (19%), 86 (75%), 84 (100%), 55 (29%); HRMS [M + H$^+$] calcd for C$_5$H$_{11}$O$_2$: 103.0759. Found: 103.0752.

10.   Activated manganese(IV) oxide was purchased from Aldrich Chemical Company, Inc. (cat. no. 217646; <5 μm, activated, 85%). Approximately 10 equivalents per oxidation are typically employed.[5]

11.   (Carbethoxymethylene)triphenylphosphorane was purchased from Aldrich Chemical Company, Inc. (cat. no. C510-6) and was used as received.

12.   The submitters purchased chloroform from Fisher Scientific UK, Ltd, while the checkers purchased chloroform from Caledon Laboratories Ltd. The solvent was distilled from potassium carbonate prior to use.

13.   Filtration was performed through a 2-cm depth of Celite® in a 14-cm diameter Büchner funnel (fitted with a 137.5 mm-diameter Fisherbrand QL-100 filter paper) attached to a 1-L Büchner flask.

14.   Pentane (Spectro grade) was purchased by the checkers from Caledon Laboratories Ltd. and used without further purification. The submitters used petroleum ether with a boiling range from 40–60 °C.

15.   60Å Ultrapure Silica gel (Silicycle) was used by the checkers, while the submitters used Fluka Silica Gel 60 (particle size, 35-70 μm).

16.   60Å Ultrapure Silica gel (Silicycle) (200 g) is placed in a 85 mm-diameter column (70 mm-depth of silica) and eluted with ~1.5-2.5 L of solvent.

17. Thin layer chromatography was carried out using Merck silica gel 60F$_{254}$ pre-coated aluminium foil plates with a thickness of 250 μm, and visualised with UV light (254 nm) and KMnO$_4$ solution.

18. The properties are as follows: mp (uncorr., pentane) 77–79 °C (Lit.[6] 73–74 °C); ^{1}H NMR (CDCl$_3$, 400 MHz) δ: 1.25 (t, 2 H, J = 6.4 Hz, overlapping signal), 1.27 (t, 6 H, J = 7.2 Hz), 1.78–1.84 (m, 2H), 4.17 (q, 4 H, J = 7.2 Hz), 5.89 (d, 2 H, J = 15.5 Hz), 6.45 (dd, 2 H, J = 9.4, 15.5 Hz); ^{13}C NMR (CDCl$_3$, 100 MHz) δ : 14.5, 17.7, 25.4, 60.5, 120.2, 149.7, 166.5; IR (film) cm^{-1}: 2979, 1712, 1645, 1144, 1047; LRMS (EI) m/z (relative intensity): 239 (MH$^+$) (35%), 193 (48%), 164 (44%), 146 (48%), 125 (50%), 119 (62%), 97 (60%), 91 (100%); HRMS [M$^+$] calcd. for C$_{13}$H$_{18}$O$_4$: 238.1205. Found: 238.1209; Anal. Calcd. for C$_{13}$H$_{18}$O$_4$: C, 65.53; H, 7.61; Found: C, 65.59; H,7.94;.

19. The submitters were able to obtain a second crop of crystals, by evaporation of the mother liquor. The solid was dissolved in 20 mL of hot pentane and the solution was cooled to room temperature. Filtration of the crystals and drying as above gave the second crop of diester.

20. 60Å Ultrapure Silica gel (Silicycle) (120 g) was placed in a 50 mm-diameter column (80 mm-depth of silica) and eluted with ~500–600 mL of solvent.

Safety and Waste Disposal Information

All hazardous materials should be handled and disposed of in accordance with "Prudent Practices in the Laboratory"; National Academy Press, Washington, DC, 1995.

3. Discussion

The importance of aldehydes as electrophilic synthetic building blocks cannot be overestimated, but their reactivity can cause difficulties. Aldehydes often undergo aerial oxidation, acid-promoted oligomerisation/polymerisation,[7] or decompose by other means; they can also be difficult to isolate due to their volatility (sometimes with lachrymatory effects) or toxicity.[8] These difficulties can be circumvented by preparing the aldehyde *in situ* and then adding the next reagent to accomplish the carbonyl addition in a one-pot oxidation-trapping process. This sequential approach was first developed by Ireland and Norbeck,[9] who employed a Swern oxidation followed by the addition of stabilized phosphorane or Grignard reagent to react with the aldehyde formed *in situ*. Similar approaches have been reported more recently by the Ley (TPAP oxidation-Wittig trapping)[10] and Bressette (PCC oxidation-Wittig trapping)[11] groups.

The first truly tandem oxidation-trapping sequence (Dess-Martin periodinane oxidation in the presence of a stabilised phosphorane), was reported by Huang in 1987,[12] with the scope of the process being expanded by Barrett *et al.* in 1997.[13] More recently, Matsuda's group have reported the use of barium permanganate for similar one-pot reactions.[14] Crich and Mo subsequently used IBX as an *in situ* oxidant with a stabilized Wittig reagent for the preparation of several 2'-deoxynucleosides[15] and more recently, Maiti and Yadav demonstrated that this IBX procedure can be employed with a range of activated and non-activated alcohols.[16]

The York group have developed a range of one-pot "tandem oxidation processes" (TOP) utilizing activated manganese(IV) oxide in combination with a variety of nucleophilic trapping agents (including phosphoranes, phosphonates, sulfuranes, amines and alcohols), and many of these procedures have found applications in the wider chemical community.[5] The low toxicity, low cost and ease of handling of activated MnO_2, combined with its commercial availability and the absence of hazardous additives/by-products, are attractive. The heterogeneous nature of the oxidant means that the work-up consists of simple filtration followed by evaporation of solvent. Although an excess of oxidant is usually required (Note 2), the spent oxidant can be reactivated/recycled.[5,17]

The manganese(IV) oxide TOP-Wittig methodology has been applied to a range of "activated" alcohols (allylic, propargylic and benzylic examples) as shown in Table 1, with the successful homologation of *Z*-allylic alcohols (*e.g.* entries 3-6) with complete retention of the pre-existing alkene geometry particularly noteworthy.[18] More surprisingly, however, is that MnO_2 (in the presence of stabilized phosphoranes) has been shown to be an efficient oxidant of "semi-" and unactivated alcohols to afford unsaturated esters in good yield (Table 2).[19]

The procedure described herein is a one-pot process for the preparation of (±)-diethyl *trans* (*E,E*)-cyclopropane-1,2-acrylate from *trans*-cyclopropane-1,2-methanol, using a MnO_2 oxidation-stabilised phosphorane olefination sequence. The process is straightforward and herein is carried

20

out on a 50-mmol scale (which could presumably be scaled up further).

Table 1 – The One-Pot Procedure using Activated Alcohols[17]

Entry	Substrate	Product	Yield (%)
1	Br⌒⌒OH	Br⌒⌒CO₂Et (CO_2Et)	81 (E,E:E,Z:Z,E:Z,Z = 18:6:3:1)
2	Cl / OH	Cl / CO_2Et	51 (E,Z = 9:1)
3	Et / OH	Et / CO_2Me	81 (E,Z;Z,Z = 9:1)
4	Pr / OH	Pr / CO_2Et	81 (E,Z;Z,Z = 9:1)
5	Br / OH	Br / $CO_2{}^tBu$	90 (E,Z;Z,Z = 5:1)
6	Br / OH	Br / O	73 (E,Z;Z,Z = 8:1)
7	≡ / OH	≡ / CO_2Et	82 (E,Z = 4:1)
8	OMe, CH₂OH, OMe	OMe, CO_2Et, OMe	80 (>98%E)
9	HO / Br, Br / OH	EtO_2C / Br, Br / CO_2Et	84 (>98% E,E,E)

Entry	Substrate	Product	Yield (%)
1			86 (>99% *E*)
2			74 (*E,Z* = 3:1)
3			58 (>98% *E*)
4			66 (*E,Z* = 6:1)
5			70 (>95% *E*)
6	C₉H₁₉—OH	C₉H₁₉—CO₂Et	80 (>95% *E*)
7	C₅H₁₁—OH	C₅H₁₁—CO₂Et	70 (>95% *E*)
8	Ph—OH	Ph—CO₂Et	86 (>95% *E*)
9			51 (>99% *E*)
10	C₉H₁₉—OH	C₉H₁₉—CON(OMe)Me	57 (*E:Z* = 11.5:1)

Table 3 – Optimization of the Synthesis

Entry	Solvent / time	Yield (%)
1	CHCl$_3$ / 18 h	72–75
2	THF / 18 h	45
3	Toluene / 18 h	35[a]

[a] The low yield may be attributed to volatility of the intermediate aldehyde

Table 3 shows the results of the optimisation of this reaction with regards the choice of solvent. The best conditions involve 10 eq. of MnO$_2$ and 1.2 eq. of phosphorane per oxidation, with chloroform as the solvent of choice.

Diethyl *trans* (*E,E*)-cyclopropane-1,2-acrylate has been reported several times in the literature, in each case by an oxidation-olefination sequence. McDonald, Verbicky and Zercher[6] reported the oxidation (by TPAP/NMO) of (*E*)-cyclopropane-1,2-methanol, followed by Horner-Wadsworth-Emmons olefination of the isolated dialdehyde. In 1996, Barrett *et al.*[20] reported the synthesis by a one-pot Dess-Martin oxidation-Wittig sequence in their approach to the transfer protein inhibitor U-106305.

In summary, the procedure presented here is an illustration of the one-pot TOP-Wittig trapping route from simple alcohols to α,β-unsaturated carbonyl compounds using MnO$_2$ as the oxidant. This procedure is operationally straightforward, and results in a reduced number of steps, giving significant time-cost benefits.

1. Department of Chemistry, University of York, Heslington, York YO10 5DD, UK.

2. Ashton, W. T.; Meurer, L. C.; Cantone, C. L.; Field, A. K.; Hannah, J.; Karkas, J. D.; Liou, R.; Patel, G. F.; Perry, H. C.; Wagner, A. F.; Walton, E.; Tolman, R. L. *J. Med. Chem.* **1988**, *31*, 2304-2315.
3. G. Zemplén, *Ber. Dtsch. Chem. Ges.*; **1927**, *60*, 1555.
4. Jakovac, I. J.; Goodbrand, H. B.; Lok, K. P.; Jones, J. B. *J. Am. Chem. Soc.* **1982**, *104*, 4659-4665.
5. Taylor, R. J. K.; Reid, M.; Foot, J.; Raw, S. A. *Acc. Chem. Res.* **2005**, *38*, 851-869, and references therein.
6. McDonald, W. S.; Verbicky, C. A.; Zercher, C. K. *J. Org. Chem.* **1997**, *62*, 1215-1222; see also, Charette, A. B.; Lebel, H. *J. Am. Chem. Soc.* **1996**, *118*, 10327-10328.
7. Buehler, C. A.; Pearson, D. E. *Survey of Organic Synthesis* Wiley Interscience, New York, 1970, 542-570.
8. (a) Gold, H. *Methoden Der Organischem Chemie* (Houbenweyl) ed. Muller, E.; Verlag, G. T.; Stuttgart, 1976, 7, 1930-1931; (b) Bayer, O. *Methoden Der Organischem Chemie* (Houbenweyl) ed. Muller, E.; Verlag, G. T.; Stuttgart, 1954, 7, 413.
9. Ireland, R. E.; Norbeck, D. W. *J. Org. Chem.* **1985**, *50*, 2198-2200.
10. MacCoss, R. N.; Balskus, E. P.; Ley, S. V. *Tetrahedron Lett.* **2003**, *44*, 7779-7781.
11. Bressette, A. R.; Glover, L. C., III *Synlett* **2004**, 738-740.
12. Huang, C. C. *J. Labelled Comp. Radiopharm.* **1987**, *24*, 675-681.
13. Barrett, A. G. M.; Hamprecht, D.; Ohkubo, M. *J. Org. Chem.* **1997**, *62*, 9376-9378.
14. Shuto, S.; Niizuma, S.; Matsuda, A. *J. Org. Chem.* **1998**, *63*, 4489-4493.
15. Crich, D.; Mo, X. *Synlett* **1999**, 67-68.
16. Maiti, A.; Yadav, J. S. *Synth. Commun.* **2001**, *31*, 1499-1506.
17. Carus Chemical Company (Peru, IL 61354, USA) offers MnO_2 recycling.
18. (a) Wei, X.; Taylor, R. J. K. *Tetrahedron Lett.* **1998**, *39*, 3815-3818; (b) Wei, X.; Taylor, R. J. K. *J. Org. Chem.* **2000**, *65*, 616-620.
19. Blackburn, L.; Wei, X.; Taylor, R. J. K. *Chem. Commun.* **1999**, 1337-1338.
20. (a) Barrett, A. G. M.; Hamprecht, D.; White, A. J. P.; Williams, D. J. *J. Am. Chem. Soc.* **1996**, *118*, 7863-7864; (b) Barrett, A. G. M.; Hamprecht, D.; White, A. J. P.; Williams, D. J. *J. Am. Chem. Soc.* **1997**, *119*, 8608-8615.

Appendix
Chemical Abstracts Nomenclature; (Registry Number)

Diethyl *trans*-1,2-cyclopropanedicarboxylate; (3999-55-1)

Bis-(Hydroxymethyl)-cyclopropane; (2345-75-7)

(±)-Diethyl (*E,E,E*)-cyclopropane-1,2-acrylate

Manganese(IV) oxide; (1313-13-9)

(Carbethoxymethylene)triphenylphosphorane; (1099-45-2)

2-Propenoic acid, 3,3'-(1,2-cyclopropanediyl)bis-, diethyl ester, [1α(E),2β(E)]-; (58273-88-4)

Richard Taylor obtained BSc and PhD (Dr. D. Neville Jones) from the University of Sheffield. Postdoctoral periods with Dr. Ian Harrison (Syntex, California) and Professor Franz Sondheimer (University College London) were followed by lectureships at the Open University and then UEA, Norwich. In 1993 he moved to a Chair at the University of York. Taylor's research interests centre on the synthesis of bioactive natural products and the development of new synthetic methodology. His awards include the Royal Society of Chemistry's Pedlar Lectureship (2007). Taylor is the immediate past-President of the RSC Organic Division and an Editor of Tetrahedron.

Leonie Campbell (nee Blackburn) was born in Leigh, Lancashire in 1975. Her undergraduate studies were undertaken at Oxford University where she received a M.Sc. (Hons) in 1998. She then moved to the University of York to join the research group of Professor Taylor for her D. Phil. working on tandem in situ oxidation-homologation procedures and their utilisation in natural product syntheses. She is currently working at AstraZeneca (Alderley Park) as a Senior Research Scientist in Medicinal Chemistry.

Graeme McAllister was born in Bellshill, Scotland in 1974. His undergraduate studies were carried out at the University of Glasgow, where he received his B.Sc. (Hons.) in 1996. He stayed at the University of Glasgow to join the research group of Richard C. Hartley, conducting research into the total synthesis of chalcomoracin, and the identification/isolation of a Diels-Alderase enzyme. In 1999, he joined the research group of Professor Taylor at the University of York, initially as a postdoctoral researcher and latterly as experimental officer, where he has carried out research, primarily into novel applications of the Ramberg-Bäcklund reaction, as well as developing new one-pot tandem methodology processes and investigating natural product syntheses.

Rob Webster was born in 1981 in Saskatoon, Saskatchewan, Canada. He obtained his BSc. from the University of Saskatchewan in 2004. During this time he had the opportunity to work in the research labs of Prof. Marek Majewski; and at both Merck-Frosst Canada Ltd. in process research under the supervision of Dr. Francis Gosselin and at Boehringer-Ingelheim Canada Ltd. in combinatorial chemistry working under Dr. Jean Rancourt and Sophie Goulet. He began his Ph.D. work in 2004 at the University of Toronto in the lab of Prof. Mark Lautens, where he is currently, in the area of natural product total synthesis and asymmetric catalysis.

FACILE SYNTHESIS OF 2-ETHYL-3-QUINOLINECARBOXYLIC ACID HYDROCHLORIDE

Submitted by Brian R. McNaughton and Benjamin L. Miller.[1]
Checked by Daniel P. Parker and John A. Ragan.

1. Procedure

A. *2-Ethyl-3-Quinolinecarboxylic Acid Methyl Ester.* A flame-dried, 1000-mL, three-necked, round-bottomed flask is equipped with a Teflon-coated magnetic stirbar, internal thermometer, glass stopper, and a reflux condenser fitted with a nitrogen inlet. The flask is charged with 10 grams of 3Å molecular sieves (Note 1), 2-nitrobenzaldehyde (10.0 g, 66.2 mmol), methyl propionylacetate (8.30 mL, 66.2 mmol) (Note 2), and zinc (II) chloride (18.0 g, 132 mmol) (Note 3). To the flask is added 175 mL of anhydrous methanol (Note 4). The flask is flushed with a stream of nitrogen, and then, with stirring, heated to an internal temperature of 67 °C for 1 h with a heating mantle. Tin(II) chloride (62.7 g, 331 mmol) is then slowly added to the flask over five min in equal portions (Note 5). The flask is again flushed with a stream of nitrogen and the reaction mixture is stirred at 67 °C for 12 h. The solution is then allowed to cool to room temperature. To a 500-mL beaker is added potassium carbonate (45.7 g, 331 mmol), which is dissolved completely in 300 mL water. The reaction solution is made alkaline (pH approximately 8, as determined by pH paper) through the slow addition of the potassium carbonate solution (ca. 225 mL) while

Org. Synth. **2008**, *85*, 27-33
Published on the Web 9/18/2007

stirring (Note 6), resulting in a light orange slurry. To this slurry is added 200 mL of diethyl ether, and the complete mixture is then filtered through a Büchner funnel (Whatman filter paper, 125 mm diameter) (Note 7). An additional 75 mL of ether is added to rinse the reaction vessel, and subsequently passed through the funnel. After all the aqueous / ethereal solution has passed through the funnel, it is washed with ethyl ether (2 x 100 mL) (Note 8). The filter cake is then transferred to a 500-mL Erlenmeyer flask and slurried in 300 mL of ethyl acetate for 30 min before being filtered through a Büchner funnel (Whatman filter paper, 125 mm diameter). The filtrates are combined, transferred to a 2000-mL separatory funnel, where the aqueous layer is removed and the organic layer is washed with brine (100 mL). The organic phase is dried by the addition of magnesium sulfate (15 g), then is filtered and and the filter cake is rinsed with ethyl acetate (100 mL). The filtrate is concentrated by rotary evaporation (40 °C, 80 mmHg) in a 1000-mL round bottom flask to provide the methyl ester **1** as a yellow oil.

B. *2-Ethyl-3-Quinolinecarboxylic Acid.*   To the 1000-mL round bottom flask containing **1** is added a mixture of 100 mL of THF and 75 mL of 2 M LiOH (aq) and a football-shaped magnetic stir bar. The flask is capped with a rubber septum, vented with a needle, and stirred at room temperature for 12 h. The THF is then removed by rotary evaporation (Note 9) to provide an aqueous solution of the lithium carboxylate.

C. *2-Ethyl-3-Quinolinecarboxylic Acid Hydrochloride.*   The above solution is made acidic, (pH of 1 as determined by pH paper) through the addition of 50 mL of 12 N hydrochloric acid which results in a dense, yellow slurry which is gently stirred for 5 min. The sususpension is chilled to 0 °C in an ice bath (internal temperature). A fritted glass funnel (60 mm) is fitted with filter paper, and the slurry is filtered with suction. The filter cake is washed with chilled (0 °C) HPLC grade ether (2 x 20 mL) (Note 10). The solids are transferred to a 100-mL, round-bottomed flask and placed on a vacuum line (22 °C, 0.2 mmHg) for 24 h to remove traces of water. The title compound is isolated as a light yellow solid 11.2–11.4 g (71–72%) (Notes 11, 12).

2. Notes

1.   The molecular sieves were washed thoroughly with methylene chloride to remove particulates and were then dried in a 200 °C oven for at least 24 h.

2. 2-Nitrobenzaldehyde (98%) and methyl propionylacetate (98%) were purchased from Aldrich Chemical Company, Inc. and were used as received.

3. Zinc(II) chloride (97%) was purchased from Strem chemicals and was used as received. Zinc(II) chloride is hygroscopic; best results are observed when using fresh material, or material stored under desiccation. The use of stoichiometric or sub-stoichiometric reagent resulted in a notable decrease in desired product formation.

4. Methanol (anhydrous, Sure-Seal) was obtained from Sigma-Aldrich.

5. Tin (II) chloride (98%) was purchased from Strem chemicals and was used as received. Complete addition of tin(II) chloride in a single step results in an exotherm of approximately 7 °C, mandating the described slow addition. It was found that 5 roughly equivalent portions (ca. 12.5 g each) of tin(II) chloride were needed to obtain high yields of the desired product.

6. Gas is evolved during this addition; care should be taken.

7. The large amounts of inorganic salts and residue from sieves produced a solid layer, thereby slowing the rate of filtration. It was found that gentle agitation of this layer by means of a spatula provided a quicker filtration.

8. Similar to Note 6, it was found that gentle agitation of the inorganic salt layer during the passing of diethyl ether provided the best extraction.

9. Rotary evaporation was conducted under a reduced pressure of 20 mmHg and heated in a 40 °C water bath.

10. HPLC grade diethyl ether (anhydrous) was obtained from J. T. Baker.

11. The product exhibits the following physicochemical properties: mp 190–192 °C; IR (thin film) 3375, 2676, 2362, 1643, 1459, 1378, 1319, 1287, 1265, 1230 cm^{-1}; ^{1}H NMR (400 MHz, D$_2$O) δ: 1.35 (t, J = 7.7 Hz, 3 H), 3.41 (q, J = 7.6 Hz, 2 H), 7.83 (t, J = 7.5 Hz, 1 H), 8.02–8.10 (m, 2 H), 8.16 (d, J = 8.3 Hz, 1 H), 9.33 (s, 1 H); ^{13}C NMR (100 MHz, D$_2$O) δ: 13.6, 27.0, 119.5, 125.7, 126.6, 130.0, 130.2, 136.9, 138.3, 149.0, 162.5, 166.9; MS (Pos. ES): 202 (M+H, 100). Product purity was established by analytical HPLC: t$_R$, 2.98 min (70% (0.1% TFA in water) / 30% (0.1% TFA in acetonitrile) for 8 min followed by a gradient to 100% (0.1% TFA in acetonitrile) at 11.0 min, (Waters Symmetry C18 column (150 mm x 4.6

mm) 220 and 254 nm). At both wavelengths, only a single peak was observed, suggesting >99% product purity.

12. Elemental analysis was problematic for establishing purity. The anhydrous formula ($C_{12}H_{12}ClNO_2$) requires: C, 60.64; H, 5.09; N, 5.89. The checkers found: C, 52.84; H, 5.52; N, 5.14. The bis-hydrate ($C_{12}H_{16}ClNO_4$) requires: C, 52.66; H, 5.89; N, 5.12. Thus, the analysis is consistent with the bis-hydrate. However, Karl-Fischer analysis indicated 10.8% water by weight, whereas the bis-hydrate requires 13.2% water. Based on this data the product is likely between a sesqui- (1.5) and bis- (2.0) hydrate.

Safety and Waste Disposal Information

All hazardous materials should be handled and disposed of in accordance with "Prudent Practices in the Laboratory"; National Academy Press; Washington, DC, 1995.

3. Discussion

The quinoline skeleton is an important motif, and has been thought of as a "privileged" structure due to its presence in a broad array of natural products and pharmaceuticals.[2] Historically, quinoline-containing small molecules such as quinine and chloroquine have most notably been recognized as anti-malarial drugs.[3] However, more recently a variety of small molecules containing the quinoline substructure have been identified targeting a wide range of conditions.[4] This variant of the Friedländer synthesis[5] described above may be used for the generation of 2- and 3-substituted quinolines, from readily available starting materials. Unlike a variety of quinoline syntheses that call for the use of harsh acidic[6] or basic[7] reactants, this procedure provides the desired quinoline skeleton through the use of mild Lewis acidic and reducing conditions, and has been shown to proceed in the presence of a variety of functional groups.[8] In addition, the carboxylic acid moiety of the title compound allows for easy coupling through solid phase synthesis,[9] as well as further synthetic elaboration. The utility of this reaction is demonstrated in the table below.

30

Table 1. One-pot quinoline synthesis using a variety of substituted aryl aldehydes and symmetric ketones.

Aryl aldehyde	Ketone	Product	% Yield
			89
			82
			94
			90
			98
			90
			70

1. Department of Dermatology, University of Rochester Medical Center, 601 Elmwood Ave., Box 697, Rochester, NY 14642, e-mail: Benjamin_Miller@urmc.rochester.edu.
2. (a) Chen, I.-S.; Chen, H.-F.; Cheng, M.-J.; Chang, Y.-L.; Teng, C.-M.; Tsutomu, I.; Chen, J.-J.; Tsai, I.-L *J. Nat. Prod.*, **2001**, *64*, 1143-1147.
3. (a) Ginsburg, H.; Krugliak, M. *Biochem. Pharm.*, **1992**, *43*, 63-70. (b) Osiadacz, J.; Majika, J.; Czarnecki, K.; Peczynska-Czoch, W.; Zakrzewska-Czerwinskia, J.; Kaczmarek, L; Sokalski, W. A. *Bioorg. & Med. Chem.*, **2000**, *8*, 937-943.
4. (a) Wang, C.-G.; Langer, T.; Kamath, P.G.; Gu, Z.-Q; Skolnick, P.; Fryer, R.I *J. Med. Chem.*, **1995**, *38*, 950-957. (b); Benard, C.; Zouhiri, F.; Normand-Bayle, M.; Danet, M.; Desmaele, D.; Leh, H.; Mouscadet, J.-F.; Mbemba, G.; Thomas, C.-M.; Bonnenfant, S.; LeBret, M.; d'Angelo, J. *Bioorg. & Med. Chem. Lett.*, **2004**, *14*, 2473-2476. (c) Jain, R.; Vaitilingam, B.; Nayyar, A.; Palde, P.B. *Bioorg. & Med. Chem. Lett.*, **2003**, *13*, 1051-1054.
5. Review articles on the Friedlander quinoline synthesis (a) Manske, R. H. F. *Chem. Rev.* **1942**, *30*, 113-144. (b) Cheng, C. C.; Yan, S. J. *Org. React.* **1982**, *28*, 37-201.
6. (a) Born, J. L. *J. Org. Chem.*, **1972**, *37*, 3952-3953. (b) Wahren, M. *Tetrahedron*, **1964**, *20*, 2773-2780.
7. (a) Riesgo, E. C.; Jin, X.; Thummel, R. P. *J. Org. Chem.* **1996**, *61*, 3017-3022. (b) Gainor, J. A.; Weinreb, *J. Org. Chem.*, **1982**, *47*, 2833-2837.
8. McNaughton, B. R.; Miller, B. L. *Org. Lett.*, **2003**, *5*, 4257-4259.
9. Lorsbach, B. A.; Kurth, M. J. *Chem. Rev.*, **1999**, *99*, 1549-1581.

Appendix
Chemical Abstracts Nomenclature; (Registry Number)

2-Nitrobenzaldehyde; (552-89-6)
Methyl propionylacetate: Pentanoic acid, 3-oxo-, methyl ester; (30414-53-0)
Zinc (II) chloride; (7646-85-7)
Tin (II) chloride; (7772-99-8)
2-Ethyl-3-Quinolinecarboxylic acid, methyl ester (119449-61-5)
2-Ethyl-3-Quinolinecarboxylic acid (888069-31-6)
2-Ethyl-3-Quinolinecarboxylic acid hydrochloride (888014-11-7)

32

Benjamin L. Miller was born in 1967 in Dayton, Ohio. He carried out his undergraduate studies at Miami University (Ohio), receiving degrees in Chemistry, Mathematics, and German in 1988. From there Ben moved to Stanford University, earning a Ph. D. in Chemistry under the direction of Paul Wender, and then to Harvard University, as an NIH postdoctoral fellow in the laboratory of Stuart Schreiber. Ben joined the faculty of the University of Rochester in 1996, where he is currently Associate Professor of Dermatology, Biochemistry and Biophysics, and Biomedical Engineering. Research in his group includes dynamic combinatorial chemistry, molecular recognition, and biomedical nanotechnology, with a particular focus on biosensing.

Brian R. McNaughton was born in Alexandria, Virginia. He attended Indiana University of Pennsylvania, graduating with a Bachelors of Science degree in 2001. He came to the University of Rochester in the Fall of 2002 and began graduate studies in Chemistry. His thesis research under the direction of Professor Benjamin L. Miller focused on the synthesis and evaluation of static and dynamic combinatorial libraries, as well as the development of novel methods in dynamic combinatorial chemistry. In 2004, he received a Master of Science degree from the University of Rochester. Following completion of his Ph.D. in 2007, he began postdoctoral research under the direction of Professor David R. Liu at Harvard University.

Daniel Parker was born in 1976 in St. Paul, Minnesota. He earned a B.S. degree in chemistry at the State University of West Georgia. He then joined Prof. Tarek Sammakia's lab at the University of Colorado, receiving his M.S. degree in 2004. He currently works as a process chemist in the Chemical Research and Development group at Pfizer in Groton, Connecticut.

PREPARATION OF A NON-SYMMETRICAL IMIDAZOLIUM SALT: 1-ADAMANTYL-3-MESITYL-4,5-DIMETHYLIMIDAZOLIUM TETRAFLUOROBORATE

Submitted by Alois Fürstner, Manuel Alcarazo, Vincent César, and Helga Krause.[1]

Checked by Chad Hopkins and Peter Wipf.[2]

1. Procedure

A. 3-(Mesitylamino)butan-2-one. A 250-mL, single-necked round-bottomed flask equipped with a Teflon-coated magnetic stir bar is charged with 3-hydroxybutan-2-one (13.2 g, 150 mmol, 1.5 equiv), mesitylamine (13.52 g, 100 mmol, 1.0 equiv) (Note 1), toluene (150 mL) (Notes 2 and 3) and a catalytic amount of concentrated HCl (approximately 0.05 mL). The flask is then fitted with a Dean-Stark trap topped with a reflux condenser fitted with a nitrogen inlet. The solution is heated at reflux in a silicon oil

Org. Synth. **2008**, *85*, 34-44
Published on the Web 9/21/2007

bath for 3 h and the water generated is collected in the Dean-Stark trap (ca. 1.5 mL). The resulting yellow solution is allowed to cool to ambient temperature and the solvent is removed by rotary evaporation (25 °C, 8 mmHg). The oily residue is transferred to a 50-mL, one-necked, round bottomed flask and is purified by short path distillation (Note 4) under high vacuum (3 x 10^{-2} mmHg) to give 16.72 g (81%) of 3-(mesitylamino)butan-2-one as a pale yellow oil, bp 94–98 °C (Notes 5, 6, and 7).

B. N-Mesityl-N-(3-oxobutan-2-yl)formamide. An oven-dried, 100-mL, three-necked, round-bottomed flask equipped with two rubber septa, a nitrogen inlet and a Teflon-coated stir bar is charged with 3-(mesitylamino)butan-2-one (16.4 g, 80 mmol, 1.0 equiv) and anhydrous THF (25 mL) (Notes 3 and 8). Acetic formic anhydride (10.6 g, 120 mmol, 1.5 equiv) (Note 9) is added in one portion to this solution, and the resulting mixture is stirred at ambient temperature for 22.5 h (Note 10). The resulting brown mixture is transferred to a 100-mL, one-necked flask and is concentrated by rotary evaporation (25 °C, 8 mmHg) to a volume of approximately 20 mL before it is added on top of a wet-packed silica gel column (diameter: 7 cm, height: 22 cm, pre-treated with hexanes/EtOAc, 4:1) (Note 11). The product is eluted with hexanes/EtOAc, 3:1) to give 14.34 g (77%) of N-mesityl-N-(3-oxobutan-2-yl)formamide as a white, crystalline solid (Notes 12 and 13).

C. 1-Adamantyl-3-mesityl-4,5-dimethylimidazolium tetrafluoroborate. A 500-mL, three-necked, round-bottomed flask equipped with a pressure-equalizing dropping funnel, a nitrogen inlet, an internal thermometer, and a Teflon-coated stir bar (length: 4 cm) is charged with N-mesityl-N-(3-oxobutan-2-yl)formamide (13.7 g, 59 mmol, 1.0 equiv) and acetic anhydride (59 mL, 622 mmol, 10.6 equiv) (Note 1). The flask is immersed in an ice/water bath before aqueous HBF$_4$ (48% w/w, 7.53 mL, 59.3 mmol, 1.01 equiv) (Note 1) is added dropwise through the addition funnel over a period of approximately 5 min to the stirred reaction mixture (Note 14). Once the addition is complete, the resulting brown solution is stirred at ambient temperature for 3 h. Anhydrous Et$_2$O (400 mL) (Notes 3 and 8) is then added, leading to the separation of 5-acetoxy-3-mesityl-4,5-dimethyl-oxazolinium tetrafluoroborate as a pale brown oil. The supernatant organic layer is carefully decanted and the residual oil is rinsed twice with Et$_2$O (2 x 200 mL). The remaining viscous oil is then suspended in toluene (100 mL) (Note

2). 1-Adamantylamine (11.54 g, 76.3 mmol, 1.3 equiv) (Note 1) is added and the resulting mixture is vigorously stirred at ambient temperature for 3 h. Addition of anhydrous Et_2O (200 mL) (Notes 3 and 8) again causes the formation of two phases. The top Et_2O layer is discarded and the viscous bottom phase is carefully rinsed with anhydrous Et_2O (2 x 200 mL) to remove excess adamantylamine. The resulting pale yellow semi-solid is suspended in toluene (150 mL) (Note 2) and then acetic anhydride (16.6 mL, 176.1 mmol, 3.0 equiv) (Note 1) and aq. HBF_4 (48% w/w, 2.24 mL, 17.6 mmol, 0.3 equiv) (Note 1) are introduced. The dropping funnel and internal thermometer are then replaced by a reflux condenser and a glass stopper and the mixture is stirred and heated to reflux in a silicon oil bath for 22 h (Note 15). After reaching ambient temperature, the mixture is transferred to a one-necked flask, the solvent is removed by rotary evaporation (25 °C, 8 mmHg), the resulting brown syrup is triturated with anhydrous Et_2O (3 x 75 mL or until the ether layer is colorless) (Notes 3 and 8), and the ether layers are discarded. The resulting white solid product is collected and dried under vacuum (0.2 mmHg) to give 14.87 g (58%) of pure 1-adamantyl-3-mesityl-4,5-dimethylimidazolium tetrafluoroborate (Notes 16, 17, and 18). Crystals for elemental analysis are obtained by dissolving a sample of the product in a minimum amount of CH_2Cl_2 (Note 2) and careful layering of the resulting solution with anhydrous Et_2O (Et_2O/CH_2Cl_2, 7:1) (Notes 3 and 8). The resulting solution is kept at ambient temperature for 3 d, resulting in the precipitation of 1-adamantyl-3-mesityl-4,5-dimethyl-imidazolium tetrafluoroborate as colorless prisms which are collected and dried under vacuum (0.2 mmHg).

2. Notes

1. 3-Hydroxybutan-2-one (97%), mesitylamine (98%), aq. HBF_4 (Alfa Aesar), 1-adamantylamine (97%, Aldrich Chemical Company, Inc.), and acetic anhydride (ACS certified, Fisher) were used as received.
2. Although the submitters took great care to ensure anhydrous conditions (Note 3), the checkers found this precaution unnecessary.
3. The submitters utilized purified solvents by distillation over the indicated drying agents prior to use and transferred them under Ar: THF, Et_2O (Mg-anthracene), hexanes, toluene (Na/K), CH_2Cl_2 (CaH_2). Reagent grade EtOAc (Fluka) was used as received. All reaction flasks were oven dried.

36

4.   The dimensions of the short path still are as follows: Jacketed, single piece construction with inlet for vacuum/inert gas, 10/18 thermometer joint at the top of the head, 14/20 joints for distillation and collection flasks, approx. 105 mm width (length of condenser) x 105 mm height (head).

5.   Working at 50% scale, the checkers obtained 8.93 g (87%).

6.   This product slowly decomposes on standing at ambient temperature and should be used in the next step without undue delay. It can be stored at –20 °C under Ar for several days.

7.   The product has the following physicochemical properties: ^{1}H NMR (600 MHz, CDCl$_3$) δ: 1.23 (d, J = 7.2 Hz, 3 H), 2.20 (s, 3 H), 2.22 (s, 3 H), 2.26 (s, 6 H), 3.96 (br s, 1 H), 4.04 (q, J = 7.2 Hz, 1 H), 6.80 (s, 2 H); ^{13}C NMR (150 MHz, CDCl$_3$) δ: 18.2, 18.8, 20.4, 27.4, 61.2, 129.0, 129.5, 130.8, 141.5, 209.9; IR (film) 3373, 2971, 2919, 1717, 1594, 1485, 1445, 1356, 1235, 1158, 1127, 1012, 855 cm^{-1}; HRMS (EI) m/z calcd for C$_{13}$H$_{19}$NO 205.1467, found 205.1469. Anal. Calcd for C$_{13}$H$_{19}$NO: C, 76.06; H, 9.33; N, 6.82. Found: C, 76.26; H, 9.51; N, 6.78.

8.   Anhydrous THF (99.9%, Acros), anhydrous Et$_2$O (ACS certified, Fisher) toluene (ACS certified, Fisher), hexanes (ACS certified, Fisher), ethyl acetate (ACS certified, Fisher), dichloromethane (ACS certified, Fisher), chloroform (stabilized with ca. 1% ethanol, Acros) were used as received.

9.   Acetic formic anhydride is conveniently prepared according to: Krimen, L. I. *Org. Synth.* **1970**, *50*, 1–3.

10.   Although TLC monitoring occasionally shows small amounts of unreacted starting material after 24 h reaction time, prolonged stirring does not improve the yields. TLC monitoring is performed with precoated plates with silica gel 60 F$_{254}$ purchased from E. Merck, Darmstadt using hexanes/EtOAc, 3:1 as the eluent; R$_f$ (substrate) = 0.73; R$_f$ (product) = 0.36.

11.   Silica gel 32-63 D (60 Å) (MP Biomedicals, Germany, purchased through Bodman Industries, Aston, PA) was used.

12.   Working at 50% scale, the checkers obtained 8.06 g (80%).

13.   The product has the following physicochemical properties: mp 100–101 °C; ^{1}H NMR (600 MHz, CDCl$_3$) δ: 0.98 (d, J = 7.2 Hz, 3 H), 2.14 (s, 3 H), 2.30 (s, 3 H), 2.38 (s, 3 H), 2.40 (s, 3 H), 4.52 (q, J = 7.2 Hz, 1 H), 6.91 (s, 1 H), 6.96 (s, 1 H), 7.99 (s, 1 H); ^{13}C NMR (150 MHz, CDCl$_3$) δ: 13.5, 18.4, 18.7, 20.9, 27.7, 59.4, 59.4, 129.0, 129.7, 133.2, 137.8, 138.7, 138.8, 163.6, 205.3; IR (KBr) 2987, 2916, 1720, 1653, 1484, 1452, 1318, 1296, 1252, 1166, 1029 cm^{-1}; HRMS (ESI+) m/z calcd for C$_{14}$H$_{19}$NO$_2$Na

256.1313, found 256.1317; Anal. Calcd for $C_{14}H_{19}NO_2$: C 72.07, H 8.21, N 6.00, found C 72.04, H 8.26, N 5.98.

14. The addition was carried out at a rate such that an internal temperature of 27 °C was not exceeded.

15. The submitters stated that the reaction requires 40–60 h for completion; however, the checkers observed completion of the reaction after 22 h based on TLC analysis. The progress of the reaction can be followed by TLC (precoated plates with silica gel 60 F_{254} purchased from E. Merck, Darmstadt) using $CHCl_3$/MeOH, 20:1 as the eluent: R_f (product) = 0.58.

16. The submitters obtained a pale grey solid (15.2–17.0 g, 60–67%) by suspending the crude product in anhydrous Et_2O (Note 3) and immersing the suspension in an ultrasound cleaning bath (Sonorex RK SIOH) for 20 min.

17. Working at 50% scale, the checkers obtained 7.99 g (62%).

18. The product has the following physicochemical properties: mp 238–241 °C (dec.); ^1H NMR (600 MHz, $CDCl_3$) δ: 1.79 (q, $J = 12.6$ Hz, 6 H), 1.94 (s, 3 H), 1.96 (s, 6 H), 2.33 (s, 6 H), 2.34 (s, 6 H), 2.61 (s, 3 H), 7.00 (s, 2 H), 8.46 (s, 1 H); ^{13}C NMR (150 MHz, $CDCl_3$) δ: 8.1, 12.8, 17.3, 21.1, 29.6, 35.2, 41.2, 63.4, 63.4, 127.2, 129.0, 129.5, 129.8, 132.9, 134.9, 141.1; IR (KBr) 3183, 2916, 2856, 1609, 1543, 1456, 1308, 1240, 1208, 1050, 903, 856, 806, 737 cm^{-1}; HRMS (ESI+) m/z calcd for $C_{24}H_{33}N_2$ 349.2644, found 349.2610; Anal. Calcd for $C_{24}H_{33}BF_4N_2$: C, 66.06; H, 7.62; N, 6.42. Found: C, 66.29; H, 7.69; N, 6.43.

Waste Disposal Information

All hazardous materials were disposed in accordance with "Prudent Practices in the Laboratories"; National Academy Press; Washington, DC, 1965.

3. Discussion

N-Heterocyclic carbenes of type **A** have gained a prominent role as ancillary ligands for a host of transition metal templates in various oxidation states[3] and as organocatalysts in their own right.[4] They are usually prepared by deprotonation of the corresponding imidazolium salts **B** with a non-nucleophilic base (Scheme 1).

Although this tremendous success story has led to a huge number of structural variants, it is surprising that several obvious and seemingly trivial substitution patterns have not been investigated in any great detail. For example, unsymmetrical imidazolium salts bearing two different aryl groups on their N-atoms are largely unknown. Likewise, non-symmetrical imidazolium salts bearing one *N*-aryl and one *N*-alkyl group are rare and essentially limited to those having primary *N*-alkyl substituents.[5] These gaps in the structural landscape reflect the inherent limitations of the established syntheses routes which largely rely on *N*-substitution reactions of appropriate imidazole precursors. While such *N*-substitutions work well with reactive halides R^2–X,[6] they are difficult to accomplish or even impossible with aryl halides, *tert*-alkyl halides and many *sec*-alkyl halides, in particular if they are chiral and non-racemic.

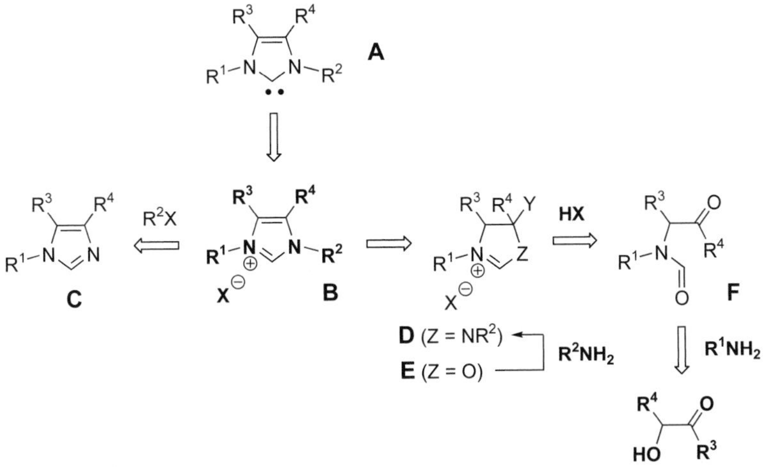

Scheme 1. Formation of imidazolium salts **B** and *N*-heterocyclic carbenes **A**: established *N*-substitution methodology (left) versus the heterocycle interconversion strategy (right).

Outlined in the procedure described above is an alternative approach based on a heterocycle interconversion strategy which provides ready access to substitution patterns beyond the reach of customary methodology.[7] Specifically, readily available *N*-formamido carbonyl derivatives **F** are cyclized to the corresponding oxazolinium salts **E** (Z = O, Y = OAc), which react with an amine of choice R^2NH_2 to give hydroxylated imidazolinium salts **D** (Z = NR^2, Y = OH) as the primary products. This condensation can be achieved with variously substituted anilines as well as secondary and tertiary amines, even if they are poorly nucleophilic and/or sterically hindered. Compounds **D** are then dehydrated to the desired imidazolium salts **B** on exposure to Ac$_2$O and catalytic amounts of a suitable mineral acid. This novel methodology is distinguished by the following characteristics:[7]

(1)   As mentioned above, a host of different amines R^{2}NH$_2$ is accommodated, thus giving access to non-symmetrical imidazolium salts **B** that are inaccessible or, at least, very difficult to make otherwise. The examples outlined in the text and in Figure 1 fall into this category.

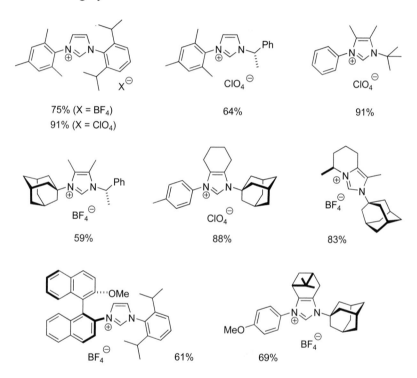

Figure 1. Selected non-symmetrical imidazolium salts prepared by the heterocycle interconversion strategy.

(2)   The intermediates **D** and **E** need not be rigorously purified; the conversion of the *N*-formamido carbonyl derivatives **F** into the desired

imidazolium salts **B** is therefore straightforward and operationally highly convenient.

(3) This largely streamlined protocol, the good to excellent overall yields, and a practical workup procedure greatly facilitate the scale-up of the method. In many cases, the final products can be readily isolated by simple precipitation with Et$_2$O, ultrasonication of the crude product, and filtration of the resulting salts.

(4) Because the imidazolium salts derive from three simple building blocks, the method is inherently flexible and allows for structural variations of every substituent on the backbone of the final products. Moreover, proper choice of the mineral acid in the cyclization step **F → E** allows different escorting counterions to be chosen.

(5) Enantiopure amines are not racemized and various chiral products, including bicyclic constructs, are also within reach (Figure 1).

This user-friendly methodology therefore nicely complements the existing routes to imidazolium salts and the *N*-heterocyclic carbenes derived thereof.[7] Because of its practicality and inherent flexibility, it is expected to foster further applications of such ligands and catalysts in advanced organic synthesis.

1. Max-Planck-Institut für Kohlenforschung, Kaiser-Wilhelm-Platz 1, D-45470 Mülheim/Ruhr, Germany; fuerstner@mpi-muelheim.mpg.de.

2. Department of Chemistry, University of Pittsburgh, Pittsburgh, PA 15260, USA; pwipf@pitt.edu.

3. (a) Arduengo, A. J. *Acc. Chem. Res.* **1999**, *32*, 913-921. (b) Herrmann, W. A. *Angew. Chem., Int. Ed.* **2002**, *41*, 1290-1309. (c) Bourissou, D.; Guerret, O.; Gabbaï, F. P.; Bertrand, G. *Chem. Rev.* **2000**, *100*, 39-91. (d) César, V. ; Bellemin-Laponnaz, S.; Gade, L. H. *Chem. Soc. Rev.* **2004**, *33*, 619-636.

4. Enders, D.; Balensiefer, T. *Acc. Chem. Res.* **2004**, *37*, 534-541.

5. In contrast, unsymmetrical imidazolinium salts (and the "saturated" *N*-heterocyclic carbenes derived thereof) are somewhat more common. For representative cases see the following and literature cited therein: (a) van Veldhuizen, J. J.; Campbell, J. E.; Giudici, R. E.; Hoveyda, A.

H. *J. Am. Chem. Soc.* **2005**, *127*, 6877-6882. (b) Waltman, A. W.; Grubbs, R. H. *Organometallics* **2004**, *23*, 3105-3107. (c) Rivas, F. M.; Riaz, U.; Giessert, A.; Smulik, J. A.; Diver, S. T. *Org. Lett.* **2001**, *3*, 2673-2676. (d) Xu, G.; Gilbertson, S. R. *Org. Lett.* **2005**, *7*, 4605-4608. (e) Bappert, E.; Helmchen, G. *Synlett* **2004**, 1789-1793. (f) Paczal, A.; Bényei, A. C.; Kotschy, A. *J. Org. Chem.* **2006**, *71*, 5969-5979.

6. For representative examples see: (a) Fürstner, A.; Ackermann, L.; Gabor, B.; Goddard, R.; Lehmann, C. W.; Mynott, R.; Stelzer, F.; Thiel, O. R. *Chem. Eur. J.* **2001**, *7*, 3236-3253. (b) Fürstner, A.; Krause, H.; Ackermann, L.; Lehmann, C. W. *Chem. Commun* **2001**, 2240-2241.

7. Fürstner, A.; Alcarazo, M.; César, V.; Lehmann, C. W. *Chem. Commun.* **2006**, 2176-2178.

Appendix
Chemical Abstracts Nomenclature; (Registry Number)

3-Hydroxybutan-2-one; (513-86-0)

Mesitylamine: Benzenamine, 2,4,6-trimethyl-; (88-05-1)

3-(Mesitylamino)butan-2-one: 2-Butanone,   3-[(2,4,6-trimethylphenyl)-amino]-; (898552-96-0)

Acetic formic anhydride; (2258-42-6)

N-Mesityl-*N*-(3-oxobutan-2-yl)formamide: Formamide,   N-(1-methyl-2-oxopropyl)-N-(2,4,6-trimethylphenyl)-; (898553-01-0)

Acetic anhydride; (108-24-7)

Aqueous HBF_4, Borate(1-), tetrafluoro-, hydrogen (1:1): (16872-11-0)

1-Adamantylamine: Tricyclo[3.3.1.13,7]decan-1-amine; (768-94-5)

1-Adamantyl-3-mesityl-4,5-dimethylimidazolium tetrafluoroborate

Alois Fürstner was born in Bruck/Mur, Austria, in 1962. He obtained his training and PhD degree from the Technical University of Graz, Austria, working on carbohydrates under the supervision of Hans Weidmann. After postdoctoral studies with the late Prof. W. Oppolzer in Geneva, he finished his Habilitation in Graz before joining the Max-Planck-Institut für Kohlenforschung, Mülheim/Ruhr, Germany, in 1993, where he is presently one of the Directors. His research interests focus on homogeneous catalysis, metathesis, and organometallic chemistry as applied to natural product synthesis.

Manuel Alcarazo was born in 1978 in Alcalá de Guadaira, Spain. After graduating from the he University of Sevilla in 2000, he obtained his PhD in 2005 at the Instituto de Investigaciones Químicas (CSIC-USe) under the supervision of Dr. José M. Lassaletta. There he worked on the synthesis and applications of N-heterocyclic carbenes derived from hydrazines. He is currently a postdoctoral associate at Prof. A. Fürstner's research group at the Max-Plank-Institut für Kohlenforschung involved in various projects ranging from ligand design to natural product total synthesis.

Vincent César was born in 1977 in Nancy, France. He was educated at the Ecole Normale Supérieure in Lyon. He joined the group of Prof. Lutz H. Gade at Université Louis Pasteur to obtain his doctorate in 2004 studying a new class of oxazolinyl-imidazolylidene ligands for catalysis. In 2004 he joined the group of Prof Dr. Alois Fürstner (Max Planck Institute für Kohlenforschung, Mülheim/Ruhr) as a postdoctoral fellow. He is currently a Chargé de Recherche at the Laboratoire de Chimie de Coordination of CNRS in Toulouse focusing on the catalytic activation and functionalization of unreactive C-H bonds.

Helga Krause was born in 1952 in Oberhausen, Germany. She started her education in 1969 as a laboratory assistant at the Max-Planck-Institute for Coal Research. After completing these studies in 1972 she worked in the group of Professor Roland Köster in boron-organic chemistry. In 1994 she joined the group of Professor Alois Fürstner working on various projects in the field of organic and organometallic chemistry.

Chad Hopkins obtained his B.Sc. (2001) from Mississippi State University. He then moved to the University of Kansas to pursue his Ph.D. with Professor Helena Malinakova, focusing on the development of novel palladium-catalyzed multi-component coupling and cascade processes for the synthesis of highly substituted homoallylic alcohols, amines, δ-lactones, and unnatural amino acids. After completion of his Ph.D. in 2007, he joined the group of Professor Peter Wipf at the University of Pittsburgh as a postdoctoral research associate and is currently working on the synthesis of new PLK1 inhibitors as well as the natural product disorazole C1.

44

PREPARATION OF A 1-ALKOXY-1-ALKYNE FROM REACTION OF A 2,2,2-TRIFLUOROMETHYL ETHER WITH AN ALKYLLITHIUM REAGENT: 1-BENZYLOXYMETHOXY-1-HEXYNE

Submitted by P. J. Kocienski and T. N. Snaddon.[1]
Checked by K. M. Brummond and T. O. Painter.

1. Procedure

Caution! Benzyl chloromethyl ether is a powerful alkylating agent and a potential carcinogen. Furthermore, it is a mild lachrymator and reacts with water and alcohols, forming hydrogen chloride. The procedure should be conducted in a well-ventilated fume hood, and inhalation and skin contact should be avoided.

A. *Benzyloxymethoxy 2,2,2-trifluoroethyl ether.* An oven-dried 500-mL, three-necked, round-bottomed flask equipped with a magnetic stir bar, a thermometer (–100 °C to 30 °C range), a rubber septum (containing nitrogen inlet and outlet needles) and a glass stopper is charged with sodium hydride (3.8 g, 95 mmol, 1.2 equiv) (Note 1) by temporary removal of the septum and 10 mL of pentane (Note 2) is added by syringe through the septum. The resulting gray suspension is stirred for 1 min and is allowed to settle. The pentane is slowly removed by syringe and the process is repeated (Note 2). Tetrahydrofuran (300 mL) (Note 3) is added and the resulting gray suspension cooled to 0 °C (internal temperature) using an ice–water–sodium chloride bath whereupon 2,2,2-trifluoroethanol (6.4 mL, 87 mmol, 1.1

equiv) (Note 4) is added drop wise *via* syringe at a rate sufficient to maintain the reaction temperature at 0 °C (approximately 10 min) [*caution:* gas evolution!]. When effervescence ceases the cooling bath is removed and the reaction is allowed to warm to 22 °C (*approximately* 15 min) giving a slightly turbid, light gray mixture. The mixture is cooled to 0 °C with an ice–water–sodium chloride bath and benzyl chloromethyl ether (11.0 mL, 79 mmol, 1.0 equiv) (Note 5) is added dropwise over 5 min by syringe. The reaction mixture is stirred at 0 °C for 30 min to produce a fine white suspension. The cooling bath is removed and the suspension is allowed to warm to 22 °C and is stirred for 1.75 h. The septum is removed and water (70 mL) is added. The mixture is transferred to a 1-L separatory funnel and the aqueous layer is separated and then extracted with dichloromethane (3 × 70 mL). The combined organic layers are dried over magnesium sulfate for 15 min, filtered and the filter cake is washed with dichloromethane (3 × 50 mL). The combined filtrates are concentrated on a rotary evaporator (Note 6) and the residual colorless oil (19.3 g) is purified by short path distillation (Note 7). The fraction boiling at 79–80 °C/15 mmHg is collected to afford 15.8 g (72 mmol, 91%) (Note 8) of pure benzyloxymethoxy 2,2,2-trifluoroethyl ether (Note 9).

B. *1-Benzyloxymethoxy-1-hexyne*. An oven-dried 500-mL, three-necked round-bottomed flask equipped with a magnetic stir bar, a thermometer (–100 °C to 50 °C range), a 125-mL pressure equalizing dropping funnel (with rubber septum) and a rubber septum (containing nitrogen inlet and outlet needles) is charged with benzyloxymethoxy 2,2,2-trifluoroethyl ether (11 g, 50 mmol, 1.0 equiv) and 150 mL of diethyl ether (Note 10) by temporarily removing the septum. The pressure-equalizing dropping funnel is charged with 98 mL of *n*-butyllithium (1.53 M in hexanes, 150 mmol, 3.0 equiv) (Note 11). The reaction flask is cooled to –75 °C (internal temperature) with a dry ice-acetone bath and the *n*-butyllithium added dropwise over 40 min at a rate sufficient to maintain the reaction temperature below –70 °C. The resulting yellow solution is allowed to warm to 0 °C slowly with stirring over 3.5 h while still in the cooling bath. Thereafter the dry ice-acetone bath is removed and the mixture is allowed to warm to 20 °C over 30 min (Note 12), and maintained at this temperature for an additional 30 min. The turbid, yellow–orange reaction mixture is transferred by cannula to a vigorously-stirred slurry of 50 g of ice and 100 mL of brine, which is cooled with an ice-water bath. After complete

46

transfer, 50 mL of diethyl ether is added and the layers are separated. The aqueous layer is extracted with 150 mL of diethyl ether–hexanes, 1:1) (Note 13). The combined organic layers are dried over magnesium sulfate for 15 min, filtered and the filter cake is washed with diethyl ether/hexanes, 1:1 (3 × 100 mL) (Note 14). The combined filtrates are concentrated on a rotary evaporator (Note 6) to afford 11.2 g of a pungent, yellow–orange oil. The crude product is purified by short path distillation (Note 15, bp 114–115 °C/0.3 mmHg) to afford 5.4 g (25 mmol, 50%) of 1-benzyloxymethoxy-1-hexyne (Note 16) as a pale-yellow oil (Note 17).

2. Notes

1.   Sodium hydride (60% dispersion in mineral oil) was purchased from Alfa Aesar and used without further purification.

2.   Reagent grade pentane was distilled prior to use from powdered calcium hydride under an atmosphere of dry nitrogen. The supernatant washes are carefully quenched by the slow addition of water.

3.   Anhydrous 99.9%, inhibitor free tetrahydrofuran was purchased from Aldrich and purified with alumina using the Sol-Tek ST-002 solvent purification system directly before use. The submitters used reagent grade tetrahydrofuran that was freshly distilled from sodium/benzophenone ketyl under an atmosphere of dry nitrogen.

4.   2,2,2-Trifluoroethanol (99%) was purchased from Alfa Aesar and distilled prior to use from anhydrous calcium sulfate and sodium hydrogen carbonate (5:1) under an atmosphere of dry nitrogen.

5.   Benzyl chloromethyl ether was prepared (0.19 mol scale) according to the procedure of Boeckman et al.[2] and was distilled (bp 70–71 °C/3.0 mmHg) from anhydrous calcium chloride. The pure benzyl chloromethyl ether can be stored over calcium chloride at –20 °C for up to two weeks prior to use. The pure material was taken directly from storage and added to the reaction mixture via a syringe equipped with a PALL Life Sciences Acrodisc CR 13 mm syringe filter (0.2 μm PTFE membrane). The submitters prepared it similarly (0.4 mol scale) according to the procedure of Boeckman et al. and freshly distilled it (bp 47–48 °C/0.05 mmHg) from anhydrous calcium chloride directly before use. Calcium chloride (fused granular, general purpose grade) was dried in an oven at 140 °C for 24 h prior to use.

6.   Bath: 20–23 °C; vacuum: approximately 12 mmHg.

7.   Short path distillation was performed with a non-jacketed, non Vigreaux still head.

8.   Yields of 88–91% were obtained on full scale and a yield of 85-88% was obtained on half scale.

9.   IR (neat): 3035, 2954, 2897, 1498, 1455, 1428, 1384, 1281, 1159, 1060, 970, 742 cm^{-1}. ^{1}H NMR (500 MHz, CDCl$_3$) δ: 3.97 (q, 2 H, J_{H-F} = 8.5 Hz), 4.66 (s, 2 H), 4.85 (s, 2 H), 7.33-7.41 (m, 5 H); ^{13}C NMR (75 MHz, CDCl$_3$) δ: 64.7 (q, J_{C-F} = 34.5 Hz), 70.1, 94.8, 124.2 (q, J_{C-F} = 276.8 Hz), 128.2, 128.7, 137.3   Anal. Calcd. for C$_{10}$H$_{11}$F$_3$O$_2$: C, 54.55; H, 5.04. Found: C, 54.65; H, 5.06.

10.   Anhydrous diethyl ether was purchased from Fisher Chemical and purified with alumina using the Sol-Tek ST-002 solvent purification system directly before use.   The submitters used diethyl ether that was freshly distilled from sodium/benzophenone ketyl under an atmosphere of dry nitrogen.

11. n-Butyllithium (1.6 M in hexanes) was purchased from Sigma–Aldrich. Its molarity was determined by titration[3] with 0.2 g of 1,3-diphenylacetone p-tosylhydrazone (Aldrich) in 5 mL of tetrahydrofuran immediately prior to use.

12.   Upon reaching 5 °C a slightly turbid yellow mixture was obtained, which became more viscous as the temperature increased.

13.   The use of diethyl ether–hexanes (1:1) aided phase separation; however, the extraction mixture should be drained of water up to the bottom of the emulsion and this separation process repeated 5 times before combining the remaining emulsion with the organic layers.

14.   Care should be exercised during the addition of magnesium sulfate to the mixture due to its high water content.   The drying agent was added in 3 portions with ice-water bath cooling of the flask to avoid an exotherm.   Once addition was complete and bubbling had ceased, the mixture was allowed to stand at room temperature.

15.   The acid-sensitivity of the product required some precautions. The vessels used to store the crude product and all components of the distillation apparatus were soaked in isopropanolic potassium hydroxide for 4–24 h, before being washed with deionized water, isopropanol, then acetone, and finally oven dried at 140 °C overnight, prior to use.   The submitters report that this glassware was soaked in methanolic potassium hydroxide for 4 h, before being washed with methanol, then acetone, and finally oven-dried before use.

16. Yields of 46–50% were obtained on full scale and a yield of 51% was obtained on half scale. A by-product fraction that appeared to contain some unreacted starting material, as determined by proton NMR, was collected at 68–72 °C as a colorless oil (approximately 2.9 g on full scale). The submitters reported a 60% yield on full scale and that the desired product was collected by distillation at 92–95 °C/0.05 mmHg as a colorless oil.

17. IR (neat): 2957, 2931, 2872, 2270, 1497, 1456, 1382, 1216, 1159, 1119, 902, 741 cm^{-1}. ^{1}H NMR (500 MHz, CDCl$_3$) δ: 0.92 (t, 3 H, J = 7.0 Hz), 1.39–1.49 (m, 4 H), 2.16 (t, 2 H, J = 7.0 Hz), 4.78 (s, 2 H), 5.03 (s, 2 H), 7.32–7.40 (m, 5 H); ^{13}C NMR (75 MHz, CDCl$_3$) δ: 13.8, 17.1, 22.0, 31.8, 39.6, 71.5, 87.6, 99.6, 128.3, 128.6, 136.6. The checkers were unable to obtain acceptable CHN analytical data. The submitters reported that CHN analysis was obtained immediately after the final distillation using an in-house service due to the sensitivity of the 1-alkoxy-1-alkyne. This afforded the following data: C, 76.95; H, 8.55. Analysis of the proton NMR indicated the presence of minor impurities, the most prevalent at 4.8–4.9 ppm, 4.6–4.7 ppm, and 3.9–4.0 ppm. These impurity peaks accounted for less than 5% of the total proton integration value.

Safety and Waste Disposal Information

All hazardous materials should be handled and disposed of in accordance with "Prudent Practices in the Laboratory"; National Academy Press; Washington, DC, 1995.

3. Discussion

The synthesis of 1-benzyloxymethoxy-1-hexyne[4] exemplifies a general method for the synthesis of 1-alkoxy-1-alkynes and 1-aryloxy-1-alkynes reported by Nakai and co-workers.[5] A 2,2,2-trifluoroethyl ether, prepared by the nucleophilic substitution of an alkyl halide by sodium 2,2,2-trifluoroethoxide, reacts with 3 equiv of an alkyllithium reagent to give the 1-alkoxy-1-alkyne in 35–80% yield (see Table). 1°-, 2°- and 3°-alkyllithiums participate equally well in the reaction according to the mechanism shown in Scheme 1. In the cases where R^1 is phenyl, the requisite 2,2,2-trifluoroethyl ether is prepared by the reaction of sodium phenolate with 2,2,2-trifluoroethyl tosylate. The sequence can also be used for the synthesis of 1-alkyl thio-and 1-arylthio-1-alkynes.[5]

Scheme 1

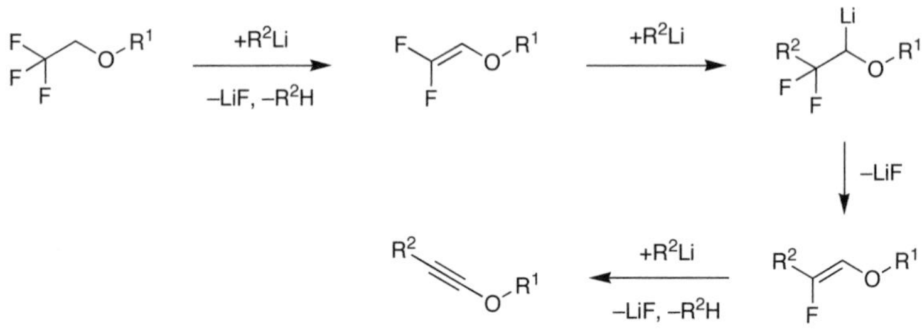

TABLE

1-ALKOXY-1-ALKYNES DERIVED FROM 2,2,2-TRIFLUOROETHYL ETHERS[4]

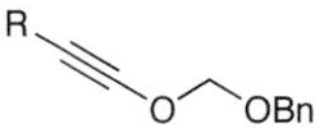

R	%		R	%		R	%
s-Bu	49		n-Bu	35		n-Bu	73
t-Bu	50		s-Bu	53		s-Bu	73
			t-Bu	44		t-Bu	80

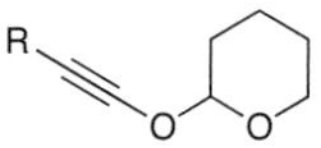

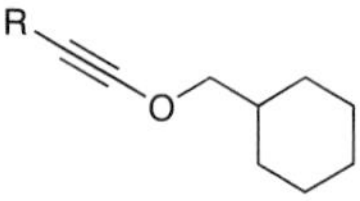

Greene and co-workers[6] developed an alternative one-pot procedure that is especially useful in the cases where the alkyllithium reagent R^2Li is inaccessible. Their procedure, depicted in Scheme 2, entails the reaction of an alcohol successively with potassium hydride (2 equivalents), trichloroethylene (1 equivalent), n-butyllithium (2.2 equivalents), and a primary iodoalkane (large excess). In this case the potassium alkoxide first generates dichloroacetylene by β-elimination of trichloroethylene and then adds to it. The resulting adduct, on treatment with n-butyllithium, undergoes elimination to the lithium acetylide whose alkylation affords the final product. Witulski and Alayric[7] have reviewed the syntheses and reactions of 1-heteroalkyl-1-alkynes.

Org. Synth. **2008**, *85*, 45-52

Scheme 2

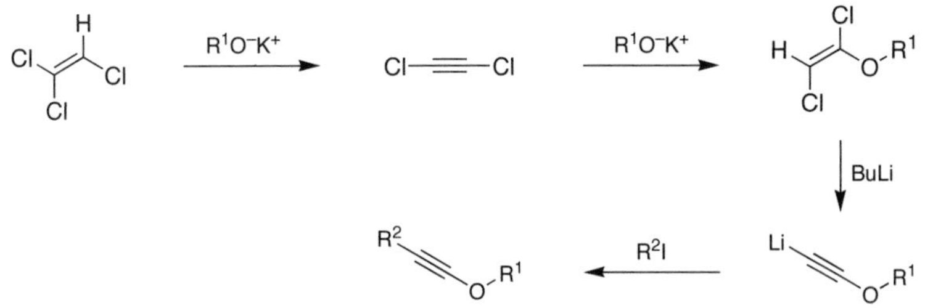

1. School of Chemistry, Leeds University, Leeds, LS2 9JT, UK
2. Connor, D. S.; Klein, G. W.; Taylor, G. N.; Boeckman, R. K.; Medwid, J. B. *Org. Synth. Coll. Vol. VI* **1988**, 101-103.
3. Lipton, M. F.; Sorenson, C. M.; Sadler, A. C.; Shapiro, R. H. *J. Organomet. Chem.* **1980**, *186*, 155-158.
4. Casson, S.; Kocienski, P. *Synthesis* **1993**, 1133-1140.
5. Tanaka, T.; Shiraishi, S.; Nakai, T.; Ishikawa, N. *Tetrahedron Lett.* **1978**, *19*, 3103-3106.
6. Moyano, A.; Charbonnier, F.; Greene, A. E. *J. Org. Chem.* **1987**, *52*, 2919-2922.
7. Witulski, B.; Alayric, C. *Science of Synthesis* **2005**, *24*, 933-956.

Appendix
Chemical Abstracts Nomenclature; (Registry Number)

Sodium hydride; (7647-69-7)

2,2,2-Trifluoroethanol (75-89-8)

Benzyl chloromethyl ether: Benzene, [(chloromethoxy)methyl]-; (3587-60-8)

Benzyloxymethoxy-2,2,2-trifluoromethyl ether: Benzene, [[(2,2,2-trifluoroethoxy)methoxy]methyl]-: (153959-88-7)

n-Butyllithium (109-72-8)

1-Benzyloxymethoxy-1-hexyne: Benzene, [[(1-hexyn-1-yloxy)methoxy]methyl]-; (162552-11-6)

1,3-Diphenylacetone *p*-tosylhydrazone: Benzenesulfonic acid, 4-methyl-, [2-phenyl-1-(phenylmethyl)ethylidene]hydrazide; (19816-88-7)

Philip Kocienski was born in Troy, New York, in 1946. His research career began under Alfred Viola while an undergraduate at Northeastern University and was further developed at Brown University, where he obtained his PhD under Joseph Ciabattoni. Postdoctoral study under George Büchi at MIT and later with Basil Lythgoe at Leeds University confirmed his interest in natural product synthesis. He has held academic appointments at Leeds University (1979-1985), Southampton University (1987-1997), and Glasgow University (1997-2000). In 2000 he returned to Leeds where his research has focussed on the applications of organometallic chemistry to natural product synthesis.

Thomas N. Snaddon was born in Stirling, Scotland in 1981. He received a B.Sc. (Hons.) in Biomolecular and Medicinal Chemistry (2003) and M.Phil. in synthetic organic chemistry (2004) from the University of Strathclyde in Glasgow. In October 2004 he moved to the University of Leeds as a Society of Chemical Industry (SCI) John Gray Scholar to pursue doctoral studies with Professor Philip J. Kocienski in the arena of natural product synthesis. In early 2008 he will move to a post-doctoral position with Professor Alois Fürstner at the Max-Planck-Institut für Kohlenforschung, Mülheim, Germany.

Thomas Painter was born in 1980 in Pittsburgh, Pennsylvania. During his undergraduate studies he interned at Valspar Corporation in 2002 and obtained his B.S. degree in chemistry from the University of Pittsburgh in 2003. He is currently pursuing graduate studies as a Bayer Fellow in the laboratory of Professor Kay Brummond at the University of Pittsburgh. His graduate research has included work on the synthesis of electron- deficient trienones and ε-lactams, and progress toward bicyclic analogs of irofulven using rhodium(I)-catalyzed cycloisomerization reactions.

DIRECT SYNTHESIS OF 2,5-DIHALOSILOLES
(2,5-Dibromo-1,1-dimethyl-3,4-diphenyl-1H-silole)

A.

B.

Submitted by Nicholas A. Morra and Brian L. Pagenkopf.[1]
Checked by Tomita Daisuke and Tsuyoshi Mita and Maskatsu Shibasaki.

1. Procedure

A. Dimethyl-bis(phenylethynyl)silane (**1**). A 250-mL, three-necked, round-bottomed flask equipped with an internal thermometer, two rubber septa and a 3-cm egg-shaped stir bar (Note 1) is charged with phenylacetylene (6.37 mL, 58.0 mmol, 2.5 equiv) (Note 2) via syringe, 60 mL of THF (Note 3) via syringe, and the solution was cooled to an internal temperature of –70 °C or colder in a dry ice/acetone bath. *n*-Butyllithium (1.58 M, 30.8 mL, 48.7 mmol, 2.1 equiv) (Note 4) is added dropwise via syringe over 5 min directly into the stirring solution of phenylacetylene, such that the internal temperature does not exceed –50 °C. The acetone/dry ice bath is replaced with an ice/water bath. When the reaction temperature reaches approximately –5 °C, dichlorodimethylsilane (2.80 mL, 23.2 mmol) (Note 5) is added dropwise via syringe over 5 min, such that the internal temperature does not exceed 10 °C. The ice bath is removed, the solution is allowed to stir at room temperature for 10 min (Note 6), and then the reaction mixture is poured slowly through an open joint into a 1-L round bottom flask containing a rapidly stirring solution of half-saturated aqueous ammonium chloride (200 mL) (Note 7). The mixture is transferred to a 500-

mL separatory funnel and the flask is rinsed with 50 mL of ethyl acetate. The organic phase is separated and the aqueous layer is extracted with ethyl acetate (2 x 50 mL). The combined organic layers are washed with 100 mL of water and 100 mL of brine, then are dried over anhydrous $MgSO_4$ (5 g, Note 8), filtered through Celite 545 (Note 9), and concentrated by rotary evaporation (35 °C, 40 mmHg). The residue is then concentrated further at 1.5 mmHg for 4 h whereupon the residue solidifies (Note 10). The yellow-white solid is dissolved in a minimal amount of boiling hexanes (Note 11), and then is cooled in a freezer (–30 °C) for 16 h to afford the pure product as white needles that are collected by vacuum filtration (5.63 g, 93% yield) (Notes 12 and 13).

B. *2,5-Dibromo-1,1-dimethyl-3,4-diphenylsilole* (**2**). *From this point forward the light sensitivity of some compounds necessitates that certain flasks be wrapped in aluminum foil to exclude light.* A 500-mL, three-necked, round-bottomed flask is equipped with two rubber septa, an internal thermometer, and 3-cm egg-shaped stir bar. Under a stream of argon (introduced through a septum with a syringe needle and balloon), lithium wire (0.53 g, 75.6 mmol, 4.5 equiv) (Note 14) is cut into (20-25) small pieces and washed with toluene (2 x 10 mL) to remove the protective oil. One septum is temporarily removed and a funnel is used to aid in the addition of the lithium pieces. The flask is then fitted with a gas adapter and evacuated (1.5 mmHg) for 10 min to ensure that all toluene is removed. The gas adapter is removed and naphthalene (10.12 g, 79.0 mmol, 4.7 equiv) (Note 15) is added with the aid of a funnel. The original septum is replaced and 50 mL of THF (Note 3) is added via syringe. The flask is sonicated for 1 h to facilitate dissolution of the lithium wire (Note 16), and the resulting dark green solution is left to stir at room temperature for 1 h. The lithium naphthalenide solution thus obtained is titrated according to a literature method[3] and found to be 1.35-1.45 M (Note 17), and the flask is wrapped in foil.

Into a 200-mL, single-necked, round-bottomed flask containing dimethyl-bis(phenylethynyl)silane (**1**) (4.37 g, 16.8 mmol) and fitted with a rubber septum (pierced with a syringe needle and argon-filled balloon) is added 120 mL of THF via syringe (Note 3). The resulting silane solution is added to the lithium naphthalenide dropwise, *over a minimum of 20 min*, via cannula (Note 18). Once the addition is complete, the solution is cooled to – 10 °C using an acetone bath with a small amount of dry ice.

Meanwhile, a solution of anhydrous $ZnCl_2$ (11.45 g, 84.0 mmol, 5.0 equiv) (Note 19) in 70 mL of THF added via syringe (Notes 3 and 20) is prepared in a 100-mL single-necked, round-bottomed flask equipped with a 2-cm stir bar and rubber septum (pierced with a syringe needle and argon-filled balloon). The $ZnCl_2$ solution is added to the naphthalenide/silane mixture via cannula over a minimum of 20 min (Note 21). The acetone/dry ice bath is removed and the resulting solution is allowed to warm to room temperature (Note 22).

Meanwhile, a 1-L, three-necked, foil-wrapped, round-bottomed flask is charged with *N*-bromosuccinimide (NBS, 7.45 g, 41.9 mmol, 2.5 equiv) (Note 23) and equipped with an internal thermometer, 3-cm egg-shaped stir bar and two rubber septa (one pierced with a syringe needle and argon-filled balloon). The flask is charged with 35 mL of THF via syringe and the resulting solution is cooled to –70 °C or colder using a dry ice/acetone bath. The room temperature silane solution is added into the cold NBS solution via cannula such that the internal temperature does not exceed –50 °C. After the addition is complete the resulting mixture is stirred for a further 30 min at –70 °C or colder. One septum is removed and the cold mixture is slowly poured into a 1-L round-bottomed flask containing a 300-mL solution of rapidly stirring half-saturated aqueous ammonium chloride (Notes 7 and 24). The heterogeneous mixture is transferred to a 1-L separatory funnel and the flask is rinsed with 50 mL of ethyl acetate. The organic phase is separated and the aqueous layer is extracted with ethyl acetate (3 x 50 mL). The combined organic layers are washed successively with 100 mL of half-saturated aqueous $Na_2S_2O_3$ (Note 25), 200 mL of water, 200 mL of brine, then are dried over anhydrous $MgSO_4$ (7 g) and quickly filtered through a thin pad of silica (Note 26) into a foil wrapped 500-mL single-necked flask. The yellow solution is concentrated by rotary evaporation (35 °C, 40 mmHg) and then at 1.5 mmHg for 1 h to give approximately 17 g of crude material (quantitative mass recovery based on both naphthalene and dibromosilole **2**). ^{1}H NMR analysis of a representative sample typically shows a 1:5.2 mixture of the dibromosilole (**2**) and naphthalene (Note 27), which corresponds to a 96% yield (6.78 g) of useable dibromosilole (**2**). It is generally unnecessary to remove the naphthalene for subsequent use, such as cross coupling reactions.

If necessary, the naphthalene can be removed by sublimation in small batches. Thus, 1.7 g of the naphthalene-silole mixture is ground into powder using a mortar and pestle, and sublimed in the dark, under vacuum (1.5

mmHg) for 3–4 hours at 35 °C (Note 28). This will typically remove 99% of the naphthalene according to ^{1}H NMR analysis. If desired, further purification can be accomplished by dissolving the residual material from the combined sublimations in hot hexanes (160 mL) followed by cooling in a freezer for 16 h. The pure 2,5-dibromosilole crystallizes as white needles, which are vacuum-filtered, washed with 10 mL of ice cold hexanes and dried in vacuo to afford 6.64 g (95% yield) of **2**. (Note 29, 30 and 31). Dibromosilole **2** stored in the dark at room temperature is stable for several weeks. Even brief exposure to sunlight during processing will lead to significant decomposition (Notes 24 and 30).

2. Notes

1.   All glassware was flame dried and allowed to cool completely to room temperature under a flow of argon before any reagents were introduced. All metal needles and cannula used throughout were acetone washed and oven dried for a minimum of 3 hours. The apparatuses were maintained under an inert atmosphere via an argon-filled balloon equipped with a needle. The balloon is inserted through a rubber septum for the duration of each reaction.

2.   Phenylacetylene, obtained from Aldrich Chemical Company, Inc. (98%), was used as received by Checkers. The submitters distilled phenylacetylene prior to use (30 mmHg, 25–30°C) and stored it in an amber glass bottle at room temperature.

3.   Anhydrous THF, purchased from Kanto Chemical Co., Inc., was used as received by Checkers. Submitters purchased anhydrous THF from VWR, further purified (dried and degassed) by an alumina column solvent dispensing system.

4.   Checkers purchased *n*-butyllithium (1.58 M in hexanes), from Kanto Chemical Co., Inc.. Submitters purchased *n*-butyllithium (2.5 M in hexanes) from Alfa Aesar and titrated according to a literature method.[2] An accurate titer is critical. Caution! *n*-Butyllithium is spontaneously flammable in air. The titrant, *N*-pivaloyl-*o*-toluidine is synthesized from *o*-toluidine and pivaloyl chloride as described by Suffert,[2] it is also commercially available from Sigma-Aldrich. To 0.373 g (1.95 mmol) of *N*-pivaloyl-*o*-toluidine in 2 mL anhydrous THF is added *n*-butyllithium slowly dropwise until the appearance of a yellow endpoint. The titration is repeated 3 times and the

closely agreeing volumes of *n*-butyllithium are averaged to calculate a concentration.

5. Dichlorodimethylsilane, purchased from Aldrich Chemical Company, Inc. (99%), was used as received by Checkers. Submitters distilled dichlorodimethylsilane over CaH_2 prior to use and stored in an amber glass bottle at room temperature.

6. The reaction takes approximately 20–30 min to warm to room temperature. Once it reaches room temperature it is stirred for an additional 10 min.

7. Ammonium chloride was purchased from Kanto Chemical Co., Inc. by Checkers (and from Caledon by Submitters), and used as received. The half saturated solution was prepared by combining equal volumes of a saturated solution and water.

8. Magnesium sulfate was purchased from Kanto Chemical Co., Inc. by Checkers (and from Caledon by Submitters), and used as received.

9. Celite 545 was obtained from Fisher Scientific and used as received. A Celite pad 6 cm in diameter and 2 cm in height was used.

10. Phenylacetylene significantly interferes with the recrystallization. Therefore it is critical that excess phenylacetylene be removed by room temperature evacuation at this point.

11. Typically 75 to 90 mL of hexanes is needed.

12. The product displayed the following physicochemical properties: mp 75 °C; R_f 0.4 (hexane/EtOAc, 9:1); ^{1}H NMR (500 MHz, CDCl$_3$) δ: 0.51 (s, 6 H), 7.30-7.37 (m, 6 H), 7.52– 7.55 (m, 4 H); ^{13}C NMR (125 MHz, CDCl$_3$) δ: 0.5, 90.6, 105.9, 122.6, 128.2, 128.8, 132.1. Anal. Calcd for $C_{18}H_{16}Si$: C, 83.02; H, 6.19. Found: C, 83.07; H, 6.23. GC: t_R (**1**), 16.53 min; t_R (naphth), 3.75 min; t_R (**2**-decomp), 7.51 min; t_R (**2**), 11.55 min (Rtx$^®$-5 GC system; 5% diphenyl-95% dimethyl polysiloxane capillary column 30 m × 0.25 mm × 0.25 μm; (50 °C (1 min), +15 °C/min to 250 °C). ^{1}H NMR and GC showed no detectable impurities.

13. Submitters obtained the product in 85–95 % yield. Checkers obtained the product in 85% yield on a half-scale reaction.

14. Lithium wire (Na content 0.5-1%), 3.2 mm diam., 98+%, in mineral oil was obtained from Sigma Aldrich. The sodium content is critical, and lithium wire without sodium fails in this reaction.

15. Naphthalene, 99.6%, was obtained from Alfa Aesar and used as received.

16. A Branson 1510 model sonicator was used. The internal reaction temperature is not allowed to exceed 30 °C.

17. An aliquot of the lithium naphthalenide solution (1.00 mL) and 1,1-diphenylethylene (0.5 mL, 2.77 mmol) are combined in a 5-mL round-bottomed flask with a 1-cm stir bar, and a 0.5 M solution of s-butanol in toluene is added slowly dropwise until the appearance of a pale yellow endpoint. If the solution is found to be less than 1.35 M, all of the lithium wire has not been dissolved and the solution should be sonicated for another hour.

18. Faster addition results in a substantial decrease in yield.

19. Anhydrous zinc chloride (99.999%: ampoule), purchased from Aldrich, was used as received by Checkers. The ampoule was opened in a glove box, and zinc chloride was measured in a glove box. Submitters purchased zinc chloride (min 97.0%) from Caledon. In their case, the zinc chloride was dried, prior to use, by flame heating the flask such that the zinc chloride fully melts while under vacuum (1.5 mmHg), allowed to cool and transferred to a glove box where it was ground to a fine power using a mortar and pestle. The Checkers observed that the above procedure was not required when using anhydrous zinc chloride (99.999%: ample), purchased from Aldrich.

20. The solution is sonicated for 10 min to facilitate dissolution of the zinc chloride.

21. There is no exotherm to this reaction, so the internal reaction temperature should not exceed 0 °C.

22. Warming to room temperature takes approximately 20 min.

23. N-Bromosuccinimide (99%) was obtained from Alfa Aesar. Prior to use, it was dried over P_2O_5 in a vacuum desiccator. Submitters recrystallized N-bromosuccinimide from hot water, and dried over P_2O_5 in a vacuum desiccator, prior to use.

24. Lengthy exposure to light during any step in part B may lead to a significant decrease in overall yield, therefore it is suggested to work in a fume hood with the lights off.

25. Sodium thiosulfate was purchased from Aldrich Chemical Company, Inc. by Checkers (and from Caledon by Submitters), and used as received. The half-saturated solution was prepared by combining equal volumes of a saturated solution and water.

26. Silica Gel was purchased from Kanto Chemical Co., Inc. by Checkers (from Silicycle by Submitters) and used as received. A silica pad 6 cm in diameter and 2 cm in height was used.

27. Naphthalene is identified by ^{1}H NMR (400 MHz, CDCl$_3$) δ: 7.50-7.54 (dd, J = 6.2, 3.0 Hz, 4 H), 7.87-7.91 (dd, J = 6.2, 3.0 Hz, 4 H). Silole is identified by ^{1}H NMR (400 MHz, CDCl$_3$) δ: 0.45 (s, 6 H), 6.93–6.96 (m, 4 H), 7.11–7.16 (m, 6 H).

28. The process was repeated ten times to purify all the naphthalene/silole mixture (1.7 g x 10 times). Submitters purified 1.0 g for each trial (1.0 g x 17 times). It is important to purify in small batches.

29. The product displays the following physicochemical properties: mp 163–164 °C; R$_f$ 0.65 (hexane/EtOAc, 9/1); ^{1}H NMR (500 MHz, CDCl$_3$) δ: 0.45 (s, 6 H), 6.93–6.96 (m, 4 H), 7.11–7.16 (m, 6 H); ^{13}C NMR (125 MHz, CDCl$_3$) δ: –6.0, 123.0, 127.6, 127.7, 129.3, 137.2, 156.2. HRMS m/z calcd for C$_{18}$H$_{17}$Br$_2$Si [M+H$^+$]: 418.9466, found: 418.9461. Anal. Calcd for C$_{18}$H$_{17}$Br$_2$Si: C, 51.45; H, 3.84. Found: C, 51.22; H, 4.04.

30. Submitters obtained the product in 85–95 % yield. Checkers obtained the product in 89% yield on a half-scale reaction.

31. Photo-decomposition leads to a brown discoloration, and decomposition is indicated by a sharp singlet at 2.78 ppm in the ^{1}H NMR spectrum (see Note 13).

Safety and Waste Disposal Information

All hazardous materials should be handled and disposed of in accordance with "Prudent Practices in the Laboratory"; National Academy Press; Washington, DC, 1995.

3. Discussion

Siloles have been extensively studied for their electronic properties,[4,5] leading to their recognition as desirable π-conjugated materials for applications such as OLEDs[6] (Organic Light Emitting Diodes) and PLEDs[7] (Polymer Light Emitting Diodes). Important contributions have been made in the areas of electro and photoluminescence,[8] nitroaromatic sensors,[9] silole-copolymers,[10] and OFETs (Organic Field-Effect Transistors).[11] We have also studied the electronic properties of donor-acceptor siloles,[12] silole oligomers,[13] siloles as precursors to substituted butadienes,[14] and the electrogenerated chemiluminescence of siloles.[15] Many of these silole chromophores have been prepared from 2,5-dihalosilole intermediates, such as that prepared here, and with the growing interest in siloles, a reliable large-scale synthesis of a 2,5-dihalosilole is becoming increasingly necessary.

In pioneering work, Tamao reported the synthesis of symmetrical 2,5-dihalo siloles via an intramolecular reductive cyclization of diethynylsilanes.[16] Yields from the original procedure ranged from 44−72%, and the reaction failed with 1,1-dimethylsiloles. The somewhat fickle and lengthy preparation of lithium naphthalenide has been shortened and now reproducibly gives essentially quantitative titers simply by using Li wire containing a small percentage of sodium and sonication of the reaction mixture. The original procedure also employed expensive t-BuPh$_2$SiCl as a stoichiometric oxidant for excess lithium naphthalenide. We,[12] and others,[17] have modified the original procedure by replacing t-BuPh$_2$SiCl with ZnCl$_2$. Besides the obvious cost savings, this modification simplifies product isolation, obviates the necessity for precise control over stoichiometry and gives higher yields. Another significant advantage of the modified ZnCl$_2$ procedure is that less reactive di-zinc silole intermediates can be selectively functionalized, and this allows for the direct preparation of asymmetrical siloles.[12] Specifically, replacing the NBS in part B of this procedure with N-chlorophthalimide followed by solid I$_2$, the asymmetrical chloroiodosilole is produced in a respectable 81% yield.[12] Alternatively, the di-zinc silole intermediate generated in part B before addition of NBS can be used directly in standard cross coupling reactions.[18]

1. Department of Chemistry, The University of Western Ontario, London, ON, Canada, N6A 5B7, bpagenko@uwo.ca.

2. Suffert, J. *J. Org. Chem.* **1989**, *54*, 509-510.

3. Screttas, C.G; Micha-Screttas, M. *J. Organomet. Chem.* **1983**, *252*, 263-265.

4. (a) Yamaguchi, S.; Limura, K.; Tamao, K. *Chem. Lett.* **1998**, 89-90. (b) Ohshita, J.; Mimura, N.; Arase, H.; Nodono, M.; Kunai, A.; Komaguchi, K.; Shiotani, M. *Macromolecules* **1998**, *31*, 7985-7987. (c) Chen, W.; Ijadi-Maghsoodi, S.; Barton, T. *Polym. Prepr. (Am. Chem. Soc., Div. Polym. Chem.)* **1997**, *38*, 189-190. (d) Stille, J. K.; Simpson, J. H.; *J. Am. Chem. Soc.* **1987**, *109*, 2138-2152.

5. For a recent review of the electronic properties of siloles: Yamaguchi, S.; Tamao, K. *Chem. Lett.* **2005**, *34*, 2-7.

6. Gerbier, P.; Aubouy, L.; Huby, N.; Hirsch, L.; Vignau, L. *Proc. SPIE* **2006**, *6192*, 61923A/1-61923A/8, DOI: 10.1117/12.667259.

7. Chen, J.; Cao, Y. *Proc. SPIE* **2006**, *6030*, 60300Q/1-60300Q/8, DOI: 10.1117/12.667639.

8. (a) Geramita, K.; McBee, J.; Shen, Y.; Radu, N.; Tilley, T. *Chem. Mater.* **2006**, *18*, 3261-3269. (b) Chen, J.; Xu, B.; Yang, K.; Cao, Y.; Sung, H.; Williams, I.; Tang, B. *J. Phys. Chem.* **2005**, *109*, 17086-17093.

9. (a) Sohn, H.; Sailor, M. J.; Magde, D.; Trogler, W. C. *J. Am. Chem. Soc.* **2003**, *125*, 3821-3830. (b) Sohn, H.; Calhoun, R. M.; Sailor, M. J.; Trogler, W. C. *Angew. Chem., Int. Ed.* **2001**, *40*, 2104-2105.

10. Wang, F.; Luo, J.; Yang, K.; Chen, J.; Huang, F.; Cao, Y. *Macromolecules* **2005**, *38*, 2253-2260.

11. Hakan, U.; Gang, L.; Antonio, F.; Tobin, J. M. *J. Am. Chem. Soc.* **2006**, *128*, 9034-9035.

12. Boydston, A. J.; Youshi, Y.; Pagenkopf, B. L. *J. Am. Chem. Soc.* **2004**, *126*, 3724-3725.

13. Boydston, A. J.; Youshi, Y.; Pagenkopf, B. L. *J. Am. Chem. Soc.* **2004**, *126*, 10350-10354.

14. Boydston, A. J.; Pagenkopf, B. L. *Angew. Chem., Int. Ed.* **2004**, *43*, 6336-6338.

15. Sartin M.; Boydston, A. J.; Pagenkopf, B. L.; Bard, A. *J. Am. Chem. Soc.* **2006**, *128*, 10163-10170.

16. Tamao, K.; Yamaguchi, S.; Shiro, M. *J. Am. Chem. Soc.* **1994**, *116*, 11715-11722.

17. Yamaguchi, S.; Endo, T.; Uchida, M.; Izumizawa, T.; Furukawa, K.; Tamao, K. *Chem. Eur. J.* **2000**, *6*, 1683-1692.

18. Tamao, K.; Uchida, M.; Izumizawa, T.; Furukawa, K.; Yamaguchi, S. *J. Am. Chem. Soc.* **1996**, *118*, 11974-11975.

Appendix
Chemical Abstracts Nomenclature; (Registry Number)

n-Butyllithium; (109-72-8)

Phenyl acetylene: Ethynylbenzene; (536-74-3)

Dichlorodimethylsilane; (75-78-5)

Dimethyl-bis-phenylethynyl silane: Benzene, 1,1'-[(dimethylsilylene)di-2,1-ethynediyl]bis-; (2170-08-3)

Naphthalene; (91-20-3)

Lithium; (7439-93-2)

Zinc chloride; (7646-85-7)

N-Bromosuccinimide: 1-Bromo-2,5-pyrrolidinedione; (128-08-5)

2,5-Dibromo-1,1-dimethyl-3,4-diphenylsilole: Silacyclopenta-2,4-diene, 2,5-dibromo-1,1-dimethyl-3,4-diphenyl-; (686290-22-2)

Brian Pagenkopf was born in 1967 in Wausau, Wisconsin. He completed his undergraduate studies at the University of Minnesota, Twin Cities, and while there worked with Takashi Okagaki (OB/GYN), Terry Davis (3M, Encapsulation) and Gary Gray (Chemistry). His graduate studies were conducted at Montana State University (Bozeman) under the direction of Tom Livinghouse. Following his graduate research, he was a NIH Postdoctoral Fellow with Erick Carreira at Caltech in 1997, and was briefly at the ETH in Zürich after the group moved there in 1998, before beginning his independent career at the University of Texas at Austin. In 2005, he moved to the University of Western Ontario. His research interests include natural product synthesis, synthetic methods, symmetric catalysis and organic materials based on siloles.

Nicholas Morra was born in 1984 in Toronto, Canada. He graduated from the University of Western Ontario in 2007 where he spent the last two years working for Prof. Brian Pagenkopf, primarily on silole based materials. He is currently working towards a PhD with his research focused on Lewis Acid catalyzed methodologies and total synthesis of natural products.

Daisuke Tomita was born in 1980 in Tokyo, Japan. He received B.S. in 2004 from the University of Chiba and M.S. in 2006 from The University of Tokyo. Presently, he is pursuing Ph.D. degree at the Graduate School of Pharmaceutical Sciences, The University of Tokyo, under the guidance of Professor Masakatsu Shibasaki. His research interests are in the area of catalytic asymmetric reaction using copper catalyst.

Tsuyoshi Mita was born in 1976 in Tokyo, Japan, and received his M.S. degree from Keio University in 2002 under the direction of Professor Tohru Yamada. In the following years, he worked as a process chemist in the pharmaceutical research laboratory of Ajinomoto Co., INC. (Kawasaki, Japan). He left his job in 2004 and entered the University of Tokyo as a Ph.D. student. After receiving his Ph.D. degree under the guidance of Professor Masakatsu Shibasaki in 2007, he joined Professor Eric N. Jacobsen's group at Harvard University as a JSPS research fellow. His current interests are the development of new asymmetric catalysis as well as the search for drug candidates based on medicinal chemistry.

ONE-POT CONVERSION OF LACTAM CARBAMATES TO CYCLIC ENECARBAMATES: PREPARATION OF 1-*TERT*-BUTOXYCARBONYL-2,3-DIHYDROPYRROLE

Submitted by Jurong Yu,[1]* Vu Truc,[1] Peter Riebel,[2] Elizabeth Hierl,[1] and Boguslaw Mudryk. [1]
Checked by Matthias Maywald and Andreas Pfaltz.

1. Procedure

1-tert-Butoxycarbonyl-2,3-dihydropyrrole.   To a 500-mL, three-necked, round-bottom flask equipped with a magnetic stirring bar, a rubber septum, a thermometer and a two-tap Schlenk adaptor connected to a bubbler and an argon/vacuum manifold (Note 1) is added N-(*tert*-butyloxycarbonyl)pyrrolidin-2-one 1 (14.1 g, 73.6 mmol) (Note 2). The flask is evacuated for 25 min at room temperature (Note 3), then is filled with argon and kept under argon atmosphere during the entire reaction. Through the rubber septum, anhydrous toluene (100 mL) is added via syringe. The flask is cooled to −70 °C (internal temperature) and a solution of lithium triethylborohydride (Super Hydride®, 81.0 mL, 1.0 M in THF, 81.0 mmol, 1.1 equiv) is added dropwise via syringe, while maintaining the temperature below −60 °C (Note 4). After complete addition, the reaction mixture is stirred for one additional hour at −70 °C to allow completion of the reaction (Note 5). Solid DMAP (90 mg, 0.736 mmol, 0.01 equiv) is added in one portion, followed by *N,N*-diisopropylethylamine (73.2 mL, 420 mmol, 5.70 equiv) and trifluoroacetic anhydride (12.3 mL, 88.3 mmol, 1.20 equiv) which are added, in sequence, dropwise by syringe through the rubber septum at rates that maintain the temperature below −55 °C (Note 6). After complete addition, the cooling bath is removed, and the reaction mixture is allowed to warm to room temperature. The mixture is then stirred for about

Org. Synth. **2008**, *85*, 64-71
Published on the Web 9/28/2007

two hours to allow completion of the reaction (Note 7). The reaction mixture is cooled to 0 °C in an ice/water bath and then is quenched by dropwise addition of water (150 mL) via syringe while keeping the temperature below 15 °C (Note 8). The mixture is transferred to a 1-L separatory funnel where the phases are separated. The organic phase is washed twice with water (150 mL each) and then is dried over anhydrous Na_2SO_4 (approximately 5 g). The solution is filtered through a fritted glass funnel with suction and the residue is washed with toluene (2 x 50 mL) and the combined filtrates are evaporated to dryness under reduced pressure (approximately 15 mmHg, water bath temperature 30 °C). The residue (14.1–16.0 g) is purified by silica gel flash chromatography (Note 9) to afford 9.05–9.75 g of product as a light-yellow oil (Note 10). Further purification by bulb-to-bulb distillation (oven temperature 50 °C, 0.06 mmHg) affords 8.71–9.10 g (70–73%) of analytically pure enecarbamate **3** as colorless oil (Note 11).

2. Notes

1.   All glassware was oven-dried, quickly assembled and heated with a heat gun under vacuum (0.05 mmHg) prior to use. The submitters used nitrogen as inert gas.

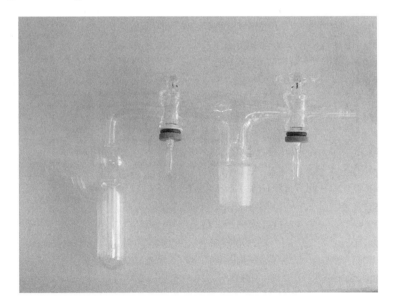

2. Submitters and Checkers used 1-(*tert*-butoxy-carbonyl)-2-pyrrolidinone **1** (97%), lithium triethylborohydride (Super-Hydride®, (1.0 M in THF,), *N*,*N*-diisopropylethylamine (99%) and trifluoracetic anhydride (>99%) as purchased from Aldrich Chemical Co. Submitters used 4-dimethylaminopyridine and toluene, p.a., as purchased from Aldrich Chemical Company, Inc. The Checkers received these chemicals from Fluka.

3. Final pressure: 0.05 mmHg.

4. A dry-ice acetone bath was used. Addition took about 40 minutes.

5. Progress of the reaction was followed by TLC on silica gel with hexane/EtOAc, 1:1. Visualization was accomplished with phosphomolybdic acid reagent (12 g in 250 mL of EtOH). The starting material has an $R_f = 0.25$ and the product has an $R_f = 0.37$.

6. Addition took about 10 min for diisopropylethylamine and 15 min for trifluoroacetic anhydride.

7. Progress of the reaction was followed by TLC analysis as described in Note 5. Enecarbamate **3** has an $R_f = 0.63$.

8. Addition took about 10 min.

9. The residue was dissolved in a minimum amount of dichloromethane, transferred to a 5 × 42 cm column packed with silica gel 60 (Merck, 0.040–0.060 mm) and eluted with 2.5 L of hexane-EtOAc 90:10 containing 0.2% (v/v) triethylamine). Enecarbamate **3** has an $R_f = 0.21$.

10. The ^{1}H NMR spectrum (400 MHz, CDCl$_3$) of the substance after chromatography showed traces of impurities with very weak signals at δ = 1.13–1.27 (0.30 H), 1.70–1.80 (0.25 H), 3.35–3.42 (0.12 H).

11. Analytical data: IR (film): 3414, 2974, 1703, 1617, 1478, 1409, 1366, 1289, 1257, 1224, 1178, 1135, 1093, 985, 882, 810, 763, 704 cm^{-1}; ^{1}H NMR (400 MHz, CDCl$_3$) δ: 1.42 (s, 9 H), 2.54–2.62 (m, 2 H), 3.59–3.70 (m, 2 H), 4.93 (d, $J = 18$ Hz, 1 H), 6.39/6.52 (br. s, 1 H) (pair of rotamers); ^{13}C NMR (101 MHz, CDCl$_3$) δ: 28.6/29.6, 44.7/45.2, 79.8/79.9, 107.4, 129.7, 151.6/152.3 (pair of rotamers); MS [EI, 70 eV] *m/z* (relative intensity): 169 (3), 113 (16), 96 (12), 68 (42), 57 (100), 41 (74); Elemental Analysis Calcd for C$_9$H$_{15}$NO$_2$: C, 63.88; H, 8.93; N, 8.28. Found: C, 63.62; H, 8.81; N, 8.08. ^{1}H NMR, ^{13}C NMR and MS data were identical to those reported in ref. 4b.

Safety and Waste Disposal Information

All hazardous materials should be handled and disposed of in accordance with "Prudent Practices in the Laboratory"; National Academy Press; Washington, DC, 1995.

3. Discussion

Cyclic enecarbamates, a class of deactivated enamines, are versatile intermediates for the synthesis of alkaloids and nitrogen-containing heterocycles. Among the available methodologies for their preparation, the reduction of lactam carbamates followed by dehydration of the lactamols is the most straightforward approach. The lactam carbamates can be reduced to lactamols by sodium borohydride (NaBH$_4$),[3] diisobutylaluminum hydride (DIBAL-H),[4] and lithium triethylborohydride (Super-Hydride®).[5] Among them, Super-Hydride® is most widely used in the laboratory, because it typically gives the cleanest reductions and highest yields. Many protocols of dehydrating lactamols to enecarbamates have been developed.[6] In comparison to the other dehydrating methods, dehydration using trifluoriacetic anhydride and a hindered base, such as 2,6-lutidine or diisopropylethylamine, was found to be the most effective. These mild conditions are compatible with a number of protecting groups and are suitable for compounds with epimerizable stereogenic centers.[7]

Recently, we simplified the two-step process by developing an efficient one-pot conversion of lactam carbamates to cyclic enecarbamates by telescoping the reduction with lithium triethylborohydride, and *in-situ* dehydration with trifluoroacetic acid and diisopropylethylamine.[8] The one-pot protocol avoids the isolation of unstable lactamol intermediate, which could undergo a ring-opening to form the tautomeric aldehyde.[9] This one-pot protocol is suitable for different azacycles (5-, 6- and 7-membered rings), and is compatible with a number of protecting groups (Boc, Cbz, Bn). Due to the mild conditions, Boc-protected *L*-pyroglutamate can be converted to its enecarbamate without any epimerization. Results are summarized in Table 1.

In general, the main advantages of this one-pot protocol are its simple operation, higher overall yields, and formation of fewer side products. This methodology has successfully been used for the preparation of enecarbamate **3c** (see Table 1) on greater than 100 kg scale.

Table 1. Examples of One-pot Conversion of Carbamate to Enecarbamate

entry	carbamate 1	enecarbamate 3	isolated yield
a	Boc–N (2-pyrrolidinone)	Boc–N (2,3-dihydropyrrole)	81%
b	Cbz–N, COOMe (5-oxopyrrolidine-2-carboxylate)	Cbz–N, COOMe	78%
c	Boc–N, EtOOC (5-oxopyrrolidine-2-carboxylate)	Boc–N, EtOOC (>99.5% ee)	95%
d	Boc–N (2-piperidinone)	Boc–N (tetrahydropyridine)	83%
e	Bn–N (glutarimide)	Bn–N, O	89%
f	Boc–N (azepanone)	Boc–N (dehydroazepane)	90%

1. Department of Process Research and Development, Bristol-Myers Squibb, New Brunswick, NJ 08903-0191, USA, E-mail: jurongyu@yahoo.com.
2. ResCom, DSM Pharma Chemicals Regensburg GmbH, D-93055 Regensburg, Germany.

3. (a) Altman, K.-H. *Tetrahedron Lett.* **1993**, *34*, 7721-7724; (b) Chamberlin, A. R.; Nguyen, H. D.; Chung, J. Y. L. *J. Org. Chem.* **1984**, *49*, 1682-1688; (c) Hubert, J. C.; Wijnberg, J. B. P. A.; Speckamp, W. N. *Tetrahedron* **1975**, *31*, 1437-1441; (d) Thomas, E. W.; Rynbrandt, R.

H.; Zimmermann, D. C.; Bell, L. T.; Muchmore, C. R.; Yankee, E. W.; *J. Org. Chem.* **1989**, *54*, 4535-4543.

4. (a) Langlois, N.; Rojas, A. *Tetrahedron* **1993**, *49*, 77-82; (b) Dieter, R. K.; Sharma, R. R. *J. Org. Chem.* **1996**, *61*, 4180-4184.

5. (a) Shono T. *Tetrahedron* **1984**, *40*, 811-850; (b) Shono, T.; Matsumura, Y.; Tsubata, K.; Sugihara, Y.; Yamane, S.; Kanazawa, T.; Aoki, T. *J. Am. Chem. Soc.* **1982**, *104*, 6697-6703; also see references 4b and 7.

6. Cossy, J.; Cases, M.; Pardo, D. G. *Synth. Commun.* **1997**, *27*, 2769-2776; also see ref 4b and 7 for discussions about dehydration protocols.

7. Oliveira, D. F.; Miranda, P. C. M. L.; Correia, C. R. D. *J. Org. Chem.* **1999**, 64, 6646-6652.

8. Yu, J.; Truc V.; Riebel, P.; Hierl, E.; Mudryk, B.; *Tetrahedron Lett.* **2005**, *46*, 4011-4013.

9. Nagasaka, T.; Tamano, H.; Maekawa, T.; Hamaguchi, F. *Heterocycles* **1987**, *26*, 617-624.

Appendix
Chemical Abstracts Nomenclature; (Registry Number)

N-(*tert*-Butyloxycarbonyl)pyrrolidin-2-one; (85909-08-6)

Super-Hydride®: Lithium triethylborohydride; (22560-16-3)

DMAP: *N, N*-Dimethyl-4-Pyridinamine: (1122-58-3)

N,*N*-Diisopropylethylamine: *N*-Ethyl-*N*-(1-methylethyl)-2-propanamine; (7087-68-5)

Trifluoracetic anhydride; (407-25-0)

1-*tert*-Butoxycarbonyl-2,3-dihydropyrrole: 1*H*-Pyrrole-1-carboxylic acid, 2,3-dihydro-, 1,1-dimethylethyl ester; (73286-71-2)

Jurong Yu received his PhD in organic chemistry from Shanghai Institute of Materia Medica, Chinese Academy of Sciences in 1988. After being a research assistant professor for two years, he moved to the United States in 1991. He did postdoctoral research for six years with Prof. J. R. Falck at the University of Texas Southwestern Medical Center at Dallas, where he worked on natural product synthesis and new synthetic methodology development. He joined Bristol-Myers Squibb Process R&D in 1997, and was a Senior Research Investigator before moving back to China in 2007. He is currently the Vice President of R&D at Nanjing Pharmatechs.

Vu Chi Truc obtained his PhD working on the syntheses and conformational analyses of 6- and 7- carbomethoxy-1 and 2-heteradecalin systems with Prof. J. A. Hirsch at Seton Hall University in 1985. He then joined the research group of Prof. M. E. Jung at UCLA for a post-Doc position for more than a year working on the syntheses of Ivermectin, an antiparasitic agent. He is currently a Principal Scientist in the Process Development Department at Bristol-Myers Squibb in New Brunswick, NJ.

Peter Riebel (born 1967 in Munich) studied chemistry at the University of Regensburg and obtained his PhD in 1996 within the workgroup of Prof. Juergen Sauer by contributing with studies on (3+2)-cycloaddition reactions. After a one year post-doc fellowship in the group of Prof. Victor Snieckus (University of Waterloo, Canada), funded by the DFG, he started his career with DSM. He is currently managing director of a site located in Regensburg which is dedicated to the early chemical development activities within DSM Pharma Chemicals (ResCom ®).

Boguslaw Mudryk earned his diploma at the Warsaw Technical University in 1978 under Professor Mieczyslaw Makosza. He completed his PhD thesis on electron transfer reactions of nitroalkanes with enolate anions in 1982 in laboratories of Professor Makosza and of Professor Glen Russell at Iowa State University. He then joined the Institute of Organic Chemistry at the Polish Academy of Sciences where he conducted research on vicarious nucleophilic substitution. In 1987 he joined Professor Thedore Cohen's group at the University of Pittsburgh where he focused his postdoctoral research on reductive lithiations of cyclic ethers. In 1997 he joined Bristol-Myers Squibb where he is now Principal Scientist responsible for developing practical synthetic routes to drug candidates.

Matthias Maywald (born 1976) received his diploma degree at the University of Göttingen in 2001 where he worked in the laboratory of Professor Armin de Meijere. He completed his Ph. D. thesis on semisynthetic enzymes and high-throughput experimentation in 2005 at the Max Planck Institute for Coal Research in Muelheim/Ruhr under Professor Manfred T. Reetz. In 2006 he began post-doctoral research with Professor Andreas Pfaltz at the University of Basel. His research interests include asymmetric synthesis and the development of new synthetic methods.

(S)-5-PYRROLIDIN-2-YL-1H-TETRAZOLE

Submitted by Valentina Aureggi,[1,2] Vilius Franckevičius,[3] Matthew O. Kitching,[3] Steven V. Ley,[3] Deborah A. Longbottom,[3] Alexander J. Oelke[3] and Gottfried Sedelmeier.[1,2]

Checked by Scott E. Denmark and Russell C. Smith.

1. Procedure

A. *(S)-2-Amido-pyrrolidine-1-carboxylic acid benzyl ester* (**2**). An oven-dried, three-necked, 1-L, round-bottomed flask equipped with an argon inlet, septum, thermocouple and a stirring bar is charged with Cbz-L-proline

Org. Synth. **2008**, *85*, 72-87
Published on the Web 10/16/2007

(1) (20.0 g, 80.2 mmol) (Note 1), di-*tert*-butyl dicarbonate (22.7 g, 104 mmol, 1.3 equiv) (Note 2), ammonium bicarbonate (7.60 g, 96.2 mmol, 1.2 equiv) (Notes 3 and 4) and acetonitrile (400 mL) (Note 5) under an argon atmosphere. After all reagents are combined, a cloudy, white mixture forms. Pyridine (3.89 mL, 48.1 mmol, 0.6 equiv) (Note 6) is added in one portion via syringe and the mixture is stirred for 5 h at room temperature (Note 7). After complete consumption of the starting material (as monitored by thin layer chromatography, Notes 8 and 9), the reaction mixture is transferred to a one-necked, 1-L, round-bottomed flask and the solvent is removed under reduced pressure with a rotary evaporator (25 °C, 15 mmHg) until approximately 100 mL remains. Ethyl acetate (200 mL) and water (200 mL) are added, the mixture is transferred to a 1-L separatory funnel and the organic phase is separated. The aqueous phase is extracted further with ethyl acetate (2 x 200 mL) and the combined organic phases are washed with brine (200 mL), then are dried (MgSO₄, 16 g), filtered, and concentrated under reduced pressure with a rotary evaporator (23-40 °C, 30 mmHg) to give a white solid (Note 10). Recrystallization of the solid from ethyl acetate (Note 11) affords 17.98-18.95 g (90-95%) of **2** as white cubes (Note 12, 13 and 14).

B. *(S)-2-Cyano-pyrrolidine-1-carboxylic acid benzyl ester* (**3**). An oven-dried, three-necked, 1–L, round-bottomed flask equipped with an argon inlet, septum, thermocouple and a stirring bar is charged with (*S*)-2-amido-pyrrolidine-1-carboxylic acid benzyl ester (**2**) (17.79 g, 71.7 mmol) and *N,N*-dimethylformamide (217 mL) (Note 15) under an argon atmosphere, which results in a clear, colorless solution. The solution is cooled in an ice/water bath for 20 min until an internal temperature of 2 °C is reached and cyanuric chloride (8.59 g, 46.6 mmol, 0.65 equiv) (Note 16) is then added in one portion. The reaction mixture is stirred at 4-6 °C (internal temperature) for 1 h, at which point the ice bath is removed and the mixture is allowed to warm to room temperature over 45 min and is stirred for an additional 2.25 h (Note 17). After complete consumption of the starting material (as monitored by thin layer chromatography, Note 18), the mixture is cooled to 5 °C using an ice-water bath and distilled water (200 mL) is added slowly (Note 19). The mixture is transferred to a 1-L separatory funnel and is extracted with ethyl acetate (3 x 200 mL). The combined organic phases are washed with lithium chloride solution (10 wt % in distilled water, 3 x 200 mL, Note 20), then are dried (MgSO₄, 18 g), filtered, and concentrated under reduced pressure with a rotary evaporator (24-40 °C, 30 mmHg) to give a colorless, viscous oil. Suction filtration through a pad of silica (Note 21) affords 16.22 g (97%) of **3** as a colorless oil, which turns to a milky, viscous oil upon storing in the refrigerator (Notes 22, 23 and 24).

C. *(S)-2-(1H-Tetrazol-5-yl)-pyrrolidin-1-carboxylic acid benzyl ester* (**4**) (Note 25). An oven-dried, single-necked, 250-mL, round-bottomed flask equipped with a gas inlet adaptor and a stirring bar is evacuated and backfilled with argon. The glass Ar adaptor is quickly removed and the flask is charged with (*S*)-2-cyano-pyrrolidine-1-carboxylic acid benzyl ester (**3**) (15.28 g, 66.4 mmol, 1.0 equiv), sodium azide (5.61 g, 86.3 mmol, 1.3 equiv) (Notes 26 and 27), triethylamine hydrochloride (11.9 g, 86.3 mmol, 1.3 equiv) (Notes 28 and 29) and toluene (65 mL) (Note 30) under an argon atmosphere, which results in a white mixture upon stirring. A reflux condenser with an Ar adaptor is fitted to the flask and the reaction mixture is heated to 95 °C (external temperature) in an oil bath for 24 h under an argon atmosphere (Note 31). After complete consumption of starting material (as monitored by thin layer chromatography, Note 32), deionized water (100 mL) is added and the mixture is transferred to a 250-mL separatory funnel. The flask is rinsed with additional toluene (25 mL) and deionized water (25 mL) and the aqueous phase is separated. The organic phase is further washed with water (50 mL) and the combined aqueous extracts are transferred to a 500-mL Erlenmeyer flask and are cooled with stirring in an ice/water bath to 0 °C (external temperature). Sodium nitrite solution (20 wt % aqueous, 21 mL, 61 mmol, Note 33) is added in one portion, followed by dropwise addition of sulfuric acid (20 wt % aqueous, 20 mL, 72 mmol, Note 34) with vigorous stirring until gas evolution ceases (Note 35), the solution is acidic (Note 36) and a sticky orange solid is formed (Note 37). The aqueous mixture is transferred to a 500-mL separatory funnel, the flask is rinsed with ethyl acetate (50 mL) and the aqueous phase is extracted with ethyl acetate (3 x 50 mL) (Note 38). The combined organic extracts are dried (MgSO$_4$, 18 g), filtered, and the solvent is removed under reduced pressure with a rotary evaporator (24-45 °C, 30 mmHg) to afford crude product **4** as a sticky, orange foam (Notes 39, 40 and 41). This material is then used directly in the next step.

D. *(S)-5-Pyrrolidin-2-yl-1H-tetrazole* (**5**). A 500-mL, single-necked, round-bottomed flask equipped with a stirring bar is charged with a solution of (*S*)-2-(1*H*-tetrazol-5-yl)-pyrrolidin-1-carboxylic acid benzyl ester (**4**) (15.33 g, 56.1 mmol) in ethanol (255 mL) (Note 42). Palladium-on-carbon

74

(10 wt %, 1.49 g) (Note 43) is added to the solution under an argon atmosphere. The flask is evacuated and purged with hydrogen gas five times on a hydrogen manifold. The mixture is then stirred under a hydrogen atmosphere at room temperature for 20-24 h (Note 44). After complete conversion (as monitored by thin layer chromatography, Note 45), the catalyst is removed by filtration through Celite using a medium-porosity fritted funnel (Notes 46 and 47), and the Celite is washed sequentially with ethanol (30 mL), acetic acid (10 mL) (Note 48) and water (50 mL), and then again with ethanol (30 mL), acetic acid (10 mL) and water (50 mL) (Note 49). The filtrate is concentrated under reduced pressure with a rotary evaporator (45 °C, 115 mmHg to remove ethanol, then 45 °C, 50 mmHg to remove water, then 45 °C, 22 mmHg to remove acetic acid), and then is dried under vacuum (0.1 mmHg) at room temperature for 13 h in a 50-mL, round-bottomed flask to afford 7.12-7.68 g (91-98%) of crude **5** as a pale brown solid. Ethanol (25 mL) is then added (Note 50) and a reflux condenser is attached to the flask. The suspension is heated in a 90 °C oil bath for 1 h, then is allowed to cool to room temperature before being cooled in the freezer (–23 °C) for 15 h. The precipitate is isolated by suction filtration, then is washed with cold ethanol (2 x 10 mL) (Notes 51 and 52) and then dried in a round bottom flask in a 40 °C oil bath (with a magnetic stirrer) under vacuum (0.1 mmHg) for 5 h, to furnish 6.09-6.94 g (78-89% over two steps) of **5** as a white solid (Notes 53 and 54).

2. Notes

1. Cbz-L-proline (**1**) (>99%) was purchased from Fluka and used as received.
2. The submitters used di-*tert*-butyl dicarbonate (>99%) purchased from Aldrich, whereas the checkers used di-*tert*-butyldicarbonate (97%) from Aldrich and each was used as received.
3. Ammonium bicarbonate (>99%) was purchased from Sigma and used as received.
4. The three solids (Cbz-L-proline (**1**), di-*tert*-butyl dicarbonate and ammonium bicarbonate) may be added in any order prior to the addition of acetonitrile.
5. The submitters used acetonitrile (HPLC grade, 99.99%) purchased from Fisher whereas the checkers used acetonitrile (HPLC grade,

99.99%) from Acros, and the solvent was distilled over calcium hydride prior to use.

6.   The submitters used pyridine (99.8%, Sureseal) purchased from Aldrich, which was used as received. The checkers used pyridine purchased from Fisher that was distilled over calcium hydride prior to use.

7.   The reaction mixture was a cloudy suspension for the duration of the reaction.

8.   The submitters reported that the R_f values of the starting material and product are 0.21 and 0.26 respectively (EtOAc). The checkers found that the use of 99/1 ethyl acetate/acetic acid improved the resolution of the spots on TLC. The R_f values of the starting material and product are 0.36 and 0.30, respectively.

9.   The submitters found that prior to work-up of the reaction, there is a UV active impurity, which does not stain in permanganate ($KMnO_4$) or molybdate ($[NH_4]_6Mo_7O_{24}.4H_2O$) dips. $R_f = 0.42$ (EtOAc). Following work-up, it is no longer present. The checkers did not observe this impurity by TLC.

10.   Crude product is usually isolated as a white crystalline solid, but is occasionally a white foam. The checkers had a difficult time solidifying the product initially, but ultimately could do so by scratching the flask and removing excess ethyl acetate under high vacuum (0.05 mmHg, 23 °C) for 6 h.

11.   The white solid (19.81 g) was transferred to a 125-mL Erlenmeyer flask and the solids were dissolved in boiling ethyl acetate (60 mL) to give a tan solution. The solution was slowly cooled to room temperature and was allowed to stand overnight. The flask was cooled in an ice/H_2O bath for 1 h prior to collection of the precipitate by suction filtration. The solids were washed with cold ethyl acetate (2 x 10 mL) and then were dried at 23 °C under high vacuum (0.05 mmHg) for 6 h to give 17.98 g (90%) of **2** as white cubes.

12.   The submitters purified **2** by chromatography using Breckland Scientific Silica Gel 60 (0.040 – 0.063 mm). The white solid was dissolved in dichloromethane (30 mL) and suction filtered through a pad of silica gel (6 cm diameter by 6 cm height), eluting with ethyl acetate (2.5 L), collecting in 10 x 50-mL round-bottomed flasks followed by 4 x 500-mL round-bottomed flasks. All fractions except fractions one to four were combined. The solvent was removed under reduced pressure with a rotary evaporator (380 mmHg, 30 °C).

13. (S)-2-Amido-pyrrolidine-1-carboxylic acid benzyl ester (2) has the following physicochemical properties: $[\alpha]_D^{25}$ −100.6 (c 0.51, CHCl₃); mp 91-93 °C; IR (film) cm⁻¹: 3329 (w), 2976 (w), 2945 (w), 1693 (s), 1674 (s), 1416 (s), 1356 (m), 1240 (w), 1117 (w), 1091 (w); ¹H NMR (500 MHz, CDCl₃) (mixture of rotamers) δ: 7.23-7.40 (br m, 5 H, PhH), 6.72 (app s), 6.13 (app s), and 5.98 (app s, 2 H, NH₂), 5.08-5.18 (m, 2 H, PhCH₂), 4.29-4.34 (m, 1 H, NCH), 3.42-3.53 (m, 2 H, NCH₂), 2.28 (app s), and 2.14 (app s, 2 H, CH₂-CH₂), 1.87-2.03 (m, 2 H, CH₂-CH₂); ¹³C NMR (125 MHz, CDCl₃) (mixture of rotamers) δ: 175.3 and 174.4 (C(O)NH₂), 155.9 and 155.0 (C(O)O), 136.3 (Ph), 128.4 (Ph), 128.0 (Ph), 127.8 (Ph), 67.2 (PhCH₂), 60.6 and 60.1 (NCH), 47.4 and 46.9 (NCH₂), 31.0 and 28.5 (NCHCH₂), 24.4 and 23.5 (NHCH₂CH₂); HRMS (ESI) m/z: calcd. for $C_{13}H_{16}N_2O_3$ [M+H]⁺, 249.1239; found [M+H]⁺, 249.1239. Anal. Calcd. for $C_{13}H_{16}N_2O_3$: C, 62.89; H, 6.50; N, 11.28. Found: C, 63.00; H, 6.49; N, 11.31.

14. The enantiomeric composition of 2 was checked by the use of CSP-HPLC: (R)-2, t_R 14.2 min (<0.1); (S)-2, t_R 19.5 min (>99.9) (Daicel Chiralpak AD-H, hexane/i-PrOH, 90:10, 1.0 mL/min, 210 and 254 nm).

15. N,N-Dimethylformamide (extra dry with molecular sieves, water <50 ppm) was purchased from Acros and used as received.

16. Cyanuric chloride (>99%) was purchased from Acros and used as received.

17. The reaction mixture changed from a pale yellow solution to a pale yellow suspension when reaction is complete.

18. The reaction was sampled by removing ~200 μL of the reaction mixture and quenching onto water (1 mL). The sample was extracted using ethyl acetate (1 mL). The R_f values of the starting material and product are 0.00 and 0.24 respectively (CH₂Cl₂). The submitters found a volatile minor impurity (R_f = 0.37, CH₂Cl₂), which is removed under vacuum (0.4 mbar) and not seen in the ¹H NMR spectrum of the product. The checkers observed a minor impurity (R_f = 0.05, CH₂Cl₂), which was removed after aqueous work-up.

19. As the water was added, the internal temperature of the mixture increased from 5 °C to 37 °C.

20. LiCl (99+%) was purchased from Sigma and used as received. LiCl (50 g) was dissolved in deionized water (450 g) to make a stock solution.

21.   The yellow oil is dissolved in dichloromethane (30 mL) and pushed through a pad of silica gel (6 cm diameter by 6 cm height, 3.5 psi), eluting with dichloromethane (2 L), collecting in 5 x 100-mL Erlenmeyer flasks followed by 3 x 500-mL Erlenmeyer flasks. All fractions except fraction one are combined on the basis of TLC analysis. The solvent is removed under reduced pressure with a rotary evaporator (15 mmHg, 24 °C).

22.   Submitters reported that the yellow oil obtained after purification by filtration through silica gel solidified in the freezer after storage for 15 h. Checkers never obtained the product as a solid, even after storage in the freezer for 5 d.

23.   (S)-2-Cyano-pyrrolidine-1-carboxylic acid benzyl ester (3) has the following physicochemical properties: $[\alpha]_D^{25}$ –91.6 (c 0.995, CHCl$_3$); IR (film) cm^{-1}: 2958 (m), 2887 (m), 2239 (w), 1709 (s), 1410 (s), 1358 (s), 1267 (s), 1180 (s); ^{1}H NMR (500 MHz, CDCl$_3$) (mixture of rotamers) δ: 7.31-7.42 (m, 5 H, PhH), 5.13-5.22 (m, 2 H, PhCH$_2$), 4.61 (dd, J_1 = 7.6 Hz, J_2 = 2.7 Hz) and 4.55 (dd, 1 H J_1 = 7.6 Hz, J_2 = 2.4 Hz, NCHCN), 3.55-3.63 (m, 1 H, NCHH'), 3.37-3.62 (m, 1 H, NCHH'), 2.00-2.30 (m, 4 H, CH$_2$CH$_2$); ^{13}C NMR (126 MHz, CDCl$_3$) (mixture of rotamers) δ: 154.2 and 153.5 (NCO), 135.9 and 135.8 (Ph), 128.4 (Ph), 128.1 (Ph), 128.0 (Ph), 118.8 and 118.6 (NCHCN), 67.7 and 67.5 (PhCH$_2$), 47.4 and 46.9 (NCH), 46.2 and 45.8 (NCH$_2$), 31.6 and 30.7 (NCH$_2$CH$_2$), 24.5 and 23.6 (NCH$_2$CH$_2$); HRMS (ESI) m/z: calcd for C$_{13}$H$_{14}$N$_2$O$_2$ [M+Na]$^+$, 253.0953; found [M+Na]$^+$, 253.0954. Anal. Calcd. for C$_{13}$H$_{14}$N$_2$O$_2$: C, 67.81; H, 6.13; N, 12.17. Found: C,67.53; H, 6.07; N, 12.35.

24.   The enantiomeric composition of 3 was checked by the use of CSP-HPLC: (R)-3, t_R 37.8 min (<0.1); (S)-3, t_R 41.7 min (>99.9) (Daicel Chiralpak AD-H, hexane/i-PrOH, 98:2, 1.0 mL/min, 254 nm).

25.   (S)-2-(1H-Tetrazol-5-yl)-pyrrolidin-1-carboxylic acid benzyl ester (4) may also be synthesized according to the method found in International Patents WO 2005/014602 A1 and WO 2007/009716.

26.   Sodium azide (>99%) was purchased from Sigma-Aldrich and used as received.

27.   Sodium azide must be weighed out using a non-metallic spatula in a fume hood.

28.   Triethylamine hydrochloride (>99%) was purchased from Fluka and used as received.

29.   The three solids ((S)-2-cyano-pyrrolidine-1-carboxylic acid benzyl ester (**4**), sodium azide and triethylamine hydrochloride) may be added in any order prior to the addition of toluene.

30.   Submitters used toluene (laboratory reagent grade, >99%), which was supplied by Fisher and distilled over calcium hydride prior to use. Checkers used toluene supplied by Fisher (ACS grade) and was dried by percolation through a column packed with neutral alumina and a column packed with Q5 reactant, a supported copper catalyst for scavenging oxygen, under a positive pressure of argon.

31.   Although no problems were encountered during the reaction, it is highly recommended to use a blast shield while heating the reaction.

32.   The R_f values of the starting material and product are 0.69 and 0.10, respectively (EtOAc).

33.   Sodium nitrite (97+%) was purchased from Sigma-Aldrich as used as received. $NaNO_2$ (40 g) was dissolved in deionized water (200 mL) to make a 2.9 M stock solution.

34.   A 3.6 M stock solution of aq. H_2SO_4 was prepared by diluting conc. H_2SO_4 (18 M, 80 mL) with deionized water (320 mL).

35.   When nitrous acid reacts with hydrazoic acid, rapid evolution of nitrogen and nitrous oxide gases occurs with concurrent production of water. The highly toxic hydrazoic acid side-product is thus completely and safely quenched under these conditions.

$$2NaNO_2 \text{ (aq)} + H_2SO_4 \text{ (aq)} \longrightarrow Na_2SO_4 \text{ (aq)} + 2HNO_2 \text{ (aq)}$$
$$HNO_2 \text{ (aq)} + HN_3 \text{ (aq)} \longrightarrow N_2 \text{ (g)} + N_2O \text{ (g)} + H_2O \text{ (l)}$$

36.   The pH of the solution is 5.0 after the addition of acid.

37.   Vigorous stirring during the quench is essential to prevent the stirring bar from becoming trapped in the sticky orange solid. An overhead stirrer is recommended for the quench of any reaction using more than 20 g of the (S)-2-cyano-pyrrolidine-1-carboxylic acid benzyl ester (**4**) starting material.

38.   Care must be taken during the extraction to make sure that the orange solid has dissolved completely.

39.   (S)-2-(1H-Tetrazol-5-yl)-pyrrolidin-1-carboxylic acid benzyl ester (**4**) has the following physicochemical properties: $[\alpha]_D^{25}$ −90.7 (c 1.29, $CHCl_3$); IR (film) cm^{-1}: 3111 (m), 2982 (m), 2889 (m), 1699 (s), 1422 (s), 1358 (s), 1125 (m), 698 (m); exothermic range: 204-291 °C (maximum: 253

°C); ^{1}H NMR (500 MHz, CDCl$_3$) (mixture of rotamers) δ: 7.31-7.35 (m, 3 H, PhH), 7.23 (app s, 1 H, PhH), 7.03 (app s, 1 H, PhH), 5.41-5.42 (m), 5.21 (d, J = 12.5 Hz), 5.11-5.16 (m), 5.07 (d, J = 12.5 Hz), and 5.02 (d, 3 H, J = 12.2 Hz, NCH, PhCH$_2$), 3.59-3.63 (m) and 3.51-3.56 (m, 2 H, NCH$_2$), 2.65-2.69 (m), 2.19-2.42 (m), 2.05-2.11 (m), and 1.86-2.00 (m, 4 H, CH$_2$CH$_2$); ^{13}C NMR (126 MHz, CDCl$_3$) (mixture of rotamers) δ: 156.5, 156.2 and 155.1 (NCO, NCN), 133.6 (Ph), 128.5 (Ph), 128.3 (Ph), 127.7 (Ph), 127.6 (Ph), 68.0 and 67.8 (Ph$\underline{C}$H$_2$), 52.5 and 51.3 (NCH), 47.2 and 47.0 (NCH$_2$), 33.0 and 29.7 (NCH$\underline{C}$H$_2$), 24.5 and 23.5 (NCH$_2\underline{C}$H$_2$); HRMS (ESI) m/z calcd. for C$_{13}$H$_{16}$N$_5$O$_2$, [M+H]$^+$, 274.1304); found: [M+H]$^+$, 274.1309.

40.   The submitters determined the enantiomeric purity through the use of CSP-SFC: (*R*)-**4**, t_R 22.4 min (<0.1); (*S*)-**4**, t_R 24.4 min (>99.9) (Daicel Chiralpak AD-H, *i*-PrOH, 10-20% gradient over 20 min followed by 20% isocratic elution, 1.0 mL/min, 200 nm). The checkers employed CSP-SFC under slightly different conditions: (*R*)-**4**, t_R 19.7 min (<0.1); (*S*)-**4**, t_R 21.7 min (>99.9) (Daicel Chiralpak AD, MeOH, 9-12 % (0.1 %/min), 2.7 mL/min, 150 bar, 210 nm).

41.   The checkers found that compound **4** solidified to a free-flowing powder (mp 81-84 °C) after concentrating an ethanolic solution and storing the residue at –20 °C for 3 days.

42.   Submitters used ethanol (>99.8%) that was purchased from Fluka or VWR and was used as received. Checkers used ethanol purchased from Aldrich and was used as received.

43.   Palladium on carbon (10 wt %) was purchased from Aldrich and used as received.

44.   Submitters used hydrogen gas contained in a triple-layered balloon and enters the flask via a three-way tap. No pressure is required for this reaction to work. Checkers used a hydrogen gas manifold that allows for the use of 1 atm of hydrogen pressure.

45.   The R$_f$ values of the starting material and product are 0.14 and 0.00 respectively (EtOAc). The starting material oxidizes blue in molybdate ([NH$_4$]$_6$Mo$_7$O$_{24}$.4H$_2$O) whereas the product does not.

46.   Submitters used Celite supplied by Aldrich (Celite 521), which was used as received. Checkers used Celite 545 supplied by Fisher, which was used as received.

47.   The Celite pad is 6.5 cm diameter by 2.5 cm height.

48.   Glacial laboratory reagent grade acetic acid (>99%) was purchased from Fisher Scientific and used as received.

49.   The checkers found problems with the filtration on smaller reaction scales. The filter cake must be thoroughly washed with water to completely remove the product from the Celite.

50.   Prior to the addition of ethanol, any lumps of crude material formed are crushed to a powder.

51.   The ethanol is cooled in an ice/water bath for 20 min, prior to being used to wash the (S)-5-pyrrolidin-2-yl-1H-tetrazole product (5).

52.   To obtain the yield quoted, it is imperative that the amount of ethanol used for washing is not exceeded from that stated.

53.   (S)-5-Pyrrolidin-2-yl-1H-tetrazole (5) has the following physicochemical properties: $[\alpha]_D^{25}$ –8.5 (c 1.04, MeOH); IR (film) cm^{-1}: 2966 (s), 2165 (s), 1927 (w), 1625 (s), 1460 (s), 1411 (s), 1325 (s), 1207 (m), 1117 (s), 1044 (s); exothermic range: 269-365 °C (maximum: 275 °C); ^{1}H NMR (500 MHz, d_6-DMSO) δ: 9.41 (br s, 1 H, NH), 4.77 (app t, 1 H, J = 7.5 Hz, NHCH), 3.23-3.35 (m, 2 H, NHCH$_2$), 2.31-2.33 (m, 1 H, NCHCHH'), 2.10-2.17 (m, 1 H, NHCHCHH'), 1.99-2.04 (m, 2 H, NHCH$_2$CH$_2$); ^{13}C NMR (126 MHz, d_6-DMSO) δ: 157.8 (NHCHC), 55.0 (NHCH), 44.7 (NHCH$_2$), 30.0 (NHCHCH$_2$), 23.2 (NHCH$_2$CH$_2$); HRMS (ESI) m/z: calcd for C$_5$H$_9$N$_5$ [M+Na]$^+$, 162.0756: found: [M+Na]$^+$, 162.0748. Anal. Calcd. for C$_5$H$_9$N$_5$: C, 43.15; H, 6.52; N, 50.33. Found: C, 42.83; H, 6.46; N, 50.01.

54.   The enantiomeric purity of the (S)-5-pyrrolidin-2-yl-1H-tetrazole (5), was checked by conversion to (S)-2-(1H-tetrazol-5-yl)-pyrrolidin-1-carboxylic acid benzyl ester (4) using the following procedure: In a flame-dried, 50-mL, 3-necked, round-bottomed flask equipped with a argon inlet, septum, thermocouple and a stir bar was charged (S)-5-pyrrolidin-2-yl-1H-tetrazole (5) (209 mg, 1.50 mmol). Dry dichloromethane chloride (11 mL) and benzyl chloroformate (222 µL, 1.58 mmol, 1.05 equiv) were added. The resulting suspension was cooled to 1 °C (internal) using an ice/water bath and pyridine (0.36 mL, 3.0 mmol, 3.0 equiv) was added dropwise (temperature was maintained below 5 °C). The resulting solution was stirred at 1-3 °C for 2 h before the ice bath was removed and the solution was allowed to room temperature and was stirred at 23 °C for 3 h. The solution was then diluted with ethyl acetate (100 mL) and was transferred to a 250-mL separatory funnel. The solution was washed with 1 N HCl (3 x 30 mL), brine (40 mL), then was dried (MgSO$_4$, 5 g) and filtered. The filtrate was concentrated in vacuo (30 mmHg, 25 °C) to give a colorless oil. Purification of the oil by flash column chromatography (silica

gel, EtOAc/MeOH, 93/7) afforded 390 mg (95%) of (*S*)-2-(1*H*-tetrazol-5-yl)-pyrrolidin-1-carboxylic acid benzyl ester (**4**) as a sticky, white foam. The enantiomeric composition was checked by the use of CSP-SFC. (*R*)-**4**, t_R 22.4 min (<0.1); (*S*)-**4**, t_R 24.4 min (>99.9); (Daicel Chiralpak AD-H, *i*-PrOH, 10-20% gradient over 20 min followed by 20% isocratic elution, 1.0 mL/min, 200 nm). The checkers employed CSP-SFC under slightly different conditions: (*R*)-**4**, t_R 19.7 min (<0.1); (*S*)-**4**, t_R 21.7 min (>99.9) (Daicel Chiralpak AD, MeOH, 9-12 % (0.1 %/min), 2.7 mL/min, 150 bar, 210 nm).

Safety and Waste Disposal Information

All hazardous materials should be handled and disposed of in accordance with "Prudent Practices in the Laboratory"; National Academy Press; Washington, DC, 1995.

3. Discussion

(*S*)-5-Pyrrolidin-2-yl-1*H*-tetrazole (**5**) is a novel proline-derived organocatalyst that was developed by the Ley group[4,5] and others[6,7] almost simultaneously. Its utility has been proven in several reaction processes, including the Mannich (A, Scheme)[4,5] and aldol reactions,[6-8] addition of ketones to nitro-olefins,[5,9] nitroalkanes/malonates to enones (B)[10-12] and α-oxyamination (C)[13-15] among others.[16-27]

The preparation of **5** shown here is a variant of those which already exist,[5,8,13,28-35] and offers the following advantages:

1) The procedures can all be carried out safely, on large scale without detriment to the yields.
2) The azide cyclization procedure avoids the generation of explosive ammonium azide during the reaction.
3) The hydrogenation protocol avoids the use of a 9/1 acetic acid:water mixture as the solvent and the concurrent extended (3 d) reaction time.
4) Product is obtained that requires very little purification.

This preparation can also be used to obtain the enantiomer of the desired product, (*R*)-5-pyrrolidin-2-yl-1*H*-tetrazole.[36]

1. Process R & D, Chemical and Analytical Development, Novartis Pharmaceuticals Corporation, 4002 Basel, Switzerland.
2. Provided the preparation method for (*S*)-5-pyrrolidin-2-yl-1*H*-tetrazole (**5**), together with the analytical data for (*S*)-2-(1*H*-Tetrazol-5-yl)-pyrrolidine-1-carboxylic acid benzyl ester (**4**) and (*S*)-5-pyrrolidin-2-yl-1*H*-tetrazole (**5**).
3. University of Cambridge, Department of Chemistry, Lensfield Road, Cambridge, CB2 1EW, United Kingdom.
4. Cobb, A. J. A.; Shaw, D. M.; Ley, S. V. *Synlett* **2004**, 558-560.
5. Cobb, A. J. A.; Shaw, D. M.; Longbottom, D. A.; Gold, J. B.; Ley, S. V. *Org. Biomol. Chem.* **2005**, *3*, 84-96.
6. Torii, H.; Nakadai, M.; Ishihara, K.; Saito, S.; Yamamoto, H. *Angew. Chem. Int. Ed.* **2004**, *43*, 1983-1986.
7. Hartikka, A.; Arvidsson, P. I. *Tetrahedron: Asymmetry* **2004**, *15*, 1831-1834.
8. Hartikka, A.; Arvidsson, P. I. *Eur. J. Org. Chem.* **2005**, 4287-4295.
9. Cobb, A. J. A.; Longbottom, D. A.; Shaw, D. M.; Ley, S. V. *Chem. Commun.* **2004**, 1808-1809.
10. Mitchell, C. E. T.; Brenner, S. E.; Ley, S. V. *Chem. Commun.* **2005**, 5346-5348.
11. Mitchell, C. E. T.; Brenner, S. E.; Garcia-Fortanet, J. *Org. Biomol. Chem.* **2006**, 2039-2039.
12. Rahbek Knudsen, K.; Mitchell, C. E. T.; Ley, S. V. *Chem. Commun.* **2006**, 66-68.

13. Momiyama, N.; Torii, H.; Saito, S.; Yamamoto, H. *Proc. Natl. Acad. Sci. USA* **2004**, *101*, 5374-5378.
14. Kim, S.-G.; Park, T.-H. *Tetrahedron Lett.* **2006**, *47*, 9067-9071.
15. Ramachary, D. B.; Barbas III, C. F. *Org. Lett.* **2005**, *7*, 1577-1580.
16. For recent reviews on the use of (*S*)-5-pyrrolidin-2-yl-1*H*-tetrazole (**5**) in organic synthesis, see: (a) Longbottom, D. A.; Franckevičius, V.; Kumarn, S.; Oelke, A. J.; Wascholowski, V.; Ley, S. V. *Aldrichimica Acta* **2007**, *in press*; (b) Longbottom, D. A.; Franckevičius, V.; Ley, S. V. *Chimia* **2007**, *5*, 247-256; (c) Limbach, M. *Chem. Biodiv.* **2006**, *2*, 119-133.
17. Chowdari, N. S.; Barbas III, C. F. *Org. Lett.* **2005**, *7*, 867-870.
18. Kumarn, S.; Shaw, D. M.; Longbottom, D. A.; Ley, S. V. *Org. Lett.* **2005**, *7*, 4189-4191.
19. Kumarn, S.; Shaw, D. M.; Ley, S. V. *Chem. Commun.* **2006**, 3211-3213.
20. Hansen, H. M.; Longbottom, D. A.; Ley, S. V. *Chem. Commun.* **2006**, 4838-4840.
21. Oelke, A. J.; Kumarn, S.; Longbottom, D. A.; Ley, S. V. *Synlett* **2006**, 2548-2552.
22. Sunden, H.; Ibrahem, I.; Eriksson, L.; Cordova, A. *Angew. Chem. Int. Ed.* **2005**, *44*, 4877-4880.
23. Yamamoto, Y.; Momiyama, N.; Yamamoto, H. *J. Am. Chem. Soc.* **2004**, *126*, 5962-5963.
24. Momiyama, N.; Yamamoto, Y.; Yamamoto, H. *J. Am. Chem. Soc.* **2007**, *129*, 1190-1195.
25. Ward, D. E.; Jheengut, V.; Beye, G. E. *J. Org. Chem.* **2006**, *71*, 8989-8992.
26. Suri, J. T.; Ramachary, D. B.; Barbas III, C. F. *Org. Lett.* **2005**, *7*, 1383-1385.
27. Thayumanavan, N. R.; Tanaka, F.; Barbas III, C. F. *Org. Lett.* **2004**, *6*, 3541-3544.
28. Grzonka, Z.; Liberek, B. *Roczniki Chem.* **1971**, *45*, 967-980.
29. Grzonka, Z.; Liberek, B. *Tetrahedron* **1971**, *27*, 1783-1787.
30. Grzonka, Z.; Gwizdala, E.; Kofluk, T. *Polish J. Chem.* **1978**, *52*, 1411-1417.
31. Almquist, R. G.; Chao, W. R.; Jennings-White, C. *J. Med. Chem.* **1985**, *28*, 1067-1071.
32. Koguro, K.; Oga, T.; Mitsui, S.; Orita, R. *Synthesis* **1998**, 910-914.

33. Bridge, A. W.; Jones, R. H.; Kabir, H.; Kee, A. A.; Lythgoe, D. J.; Nakach, M.; Pemberton, C.; Wrightman, J. A. *Org. Proc. Res. Dev.* **2001**, *5*, 9-15.
34. Demko, Z. P.; Sharpless, K. B. *Org. Lett.* **2002**, *4*, 2525-2527.
35. Franckevičius, V.; Rahbek Knudsen, K.; Ladlow, M.; Longbottom, D. A.; Ley, S. V. *Synlett* **2006**, 889-892.
36. (*R*)-5-Pyrrolidin-2-yl-1*H*-tetrazole has the same physical properties as the (*S*)-enantiomer, aside from the optical rotation, which is equal and opposite: $[\alpha]_D^{25}$ +9 (*c* 1.0, MeOH). The enantiomeric composition was checked by the use of CSP-SFC. (*R*)-**4**, t_R 22.4 min (>99.9); (*S*)-**4**, t_R 24.4 min (<0.1); (Daicel Chiralpak AD-H, *i*-PrOH, 10-20% gradient over 20 min followed by 20% isocratic elution, 1.0 mL/min, 200 nm)

Appendix
Chemical Abstracts Nomenclature; (Registry Number)

(*S*)-5-Pyrrolidin-2-yl-1*H*-tetrazole: 2*H*-Tetrazole, 5-[(2*S*)-2-pyrrolidinyl]-;
    (33878-70-5)
(*S*)-2-Amido-pyrrolidine-1-carboxylic acid benzyl ester: 1-
    Pyrrolidinecarboxylic acid, 2-(aminocarbonyl)-, phenylmethyl ester,
    (2*S*)-; (34079-31-7)
Cbz-L-proline: 1,2-Pyrrolidinedicarboxylic acid, 1-(phenylmethyl) ester,
    (2*S*)-; (1148-11-4)
Di-*tert*-butyl dicarbonate: Dicarbonic acid, C,C'-bis(1,1-dimethylethyl)
    ester; (24424-99-5)
(*S*)-2-Cyano-pyrrolidine-1-carboxylic acid benzyl ester: 1-
    Pyrrolidinecarboxylic acid, 2-cyano-, phenylmethyl ester, (2*S*)-;
    (63808-36-6)
Cyanuric chloride: 2,4,6-Trichloro-1,3,5-triazine; (108-77-0)
(*S*)-2-(1*H*-Tetrazol-5-yl)-pyrrolidin-1-carboxylic acid benzyl ester: 1-
    Pyrrolidinecarboxylic acid, 2-(2*H*-tetrazol-5-yl)-, phenylmethyl ester,
    (2*S*)-; (33876-20-9)
Sodium azide; (26628-22-8)
Triethylamine hydrochloride: Ethanamine, *N,N*-diethyl-, hydrochloride
    (1:1); (554-68-7)

Steven V. Ley received his PhD from Loughborough University in 1972, after which he carried out post-doctoral research with Professor Leo Paquette at Ohio State University, followed by Professor Derek Barton at Imperial College London. In 1975, he joined that Department as a lecturer and became Head of Department in 1989. In 1992, he moved to the 1702 BP Chair of Organic Chemistry at the University of Cambridge and became a Fellow of Trinity College. He was elected to the Royal Society in 1990 and was President of the Royal Society of Chemistry (RSC) 2000-02. Steve has been the recipient of many prizes and awards including the Yamada-Koga Prize, Nagoya Gold Medal, ACS Award for Creative Work in Synthetic Organic Chemistry and the Paul Karrer Medal.

Valentina Aureggi was born 1977 in Como, Italy. She obtained her Diploma in 1999 and MSci degree in 2003 at Insubria University. During this five-year course of study, her final year project was carried out at the University of Neuchâtel, investigating the synthesis of amido silyloxy dienes under the supervision of Professor Reinhard Neier. Following this, she obtained her PhD in 2007 under the supervision of Professor Gottfried Sedelmeier in the Department of Process Research and Development of Novartis Pharma in Basel. She is currently pursuing post-doctoral research in Professor Mark Lautens's group at the University of Toronto.

Vilius Franckevičius was born in 1983 in Kaunas, Lithuania. He studied Natural Sciences at the University of Cambridge, where he undertook his final year project on the development of new organocatalysts under the supervision of Professor Steven V. Ley, and subsequently obtained his MSci degree in Natural Sciences (Chemistry) in 2005 (Fitzwilliam College). He is currently a PhD student in the Ley group where he is involved in the application of organocatalytic methodology in natural product synthesis.

Matthew O. Kitching was born in 1983 in Wolverhampton, England. He completed his undergraduate degree in Natural Science at the University of Cambridge in 2005. Following this, he joined Professor Steven V. Ley's group as a summer project student where he worked on development of new organocatalytic species, including the pyrrolidinyl tetrazole. After completing his MSci degree in Natural Sciences (Chemistry) at the University of Cambridge, working with Dr. Jane Clarke on protein thermodynamic stability and its relationship to disease, he returned to Professor Steven V. Ley's group for his PhD as part of the Innovative Technology Centre studying microencapsulated palladium catalysts.

86

Deborah A. Longbottom received her undergraduate degree from the University of Durham in 1997 and, following a year working in the pharmaceutical industry, came to Cambridge to carry out her PhD under the guidance of Professor Steven V. Ley. In 2002, she pursued post-doctoral research with K. C. Nicolaou and returned to the Ley Group early in 2004. Her research interests have encompassed both natural product synthesis, e.g. polyenoyltetramic acids and depsipeptides, and method development e.g. novel uses of the Burgess reagent and organocatalytic methodologies. Currently, she is a senior research associate in the Ley group. As of October 2007, she will hold a joint position between the Department of Chemistry and Homerton College, Cambridge, with an additional By-fellowship at Churchill College.

Alexander J. Oelke was born in 1980 in Reinbek, Germany. He studied chemistry at the University of Hamburg, where he obtained his Diploma in 2006 under the supervision of Professor Chris Meier and in collaboration with Professor Steven V. Ley for the development of an organocatalytic tandem procedure for the synthesis of chiral pyridazine derivatives. He is currently a PhD student in the Ley group at the University of Cambridge, where he is involved in the application of organocatalytic methodology in natural product synthesis.

Gottfried Sedelmeier was born in 1948 in Schallstadt, Germany. He studied Chemistry at the University of Freiburg, where he obtained his PhD in 1979 under the supervision of Professor Horst Prinzbach. He then joined the Process Research Department of Ciba-Geigy Pharma (now Novartis) in Basel, where he has worked until the present. Since 1991, he has also held a Lectureship position at the University of Freiburg, where he teaches on the subject of Industrial Pharmaceutical Chemistry and in 2003, he became an honorary Professor at that same institution.

As an undergraduate student at Illinois Wesleyan University, Russell Smith performed organic research for four years. His undergraduate advisor, Dr. Ram S. Mohan, sparked his love for organic chemistry and experimentation and he graduated with a BS degree with Research Honors in 2004. He joined the research group of Scott Denmark at the University of Illinois, where his current project is studying the use of aryl silanolates in cross-coupling reactions to form biaryl compounds. When not in lab, he enjoys spending time with his wife, Lindsay, and dog, Lucky. He is a football fanatic and cannot wait until fall every year.

SINGLE-STEP SYNTHESIS OF ALKYNYL IMINES FROM N-VINYL AND N-ARYL AMIDES. SYNTHESIS OF N-[1-PHENYL-3-(TRIMETHYLSILYL)-2-PROPYNYLIDENE]-BENZENAMINE.

Submitted by Mohammad Movassaghi and Matthew D. Hill.[1]
Checked by Rie Motoki and Masakatsu Shibasaki.

1. Procedure

N-[1-phenyl-3-(trimethylsilyl)-prop-2-ynylidene]-benzenamine.   A flame-dried, 200-mL, single-necked, round-bottomed flask is equipped with a 2.5-cm football-shaped stir bar, sealed with a rubber septum under an argon atmosphere, and fitted with an argon-filled balloon through a syringe needle. The flask is charged with anhydrous tetrahydrofuran (80 mL, Note 1) and cooled to 0 °C (external temperature) in an ice-bath. Vigorous stirring is commenced and (trimethylsilyl)acetylene (13.3 mL, 9.41 g, 95.8 mmol, 2.70 equiv, Note 2) is added to the flask via syringe. Butyllithium (40.6 mL, 95.8 mmol, 2.36 M, 2.70 equiv, Note 3) is added to the THF solution of (trimethylsilyl)acetylene over 10 min via syringe as the mixture was kept at 0 °C.   After 5 min, the colorless solution of the lithium (trimethylsilyl)acetylide is removed from the ice-bath and allowed to warm to ambient temperature and kept at this temperature for an additional 40 min. During this time, another flame-dried, 300-mL, single-neck round-bottomed flask is equipped with a 2.5-cm football-shaped stir bar, charged with solid copper(I) bromide–dimethyl sulfide complex (19.71 g, 95.84 mmol, 2.700 equiv, Note 2), sealed with a rubber septum under an argon atmosphere, and fitted with an argon-filled balloon through a syringe needle. The flask containing copper bromide is charged with tetrahydrofuran (80 mL, Note 1)

Org. Synth. **2008**, *85*, 88-95
Published on the Web 10/11/2007

and the mixture is cooled in a dry ice-acetone bath (–78 °C, external temperature). The lithium (trimethylsilyl)acetylide solution at ambient temperature is slowly transferred via cannula over 5 min to the cold (–78 °C, external temperature) flask containing copper bromide. The bright-yellow solution is maintained at –78 °C for an additional 5 min. The flask containing the yellow copper acetylide solution is moved from the dry ice-acetone bath to an ice-water bath and maintained for 20 min prior to use.

A flame-dried, 500-mL single-necked, round-bottomed flask equipped with a 2.5-cm stir bar is charged with benzanilide (1, 7.00 g, 35.5 mmol, 1 equiv, Note 2), sealed with a rubber septum under an argon atmosphere, and fitted with an argon-filled balloon through a syringe needle. Anhydrous dichloromethane (71 mL, Note 1) followed by 2-chloropyridine (13.4 mL, 16.1 g, 142 mmol, 4.00 equiv, Note 4) are added via syringe and the heterogeneous mixture is vigorously stirred and cooled by placement in a dry ice-acetone bath (–78 °C, external temperature). After 10 min, trifluoromethanesulfonic anhydride (Tf$_2$O, 7.03 mL, 12.0 g, 42.6 mmol, 1.20 equiv, Note 2) is added via syringe over 2 min. After five min, the reaction flask is moved to an ice-water bath for a period of five min, and then returned to the dry ice-acetone bath. After five min, the copper (trimethylsilyl)acetylide solution prepared above is transferred via cannula over 5 min to the cold (–78 °C) solution of the activated amide and the reaction mixture is kept in a dry ice-acetone bath (Note 5). After five min, the reaction flask is moved from the dry ice-acetone bath to an ice-water bath. After five min, the cold bath is removed and the reaction mixture is allowed to warm to ambient temperature. After 20 min, the reaction mixture is filtered through Celite into a 500-mL, side-armed Erlenmeyer flask with vacuum filtration (Note 6). The combined filtrate is transferred to a 500-mL round-bottomed flask and concentrated with a rotary evaporator (30 °C, 70 mmHg). The remaining brown oil is purified by flash column chromatography (Note 7) to afford 8.99 g (91%) (Note 8) of imine 2 as a bright yellow oil (Note 9).

2. Notes

1. Submitters purchased tetrahydrofuran and dichloromethane from J.T. Baker (Cycletainer[TM]), which was dried by the method of Grubbs et al. under positive argon pressure.[2] Checkers purchased dehydrated

tetrahydrofuran and dichloromethane from Kanto Chemical Co., Inc. and used without further purification.

2. Trimethylsilylacetylene (98%), copper(I) bromide-dimethyl sulfide complex (98%), benzanilide (98%), and trifluoromethanesulfonic anhydride (99+%) were purchased from Aldrich Chemical Company, Inc. and were used as received.

3. Butyllithium (2.36 M) in hexanes was purchased from Aldrich Chemical Company, Inc. and the molarity was determined by titration using diphenylacetic acid as an indicator (average of three determinations).[3]

4. 2-Chloropyridine (Aldrich Chemical Company, Inc.) was distilled from calcium hydride and stored under an argon atmosphere. The yield of the desired product dropped when less 2-chloropyridine was used (see the Supporting Information in reference 7). Prudent experimental practices should be followed in handling 2-chloropyridine and any inhalation, ingestion, skin contact, or eye contact should be avoided.

5. The reaction mixture turned brown as the acetylide was added and gradually regained a bright yellow color upon completion of the addition.

6. At ambient temperature, the yellow solution became heterogeneous and turned dark brown as white salts precipitated. The solids were removed by vacuum filtration (20 mmHg) through a plug of Celite (50 g) in a sintered glass funnel (9 cm diameter). The Celite was rinsed thoroughly with dichloromethane (100 mL). The combined filtrates were concentrated under reduced atmosphere (75 mmHg, 30 °C), at which time additional salts precipitated. This crude residue was purified by flash column chromatography.

7. Flash column chromatography (7 cm diameter, 16 cm height) was performed using silica gel (60-Å pore size, 32–63 μm, standard grade, Sorbent Technologies). Hexanes (300 mL) were used as the first eluent to remove the bis(trimethylsilylacetylene) (R_f = 0.75, hexane/ethyl acetate, 9:1, visualized by $KMnO_4$ stain). The eluent was changed to hexane/ethyl acetate, 13:1 until alkynylimine **2** (R_f = 0.50, hexane/ethyl acetate, 9:1, visualized by UV and $KMnO_4$ stain) was passed through the column (fractions 27–61 collected, ~20-mL fractions) and the combined fractions were concentrated under reduced pressure (30 °C, 75 mmHg). Any residual 2-chloropyridine (R_f = 0.33, hexane/ethyl acetate, 9:1, visualized by UV) can be readily removed from the imine **2** with a second flash column chromatography on silica gel (7 cm diameter, 16 cm height and hexane/ethyl

90

acetate, 29:1 as eluent). TLC was performed on EMD Silica Gel 60F plates. Checkers used Merck silica gel 60 (0.040-0.063 mm, for column chromatography). Checkers used TLC plates purchased from Merck.

8. Submitters obtained the product in 91–98 % yield. Checkers obtained the product in 96% yield on a half-scale reaction.

9. The analytical data for imine 2 is as follows: ^{1}H NMR (500 MHz, CDCl$_3$) δ: 0.14 (s, 9 H), 7.11-7.14 (m, 2 H), 7.17 (tt, 1 H, J = 7.5, 1.1 Hz), 7.36-7.40 (m, 2 H), 7.45-7.51 (m, 3 H), 8.18-8.21 (m, 2 H); ^{13}C NMR (125 MHz, CDCl$_3$) δ: –0.5, 97.5, 105.4, 120.9, 125.0, 128.3, 128.5, 128.6, 131.4, 137.0, 150.1, 151.7; IR (neat): 3062, 3030, 2960, 2152, 1587, 1563, 1251, 844, 748, 690 cm^{-1}; EI-HRMS calcd for C$_{18}$H$_{19}$NSi$^+$ (M$^+$): 277.1281, found 277.1281; Anal. Calcd for C$_{18}$H$_{19}$NSi: C, 77.93; H, 6.90; N, 5.05. Found: C, 78.23; H, 6.79; N, 5.06. Imine 2 is air stable at ambient temperature for weeks. For prolonged storage (several months) samples of 2 were sealed under an argon atmosphere.

Waste Disposal Information

All hazardous materials should be handled and disposed of in accordance with "Prudent Practices in the Laboratory"; National Academy Press; Washington, DC, 1995.

3. Discussion

Azaheterocycles are present in natural products, pharmaceuticals, and functional materials.[4] The cycloisomerization of readily assembled acyclic precursors offers an attractive strategy for the synthesis of heterocycles.[5,6] We reported[7] a two-step procedure for the preparation of pyridine and quinoline derivatives that takes advantage of readily available N-vinyl[8] and N-aryl amides as starting materials. This methodology required the development of the herein discussed convergent and mild synthesis of alkynyl imines, versatile precursors for a variety of azaheterocycles.[9,10] The unique activation of amides with trifluoromethanesulfonic anhydride[11] in the presence of 2-chloropyridine[12] as an acid scavenger enables the single-step synthesis of a wide range of alkynyl imines (Table 1), including sensitive N-vinyl and N-heterocyclic imines. For comparison, application of existing multi-step methods toward the synthesis of the highly sensitive N-2-thienyl and N-dihydropyranyl alkynyl imines shown in Table 1 (entries 12 and 14) gave none and <10% yield of the desired imines, respectively. Importantly, these imines were obtained using the herein described single-step procedure.

TABLE 1

Entry	Amides	Alkynyl Imines	Yield (%)[a]	Entry	Amides	Alkynyl Imines	Yield (%)[a]
	(Ph–C(O)–NH–aryl(R,R',R"))	(Ph–C≡C–SiMe$_3$ imine of aryl(R,R',R"))		9	Ar = Ph	(Ph–N=CH–CH=CH–Ar ; C≡C–SiMe$_3$)	78
				10	Ar = 1-naphthyl		86
				11	Ar = 3,4-dimethoxyphenyl		92
1	R = H R' = H R" = OMe		89				
2	R = OMe R' = H R" = H		96	12	(Ph–C(O)–NH–thienyl)	(thiophene imine, Ph–C≡C–SiMe$_3$)	63[b]
3	R = H R' = CF$_3$ R" = H		73				
	(R–C(O)–N(Ph)H)	(R–C(=N–Ph)–C≡C–SiMe$_3$)		13	(Ph–C(O)–NH–thienyl)	(thiophene imine, C≡C–SiMe$_3$)	98
4	R = ^{c}C$_6$H$_{11}$		85				
5	R = ^{t}Bu		83	14	(Ph–C(O)–NH–dihydropyranyl)	(dihydropyran imine, Ph–C≡C–SiMe$_3$)	99
6	R = N(CH$_2$CH$_2$)$_2$O		80				
7	(thienyl–C(O)–N(Ph)H)	(thienyl–C(=N–Ph)–C≡C–SiMe$_3$)	81	15	(Ph–C(O)–NH–pyrrolyl–NSi^{i}Pr$_3$)	(pyrrole NSi^{i}Pr$_3$ imine, Ph–C≡C–SiMe$_3$)	82
8	(Ph–C(O)–NH–cyclohexenyl)	(cyclohexenyl imine, Ph–C≡C–SiMe$_3$)	75				

a. Isolated yields: all entries are average of two experiments. Optimum conditions used uniformly. b. No warming prior to isolation.

1. Department of Chemistry, Massachusetts Institute of Technology, Cambridge, MA 02139, E-mail: movassag@mit.edu
2. Pangborn, A. B.; Giardello, M. A.; Grubbs, R. H.; Rosen, R. K.; Timmers, F. J. *Organometallics* **1996**, *15*, 1518-1520.
3. Kofron, W. G.; Baclawski, L. M. *J. Org. Chem.* **1976**, *41*, 1879-1880.
4. For reviews, see: (a) Undheim, K.; Benneche, T. In *Comprehensive Heterocyclic Chemistry II*; Katritzky, A. R., Rees, C. W., Scriven, E. F. V., McKillop, A., Eds.; Pergamon: Oxford, 1996; Vol. 6; p 93. (b) Jones, G. In *Comprehensive Heterocyclic Chemistry II*; Katritzky, A. R., Rees, C. W., Scriven, E. F. V., McKillop, A., Eds; Pergamon: Oxford, 1996; Vol. 5; p 167. (c) Joule, J. A.; Mills, K. In *Heterocyclic Chemistry*, 4th ed.; Blackwell Science Ltd.: Cambridge, MA, 2000; p

194. (d) Henry, G. D. *Tetrahedron* **2004**, *60*, 6043-6061. (e) Michael, J. P. *Nat. Prod. Rep.* **2005**, *22*, 627-646. (f) Abass, M. *Heterocycles* **2005**, *65*, 901-965.

5. For reviews on metal-catalyzed heterocycle synthesis, see: (a) Bönnemann, H.; Brijoux, W. *Adv. Heterocycl. Chem.* **1990**, *48*, 177-222. (b) Nakamura, I.; Yamamoto, Y. *Chem. Rev.* **2004**, *104*, 2127-2198. (c) Zeni G.; Larock, R. C. *Chem. Rev.* **2004**, *104*, 2285-2310. (d) Varela, J. A.; Saá, C. *Chem. Rev.* **2003**, *103*, 3787-3802.

6. For recent metal-catalyzed azaheterocycle syntheses, see: (a) Roesch, K. R.; Larock, R. C. *Org. Lett.* **1999**, *1*, 553-556. (b) Varela, J. A.; Castedo, L.; Saá, C. *J. Org. Chem.* **2003**, *68*, 8595-8598. (c) Zhang, X.; Campo, M. A.; Yao, T.; Larock, R. C. *Org. Lett.* **2005**, *7*, 763766. (d) McCormick, M. M.; Duong, H. A.; Zuo, G.; Louie, J. *J. Am. Chem. Soc.* **2005**, *127*, 5030-5031 and references cited therein.

7. Movassaghi, M.; Hill. M. D. *J. Am. Chem. Soc.* **2006**, *128*, 4592-4593.

8. (a) Muci, A. R.; Buchwald, S. L. *Top. Curr. Chem.* **2002**, *219*, 131-209. (b) Hartwig, J. F. In *Handbook of Organopalladium Chemistry for Organic Synthesis*; Negishi, E., Ed.; Wiley-Interscience: New York, 2002; p. 1051. (c) Beletskaya, I. P.; Cheprakov, A. V. *Coordin. Chem. Rev.* **2004**, *248*, 2337-2364. (d) Dehli, J. R.; Legros, J.; Bolm, C. *Chem. Commun.* **2005**, 973-986.

9. For representative reports on the derivatization of alkynyl imines to azaheterocycles, see: (a) Kel'in, A. V.; Sromek, A. W.; Gevorgyan, V. *J. Am. Chem. Soc.* **2001**, *123*, 2074-2075. (b) Van den Hoven, B. G.; Alper, H. *J. Am. Chem. Soc.* **2001**, *123*, 10214-10220. (c) Sangu, K.; Fuchibe, K.; Akiyama, T. *Org. Lett.* **2004**, *6*, 353-355. (d) Sromek, A. W.; Rheingold, A. L.; Wink, D. J.; Gevorgyan, V. *Synlett* **2006**, *14*, 2325-2328.

10. For reports on the synthesis of alkynyl imines via a two-step procedure involving imidoyl chlorides, see reference 9 and: (a) Ried, W.; Erle, H.-E. *Chem. Ber.* **1979**, *112*, 640-647. (b) Austin, W. B.; Bilow, N.; Kelleghan, W. J.; Lau, K. S. Y. *J. Org. Chem.* **1981**, 46, 2280-2286. (c) Lin, S.-Y.; Sheng, H.-Y.; Huang, Y.-Z. *Synthesis* **1991**, 235-236.

11. (a) Baraznenok, I. L.; Nenajdenko, V. G.; Balenkova, E. S. *Tetrahedron* **2000**, *56*, 3077-3119. (b) Charette, A. B.; Grenon, M. *Can. J. Chem.* **2001**, *79*, 1694-1703.

12. For a related discussion of amide activation using this reagent combination, see: Movassaghi, M.; Hill, M. *J. Am. Chem. Soc.* **2006**, *128*, 14254-14255.

Appendix
Chemical Abstracts Nomenclature; (Registry Number)

N-Phenylbenzenecarboxamide (benzanilide); (93-98-1)
2-Chloropyridine; (109-09-1)
Trifluoromethanesulfonic acid anhydride; (358-23-6)
1-(Trimethylsilyl)acetylene; (1066-54-2)
N-[1-Phenyl-3-(trimethylsilyl)-2-propyn-1-ylidene]-benzeneamine; (77123-64-9)

Mohammad Movassaghi carried out his undergraduate research with Professor Paul A. Bartlett at U.C. Berkeley, where he received his BS degree in chemistry in 1995. Mo then joined Professor Andrew G. Myers' group for his graduate studies and was a Roche Predoctoral Fellow at Harvard University. In 2001, Mo joined Professor Eric N. Jacobsen's group at Harvard University as a Damon Runyon–Walter Winchell Cancer Research Foundation postdoctoral fellow. In 2003, he joined the chemistry faculty at MIT where his research program has focused on the total synthesis of alkaloids in concert with the discovery and development of new reactions for organic synthesis.

Matthew D. Hill was born in 1980 and grew up in Cleveland, Ohio. Matthew pursued his undergraduate degrees at Ohio University where he studied biochemistry, molecular biology, and legal communication. While an undergraduate he worked in the labs of Professor Mark McMills at Ohio University and Professor Koji Nakanishi at Columbia University. In 2003, Matthew joined Professor Movassaghi's group at MIT where his research has focused on development of new methodologies for azaheterocycle synthesis. In his spare time, Matthew enjoys fishing, boating, volleyball, and most other outdoor recreational activities.

Rie Motoki was born in 1981 in Tokushima, Japan. She graduated in 2004 and received her M.S. degree in 2006 from the University of Tokyo under the direction of Professor Masakatsu Shibasaki. The same year she started her Ph.D. study under the supervision of Professor Shibasaki. Her research interest is design and synthesis of biologically active compounds.

PALLADIUM-CATALYZED DEHYDRATIVE ALLYLATION OF HYPOPHOSPHOROUS ACID WITH ALLYLIC ALCOHOLS. PREPARATION OF CINNAMYL-*H*-PHOSPHINIC ACID

Submitted by Karla Bravo-Altamirano and Jean-Luc Montchamp.[1]
Checked by Alena Rudolph and Mark Lautens.

1. Procedure

A 1-L, round-bottomed flask (Note 1) equipped with a magnetic stirring bar is charged under air, with a solution of concentrated hypophosphorous acid (19.80 g, 300 mmol, 2.0 equiv) (Note 2) in *N,N*-dimethylformamide (300 mL, via a graduated cylinder) (Note 3). Mesitylene (20.9 mL, 18.03 g, 150 mmol, 1.0 equiv) (Note 4) and cinnamyl alcohol (19.7 mL, 20.57 g, 153 mmol, 1.0 equiv) (Note 5) are added via syringe (Note 6). The flask is then fitted with a rubber septum and placed on a magnetic stir plate. After stirring for 5 min, Pd(OAc)$_2$ (0.067 g, 0.300 mmol, 0.002 equiv) (Note 7), and 9,9-dimethyl-4,5-bis(diphenylphosphino)xanthene (0.190 g, 0.330 mmol, 0.0022 equiv) (Note 8) are added by temporarily removing the septum. Material adhering to the sides of the reaction flask is rinsed into the reaction mixture with 5 mL of *N,N*-dimethylformamide, resulting in a clear, brown solution. Stirring is maintained and the reaction flask is equipped with a Claisen adapter fitted with a reflux condenser with nitrogen inlet, and a thermocouple temperature probe adapter. Under a nitrogen atmosphere, the system is placed in a heating mantle filled with sand (Note 9) and the thermocouple is inserted through the adapter. The solution is heated at 85 °C (internal temperature) for 7 h (Notes 10, 11). Heating and stirring are then interrupted, and the resulting solution is allowed to cool to room temperature (Notes 12, 13). After removing the nitrogen inlet and the water condenser, the reaction mixture is concentrated for 1 h by rotary evaporation (50 °C, 0.5 mmHg). The residue is diluted with ethyl acetate (150 mL) and treated with activated charcoal (3.0 g) (Note 14). The resulting heterogeneous mixture is stirred

96

for 30 min and filtered in vacuo through a Celite pad (Note 15) in a Büchner funnel. The Celite is carefully washed with three 100-mL portions of ethyl acetate (Note 16) and the combined washings are transferred to a 1-L separatory funnel. The organic layer is washed with aqueous HCl (2 M, 250 mL) (Note 17) and the aqueous phase is separated and extracted with two 125 mL portions of ethyl acetate. The combined organic layers are washed with 200 mL of brine, then are retreated with charcoal (1.0 g) and MgSO$_4$ (20 g) (Notes 18, 19), then are filtered through a second Celite pad in a Büchner funnel (Notes 15, 20), and concentrated under reduced pressure (45 °C, 150 mmHg). The resulting pale-yellow solid (around 26 g, 95% yield) (Note 21) is dissolved in 80 mL of hot dichloromethane (35–38 °C), and about 140 mL of hexane (38 °C) is added until a light-yellow, homogeneous solution is obtained. The solution is cooled at –15 °C for 2 h and the resulting white crystals (21.8 g) are collected by suction filtration on a Büchner funnel, and are washed with ice-cold hexane (100 mL). The filtrate is concentrated under reduced pressure (45 °C, 150 mmHg) and the residue is dissolved in 20 mL of dichloromethane (35 °C), and 30 mL of hexane (35 °C). The solution is cooled at –15 °C overnight and a second crop of crystals is collected by suction filtration, and are washed with ice-cold hexane (20 mL). The two crops of crystals are combined and dried overnight at 0.1 mmHg to provide 23.2 g of cinnamyl-*H*-phosphinic acid (83%) as white crystals (Note 22).

2. Notes

1.   The success of the reaction does not depend on having previously dried the glassware, or on adding the reagents under a nitrogen atmosphere.

2.   Aqueous hypophosphorous acid (50 wt.%) was purchased from Aldrich Chemical Company, Inc. and concentrated before reaction, according to the following procedure. The 1-L round-bottomed reaction flask was charged with 39.6 g of 50% aqueous hypophosphorous acid. The acid was concentrated for 30 min by rotary evaporation (40 °C, 0.3 mmHg).

3.   Reagent grade *N,N*-dimethylformamide (≥99.8%) was purchased from Aldrich Chemical Company, Inc. and used as received.

4.   Mesitylene (standard for GC, ≥99.8%) was purchased from Fluka and used as received. This reagent does not interfere with the reaction; it works only as internal standard for GC-monitoring of the reaction progress and can be omitted. The checkers omitted the use of mesitylene.

5.   Cinnamyl alcohol (98%) was purchased from Aldrich Chemical Company, Inc. and used without further purification.

6.   Due to the low melting point of cinnamyl alcohol (30–33 °C), the reagent was immersed in a water bath at 45 °C for 30 min before use to facilitate its addition via syringe. A preheated (45–50 °C) 20–mL glass syringe fitted with a short needle (50 mm) was used in order to avoid solidification of the reagent during the addition.

7.   Palladium (II) acetate, min. 98% (99.9+%-Pd) was purchased from Strem Chemicals, Inc. and used as received.

8.   9,9-Dimethyl-4,5-bis(diphenylphosphino)xanthene   (Xantphos) (97%) was purchased from Aldrich Chemical Company, Inc. and used as received.   The checkers purchased Xantphos (min 98%) from Strem Chemicals, Inc. and used it as received.

9.   The surface of the solution was below the sand level and good stirring was maintained along the process.

10.   A J-KEM Scientific, Inc. temperature controller model 150 with a Teflon-coated thermocouple was used with the heating mantle. The thermocouple was placed inside the solution (1-2 inches) and the temperature was set to 85 °C. The reaction time was measured once the internal temperature of the solution was stabilized at 85±3 °C, which took about 20-30 min. The checkers used a mercury thermometer to monitor the internal temperature of the reaction.

11.   The reaction was terminated when TLC analysis indicated that all the cinammyl alcohol was consumed. TLC was conducted using Merck silica gel 60 F-254 plates (elution with hexanes/ethyl acetate, 7:1; visualization by UV, and by immersion in anisaldehyde stain (by volume: 93% ethanol, 3.5% sulfuric acid, 1% acetic acid, and 2.5% anisaldehyde) followed by heating; R_f cinnamyl alcohol = 0.13, blue spot on anisaldehyde. The progress of the reaction was also monitored by gas chromatography. GC analysis was performed on a HP5890 Gas Chromatograph equipped with a HP5 capillary column (30-m x 0.32-mm x 0.25-µm) and FID detector, under the following conditions: flow(He) = 0.9 mL/min at 60 °C, under constant pressure at 5 psi; inlet temp 200 °C; oven temp 60 °C, 1 min; ramp1 5 °C/min; final temp1: 160 °C, final time1: 0 min; then ramp2: 25 °C/min; final temp2: 280 °C, final time2: 20 min; detector 280 °C; split mode with constant make-up. For GC analysis, 3 drops of sample was diluted in 1 mL of diethyl ether. The solution was washed with 1 mL of saturated aqueous $NaHCO_3$ solution and 1-µL of the organic solution was injected into the GC;

t_R (cinnamyl alcohol), 18.72 min; t_R (mesitylene). 8.69 min. The checkers monitored the reaction by TLC only.

12. The submitters removed the sand bath and replaced it with a water bath. The checker removed the reaction mixture from the sand bath, continued stirring for 30 min and subsequently removed the solvent under reduced pressure as described in the procedure. The reaction mixture was not fully cooled to room temperature in a water bath prior to evaporation of the solvent.

13. One mL of the reaction mixture at room temperature was placed in an NMR tube for analysis. ^{31}P NMR (121.47 MHz, DMF) δ: 28.35, ~ 118% (dt, J_{HP} = 529 Hz, J = 19 Hz, Product), 4.70, ~ 48% (t, J_{HP} = 526 Hz, H_3PO_2), 3.02, ~ 34% (d, J_{HP} = 641 Hz, H_3PO_3). The ^{31}P NMR yields were determined by integration of all the resonances in the ^{31}P NMR spectra. The checkers did not perform this analysis.

14. Activated charcoal (Purum p.a.) was purchased from Fluka and used as received.

15. Celite 545 was purchased from Fischer Scientific Co. A slurry mixture of 40 g of Celite in ethyl acetate (about 50 mL) was placed in a Büchner funnel (7 cm diameter, 10-15 μ). The checkers used technical grade Celite from ACP Chemicals Inc.

16. A milky suspension was obtained with some dark, gel-like precipitate.

17. The submitters observed an emulsion with some black precipitate is formed at the interphase, which can be broken by using a stirring rod. Some black precipitate goes into the organic phase. The checkers did not observe the formation of an emulsion with black precipitate.

18. Magnesium sulfate (anhydrous, ≥97%) was purchased from Aldrich Chemical Company, Inc.

19. The mixture was stirred with activated charcoal (Note 15) for 15 min, then $MgSO_4$ was added and stirring was continued for another 15 min.

20. The Celite pad was washed with two 50-mL portions of ethyl acetate.

21. Cinnamyl-H-phosphinic acid was pure according to the melting point (84–85 °C), and NMR analysis (^{31}P, ^{1}H, ^{13}C). The material was recrystallized to remove any traces of Pd.

22. Full characterization of the product was as follows: mp 83–85 °C; ^{1}H NMR (400 MHz, CDCl$_3$) δ: 2.75 (dd, J_{HP} = 19.4, 7.6 Hz, 2 H), 6.04–6.13 (m, 1 H), 6.51 (dd, J = 15.8, 5.8 Hz, 1 H), 7.01 (d, J_{HP} = 558.7 Hz, 1

H), 7.20–7.35 (m, 5 H), 11.90 (bs, 1 H); ^{13}C NMR (100 MHz, CDCl$_3$) δ: 34.7 (d, J_{PC} = 90.9 Hz, CH$_2$), 116.8 (d, J_{PCC} = 10.1 Hz, CH), 126.5 (d, J_{PCCCC} = 2.2 Hz, CH), 128.0 (CH), 128.8 (CH), 136.2 (d, J_{PCCC} = 14.6 Hz, CH), 136.7 (d, J_{PCCCC} = 4.1 Hz, C); ^{31}P NMR (161.82 MHz, CDCl$_3$) δ: 35.51 (d, J_{PH} = 557.9 Hz); IR (thin film, NaCl), cm^{-1}: 2619 and 1682 (P-O-H); 2422, 2326 and 2177 (P-H); 1217 (P=O); 970 and 728; UV (EtOH, c≈8 μM) λ_{max} = 255 nm; MS m/z (relative intensity): 182 (28), 118 (19), 117 (100), 116 (13), 115 (41), 91 (19). HRMS (EI) m/z Calcd for C$_9$H$_{11}$O$_2$P: 182.0497. Found: 182.0497. Anal. Calcd. for C$_9$H$_{11}$O$_2$P: C, 59.34; H, 6.09. Found: C, 58.89; H, 6.10. Analysis by Reverse Phase Ion-Pairing HPLC:[2] t_R 1.43 min. Agilent Zorbax® Eclipse XDB-C8 column (4.6 x 150 mm, 5μm) with a guard column (Agilent Zorbax® ODS, 4.6 x 12.5 mm, 5μm), 1 mL/min flow (isocratic), using as mobile phase a buffer (5 mM hexadecyl-trimethylammonium bromide, 50 mM ammonium acetate, and 2% MeOH. pH 4.85, adjusted with acetic acid). Injection volume: 5 μL, C≈0.24 mg/mL (EtOH). The checkers did not verify the HPLC data.

Safety and Waste Disposal Information

All hazardous materials should be handled and disposed of in accordance with "Prudent Practices in the Laboratory"; National Academy Press; Washington, DC, 1995.

3. Discussion

Palladium-catalyzed dehydrative substitutions of allylic alcohols with carbon, oxygen, nitrogen, and sulfur nucleophiles have started to emerge as efficient and atom-economical processes in organic synthesis.[3,4] These reactions directly use allylic alcohols instead of activated esters or halides, which are prepared from the corresponding allylic alcohols. Though attractive, this type of allylic substitution is generally slow, and often requires the addition of catalytic or stoichiometric amounts of activating agents.[3] Metal-catalyzed C-P bond formation through allylation processes are by far less developed and just a couple of examples involving the use of allylic acetates or carbonates as electrophiles, in presence of stoichiometric amounts of base or silylating agent, are known.[5] This approach is not practical due to the formation (P(III) species) and/or manipulation of air sensitive compounds (nickel(0) catalysts). Hypophosphorous compounds

100

(ROP(O)H$_2$) exist in a tautomeric equilibrium between the P(V) and P(III) forms due to the presence of labile hydrogen atoms. Under the influence of transition metals they are known to undergo transfer hydrogenation reactions, but their reactivity can be harnessed through adequate selection of catalysts.[6] The challenge to overcome the reductive pathway in palladium-catalyzed allylation reactions was initially addressed by the development of a cross-coupling reaction of allylic halides, acetates, benzoates and carbonates with amine salts of esters of hypophosphorous acid.[7,8] Subsequently, a palladium-catalyzed dehydrative allylation of H$_3$PO$_2$ with allylic alcohols *in the absence of any additives* was successfully achieved.[9] The present procedure describes a sound and environmentally friendly approach for the preparation of allylic-*H*-phosphinic acids, having water as byproduct (Table 1). In most cases, primary *H*-phosphinic acids can be isolated in moderate to good yields after a simple extractive workup (>95% purity). Since substrates that possess a terminal double bond and secondary or tertiary alcohols in the allylic position undergo rearrangement to form a primary C-P bond, the reaction is considered to proceed via π-allylpalladium intermediates. When using low molecular weight allylic alcohols (3-4 carbons), an *in situ* esterification with alkoxysilanes to the corresponding *H*-phosphinate esters improves the yield significantly. The reaction is highly selective towards formation of the *E*-isomers. Secondary allylic alcohols react successfully in this reaction, although they require more concentrated conditions and slightly higher catalyst loading. The reaction is water and air tolerant, and requires as little as 0.2 mol% of Pd (Pd$_2$dba$_3$, Pd(OAc)$_2$ or PdCl$_2$), with Xantphos as a ligand. *N,N*-Dimethylformamide (85 °C) is required as a solvent. Good yields of products are obtained even with equimolar amounts of H$_3$PO$_2$ and allylic alcohols, but 2 equivalents of H$_3$PO$_2$ appear to improve the yield.

Allylic *H*-phosphinic acids have been prepared previously from the reaction of an allylic halide with (TMSO)$_2$PH.[10] However, this method requires wasteful silylation, a halide-containing electrophile, and it is difficult to prevent the formation of symmetrically disubstituted products.[11]

Another synthetic approach to allylic-*H*-phosphinate esters is a base-promoted direct alkylation of alkyl phosphinates,[12] or a Michaelis-Becker reaction of masked hypophosphorous synthons.[13]

This novel catalytic phosphorous-carbon bond formation reaction is a powerful and atom-economical entry into allylic organophosphorus compounds from readily available allylic alcohols.

Table 1. Scope of the Pd-Catalyzed Allylation[a]

Entry	Alcohol	H-phosphinic acid	isolated yield %
1a	R = E-Pr		62
1b	R = E-Pr		92
1c	R = Me		50 (88)[d]
1d	R = CH₂OBn		67[d]
1e	R = CO₂Et		77[d]
2a	R = E-Pr		52
2b	R = E-Pr		100
2c	R = Me		86
3			68
4			74
5a	R = Et		92
5b	R = p-FC₆H₄		96
6a	R = CH₂		78
6b	R = NCO₂Et		73
7a	R = H		82
7b	R = prenyl		93[c]
7c	R = geranyl		98[c]
8			43[c,d]
9a	R = H		45[d,e]
9b	R = Me		52[e]

a See Reference 9 for details. Reactions were conducted in DMF over 4Å sieves (0.2 M) at 85 °C, with 0.5 mol% Pd/xantphos, and n equiv H₃PO₂. n = 1, entries 1a-c, 2-3, 7; n = 2, entries 4-6; n = 2.5, entries 1d-e; n = 3, entries 8-9. Reaction times: 2-8 h. b Isolated yield after extractive work-up. c 1:1 mixture of isomers. d After esterification and chromatographic purification. e Conditions: 2 mol% Pd/xantphos, 3 equiv H₃PO₂, 85 °C, DMF (2 M).

1. Department of Chemistry, Texas Christian University, Fort Worth, TX 76129. Financial support by the National Science Foundation (CHE-0242898) is gratefully acknowledged. K.B.A. was supported in part by a Texas Christian University Graduate Fellowship.

2. Richardson, D.; Sadi, B. B. M.; Caruso, J. A. in *Ultra-Trace Analysis of Organophosphorus Chemical Warfare Agent Degradation Products by HPLC-ICP-MS.* Agilent Technologies, Online Literature at www.agilent.com/chem.

3. Reviews: (a) Muzart, J. *Tetrahedron* **2005**, *61*, 4179-4212. (b) Tamaru, Y. *Eur. J. Org. Chem.* **2005**, 2647-2656.

4. Examples of direct allylation without additives: (a) Olsson, V. J.; Sebelius, S.; Selander, N.; Szabó, K. J. *J. Am. Chem. Soc.* **2006**, *128*, 4588-4589. (b) Kayaki, Y.; Koda, T.; Ikariya, T. *Eur. J. Org. Chem.* **2004**, 4989-4993. (c) Kayaki, Y.; Koda, T.; Ikariya, T. *J. Org. Chem.* **2004**, *69*, 2595-2597. (d) Trost, B. M.; Spagnol, M. D. *J. Chem. Soc., Perkin Trans. 1* **1995**, 2083-2096. (e) Vedejs, E.; MacKay, J. A. *Org. Lett.* **2001**, *3*, 535-536.

5. (a) Lu, X.; Zhu, J. *Synthesis* **1986**, 563-564. (b) Lu, X.; Zhu, J.; Huang, J.; Tao, X. *J. Mol. Catal.* **1987**, *41*, 235-243. (c) Lu, X.; Tao, X.; Zhu, J.; Sun, X.; Xu, J. *Synthesis* **1989**, 848-850. (d) Lu, X.; Zhu, J. *J. Organomet. Chem.* **1986**, *304*, 239. (e) Fiaud, J. C. *J. Chem. Soc. Chem. Commun.* **1983**, 1055-1056.

6. Reviews: (a) Montchamp, J.-L. *J. Organomet. Chem.* **2005**, *690*, 2388-2406. (b) Montchamp, J.-L. *Specialty Chemicals Magazine* **2006**, *26*, 44-46. (c) Bravo-Altamirano, K.; Montchamp, J.-L., "Phosphinic Acid, Alkyl Esters" in Encyclopedia of Reagents for Organic Synthesis (eEROS). *In press.*

7. Bravo-Altamirano, K.; Huang, Z.; Montchamp, J.-L. *Tetrahedron* **2005**, *61,* 6315-6329.

8. Bravo-Altamirano, K.; Montchamp, J.-L. *Poster 313*, 231st National Meeting of the American Chemical Society, Atlanta, March, 2006.

9. Bravo-Altamirano, K.; Montchamp, J.-L. *Org. Lett.* **2006**, *8*, 4169-4171.

10. Boyd, E. A.; Regan, A. C.; James, K. *Tetrahedron Lett.* **1994**, *35*, 4223-4226.

11. (a) Bujard, M.; Gouverneur, V.; Mioskowski, C. *J. Org. Chem.* **1999**, *64*, 2119-2123. (b) Majewski, P. *Phosphorus, Sulfur Silicon* **1989**, *45*,

151-154. (c) Kurdyumova, N. R.; Ragulin, V. V.; Tsvetkov, E. N. *Russ. J. Gen. Chem.* **1994**, *64*, 380-383.

12. Abrunhosa-Thomas, I.; Ribière, P.; Adcock, A. C.; Montchamp, J.-L. *Synthesis* **2006**, *2*, 325-331. (b) Gallagher, M. J.; Ranasinghe, M. G.; Jenkins, I. D. *Phosphorus, Sulfur, Silicon* **1996**, *115*, 255-259.

13. Baylis, E. K. *Tetrahedron Lett.* **1995**, *36*, 9385-9388.

Appendix
Chemical Abstracts Nomenclature; (Registry Number)

Hypophosphorous acid: Phosphinic acid; (6303-21-5)

Cinnamyl alcohol: 3-Phenyl-2-propen-1-ol; (104-54-1)

Palladium(II) acetate: Pd(OAc)$_2$; (3375-31-3)

9,9-Dimethyl-4,5-bis(diphenylphosphino)xanthene: Xantphos; (161265-03-8)

Cinnamyl-H-phosphinic acid: [(2*E*)-3-phenyl-2-propenyl]-Phosphinic acid; (911128-46-6)

Jean-Luc Montchamp was born in Lyon, France. He completed his undergraduate studies at the Ecole Superieure de Chimie Industrielle de Lyon (ESCIL), now known as CPE. He obtained his Ph.D. from Purdue University in 1992, under the direction of Professor John W. Frost. After postdoctoral experiences at Michigan State University and at the Scripps Research Institute, he returned to Purdue University for a postdoctoral stay with Professor Ei-ichi Negishi. He became Assistant Professor at Texas Christian University in 1998, and was promoted to Associate Professor in 2004. His research interests include the development of methodology for phosphorus-carbon bond formation, especially using hypophosphorous derivatives and the medicinal chemistry of phosphorus-containing analogs of natural products.

Karla Bravo-Altamirano was born in 1979, in Oaxaca, Mexico. She received a B.S. degree in Chemistry in 2002 from Universidad de las Américas Puebla, Mexico, where she conducted research for the group of Prof. Cecilia Anaya Berrios. She is currently pursuing graduate studies at Texas Christian University, under the guidance of Prof. Jean-Luc Montchamp. Her research focuses on the development of new methodologies for the synthesis of *H*-phosphinic acid derivatives and their *P*-chiral counterparts.

Alena Rudolph was born in Ottawa, Canada in 1978. She received her Bachelor of Science degree (Honors, Cooperative) in Chemistry from the University of Waterloo in 2002, where she conducted research in the group of Prof. Eric Fillion. After working in the department of Medicinal Chemistry at Abbott Bioresearch Center in Worcester, MA, she returned to Canada and is currently pursuing her Ph.D. under the supervision of Prof. Mark Lautens at the University of Toronto. Her research is currently focused on the development of tandem processes to generate functionalized heterocycles via a C–H activation pathway.

SYNTHESIS AND RESOLUTION OF RACEMIC *TRANS*-2-(*N*-BENZYL)AMINO-1-CYCLOHEXANOL: ENANTIOMER SEPARATION BY SEQUENTIAL USE OF (*R*)- AND (*S*)-MANDELIC ACID

[Cyclohexanol, 2-(*N*-benzyl)amino, (1*S*,2*S*)- and (1*R*,2*R*)-]

Submitted by Ingo Schiffers and Carsten Bolm.[1]
Checked by Scott E. Denmark, Eric Woerly, Aurélie Toussaint, and Andreas Pfaltz.

1. Procedure

A.   Racemic   trans-2-(N-benzyl)amino-1-cyclohexanol   (rac-3).   An autoclave (Note 1) with a properly fitted 400-mL glass insert (Note 2) is charged with cyclohexene oxide (**1**) (61.5 mL, 59.7 g, 0.596 mol, 1.1 equiv) and benzylamine (**2**) (60.0 mL, 58.8 g, 0.543 mol, 1 equiv) (Note 3), equipped with a magnetic stirring bar (Note 4), sealed and flushed with

Org. Synth. **2008**, *85*, 106-117
Published on the Web 10/16/2007

nitrogen. The reaction mixture is placed in a 250 °C preheated oven for 6 h (Note 5), then cooled to ambient temperature, diluted with dichloromethane (60 mL) (Note 6) and transferred into a 1000-mL single-necked, round-bottomed flask. The glass inlay is rinsed with dichloromethane (3 × 50 mL) and the combined organic phases are concentrated using a rotary evaporator (30 mmHg, ambient temperature). The residual cyclohexene oxide is removed under reduced pressure (1 mmHg) at room temperature over 11 h to yield 110.68 g (99%) amino alcohol *rac*-**3** as a light yellow solid, which is suitable for use in the next step without further purification (Note 7).

B. *(S)-Mandelic acid salt of (1R,2R)-trans-2-(N-benzyl)amino-1-cyclohexanol (4) and (R)-mandelic acid salt of (1S,2S)-trans-2-(N-benzyl)amino-1-cyclohexanol (ent-4).* A 1-L single-necked, round-bottomed flask containing a magnetic stirring bar is equipped with a pressure-equalizing addition funnel fitted with an argon inlet (Note 8). The flask is charged with amino alcohol *rac*-**3** (82.12 g, 0.40 mol, 1.0 equiv) dissolved in ethyl acetate (600 mL) and a solution of (*S*)-mandelic acid (30.43 g, 0.20 mol, 0.5 equiv) (Note 9) in ethyl acetate (200 mL) and diethyl ether (100 mL) (Note 10) is added via the addition funnel over a period of 5 h at room temperature (Note 11). After the addition is complete the dropping funnel is rinsed with diethyl ether (2 × 5 mL) and the reaction mixture is stirred overnight at ambient temperature, followed by 5 h at 0 °C. The precipitated ammonium salt is collected by suction filtration, then is washed with ethyl acetate (100 mL), followed by diethyl ether (2 × 100 mL), and dried *in vacuo* at 0.05 mmHg at room temperature over 1 h (Note 12) to afford 52.86 g (0.15 mol, 74% based on the amount of mandelic acid) of the (*S*)-mandelic acid salt of (1*R*,2*R*)-*trans*-2-(*N*-benzyl)amino-1-cyclohexanol (**4**) as a colorless solid (Note 13). The filtrate from the above procedure is transferred to a 2-L separatory funnel, then is washed with 1 N aq. NaOH solution (3 × 50 mL) and the aqueous layer is back-extracted with diethyl ether (3 × 50 mL). The combined organic layers are dried (MgSO₄, approx. 50 g), filtered and concentrated under reduced pressure (40 °C, 100 mbar) to give 49.20 g of the crude amino alcohol (1*S*,2*S*)-**3** as a pale yellow oil. The oily residue is dissolved in ethyl acetate (400 mL), transferred into a 1-L flask, and treated with a solution of (*R*)-mandelic acid (30.43 g, 0.20 mol, 0.5 equiv) in ethyl acetate (200 mL) and diethyl ether (100 mL) analogously to the above described procedure, to deliver 55.50 g (0.15 mol, 78% based on the amount of mandelic acid) of the (*R*)-mandelic acid salt of (1*S*,2*S*)-

trans-2-(*N*-benzyl)amino-1-cyclohexanol (*ent*-**4**) as a colorless solid (Note 14).

C. Liberation of the amino alcohols and recovery of mandelic acid. In a 1-L separatory funnel, the mandelic acid ammonium salt **4** or *ent*-**4** (50.04 g, 0.14 mol) is partitioned between ethyl acetate (500 mL) (Note 10) and 2 N aq. HCl solution (200 mL). Then, the mixture is manually and vigorously shaken until the salt is completely dissolved. The organic layer is additionally washed with 2 N aq. HCl solution (2 × 25 mL) and the combined aqueous phases are back-extracted with ethyl acetate (3 × 100 mL). The combined organic phases are dried (MgSO$_4$, approx. 50 g), filtered and concentrated under reduced pressure to provide 19.80–20.02 g (93–94%) of the corresponding mandelic acid enantiomer (Note 17). To a mixture of the acidic aqueous phase and diethyl ether (200 mL) in the same separatory funnel, 5 N NaOH (280 mL) is added carefully in small portions over a period of 45–60 minutes. After separation, the aqueous layer is extracted with diethyl ether (4 × 100 mL) and the combined organic phases are dried (MgSO$_4$, approx. 30 g), filtered, and concentrated under reduced pressure (40 °C, 100 mbar) to yield 25.80–26.74 g (90–93%) (Note 18) of the corresponding *trans*-2-(*N*-benzyl)amino-1-cyclohexanol enantiomer (1*R*,2*R*)-**3** or (1*S*,2*S*)-**3** as a white solid (Note 19).

2. Notes

1. A custom made high-pressure autoclave was used. See Figures 1 and 2 for autoclave dimensions. The dimensions for the Teflon and rubber O-rings are: Teflon ring 4 ¼" (ID) x 4 7/16" (OD) x 3/32" thick; rubber O-ring 3" (ID) x 3 3/16" (OD) x 3/32" thick.

2. Glass inlay dimensions: outer diameter = 7 cm; inner diameter = 6.5 cm; height = 13 cm.

3. Benzylamine (99%) and cyclohexene oxide (98%) were purchased from Acros Organics and used as received.

4. Magnetic stirring was used when purging the system with nitrogen. In the original procedure, magnetic stirring was used during the reaction. Magnetic stirring was not used in the Checker's procedure.

5. A Stabil-Therm Oven (model #OV-12A) at a high power setting of 5 was used as a heat source.

6.   Dichloromethane (>99.5%) was purchased from Sigma-Aldrich and used as supplied.   The Submitters used dichloromethane (>99%) purchased from Merck as received.

7.   The NMR spectra indicated a purity of 91–94% for *rac*-**3**. It is also available from Acros Organics (>99%). mp 57.6 – 58.6 °C. ^{1}H NMR (500 MHz, CDCl$_3$) δ: 1.00 (q, J = 10 Hz, 1 H), 1.06–1.56 (m, 5 H), 1.60–1.94 (m, 3 H), 1.94–2.12 (m, 1 H), 2.19 (d, J = 10 Hz, 1 H), 2.26–2.50 (m, 1 H), 3.21–3.30 (m, 1 H), 3.69 (d, J = 15 Hz, 1 H), 3.96 (d, J = 15 Hz, 1 H), 7.10–7.58 (m, 6 H); ^{13}C NMR (125 MHz, CDCl$_3$) δ: 24.3, 25.1, 30.5, 33.2, 50.7, 63.1, 73.8, 127.0, 128.1, 128.4, 140.5; IR (thin film): 3305 (m), 3062 (m), 3028 (m), 2930 (s), 2857 (s), 1604 (w), 1496 (w), 1452 (s), 1354 (w), 1199 (w), 1074 (m), 1029 (m), 977 (w), 845 (w), 747 (m), 699 (s) cm^{-1}. MS (EI, 70eV) m/z : 205 (M$^+$ (11%)), 146 (27%), 121 (20%), 115 (16%), 106 (27%), 92 (17%), 91 (100%), 56 (58%).

8.   In the original procedure, a gas inlet adapter with stopcock connected to a nitrogen-filled balloon was used.

9.   (*S*)-Mandelic acid (>99%) and (*R*)-mandelic acid (99%) were purchased from Acros Organics and used as received.

10.   Ethyl acetate (>99%) and diethyl ether (>99%) were purchased from Fluka and used as received.

11.   After addition of one-third of the mandelic acid solution, the ammonium salt began to precipitate.

12.   A sintered glass funnel (100 mm diameter) was used. After washing the filter cake, suction filtration was continued for 5 min to dry the ammonium salt by a stream of air. Then, the damp solid was transferred into a 250-mL single-necked, round-bottomed flask to remove residual solvents under vacuum.

13.   In a run carried out on half-scale, 25.70 g of product was obtained (72% yield). The product has the following characteristics: mp 146 °C; $[\alpha]_D^{25}$ +16.0 (*c* 2.15, CHCl$_3$). ^{1}H NMR (400 MHz, CDCl$_3$) δ: 0.96–1.35 (m, 4 H), 1.58–1.73 (m, 3 H), 1.90 (d, J = 12.6 Hz, 1 H), 2.53 (dt, J = 4.0 Hz, J = 12.0 Hz, 1 H), 3.03 (dt, J = 4.3 Hz, J = 10.6 Hz, 1 H), 3.46 (d, J = 12.9 Hz, 1 H), 3.89 (d, J = 12.6 Hz, 1 H), 4.90 (s, 1 H), 7.15–7.32 (m, 8 H), 7.49 (s, 1 H), 7.52 (s, 1 H); ^{13}C NMR (100 MHz, CDCl$_3$) δ: 23.9, 24.2, 26.6, 34.0, 48.4, 62.5, 70.4, 74.4, 76.8, 126.7, 127.3, 128.1, 128.9, 129.1, 130.0, 131.4, 142.3, 178.6. IR (KBr) 3362, 3036, 2937, 2859, 1606, 1554, 1490, 1447, 1370, 1086, 1061, 1041 cm^{-1}; MS (EI, 70 eV) m/z : 205 (34), 146 (88),

114 (12), 107 (51), 91 (100), 79 (29), 77 (22); Anal. Calcd for $C_{21}H_{27}NO_4$ (357.44): C, 70.56; H, 7.61; N, 3.92. Found: C, 70.54; H, 7.50; N, 3.75.

14. In a run carried out on half-scale, 27.92 g of product *ent*-**4** was obtained (78% yield); $[\alpha]_D^{25}$ –14.3 (*c* 2.15, $CHCl_3$). Both ammonium salts, **4** and *ent*-**4**, were diastereomerically pure according to NMR analysis. To determine the enantiomeric excess (ee) of the free amino alcohol, a 50 mg sample of the mandelic acid salt was partitioned between 1 N aq. NaOH solution (4 mL) and diethyl ether (3 × 10 mL). The combined organic phases were dried ($MgSO_4$, 50 mg), filtered, concentrated under reduced pressure (40 °C, 300 mbar), and analyzed by HPLC analysis using a chiral stationary phase (for conditions, see Note 19). The described procedure yields (1*R*,2*R*)-**3** with 99% ee and (1*S*,2*S*)-**3** with >99% ee (no signal detected for the minor enantiomer in HPLC analysis). One recrystallization (Note 15) of (1*R*,2*R*)-**3** followed by liberation of the amino alcohol as described in Step C, provides product that showed no signal for the minor enantiomer in the CSP-HPLC analysis.

15. General procedure for the recrystallization: A 1-L, three-necked, round-bottomed flask equipped with a magnetic stirring bar, two rubber septa, and a reflux condenser fitted with an argon inlet (Note 8) is charged with mandelic acid salt **4** (20 g, 56 mmol) and ethyl acetate (600 mL). The suspension is heated to reflux before ethanol (Note 16) is added in portions via syringe until complete dissolution occurs (25 mL in total). Then the stirring is stopped, the oil bath is removed, and the clear solution is allowed to cool to room temperature overnight. The crystallized ammonium salt is collected by suction filtration (Note 12), washed with ethyl acetate (15 mL), followed by diethyl ether (2 × 30 mL), and dried *in vacuo* affording 18.30 g (91%) of enantiomerically pure **4**.

16. Ethanol (≥ 99.9%) was purchased from Fluka and used as received.

17. The recovered mandelic acid showed an identical value for optical rotation in comparison to the starting material; (*R*)-mandelic acid $[\alpha]_D^{25}$ –152.4 (*c* 2.5, H_2O), (*S*)-mandelic acid $[\alpha]_D^{25}$ +153.5 (c 2.8, H_2O). These values are in accord with the literature (–150 to –155 for (*R*)-mandelic acid and +153 to +155 for (*S*)-mandelic acid).

18. In the runs carried out on half-scale, 12.86–12.97 g of products were obtained (89–90% yield). CSP-HPLC separation conditions: t_R (1*S*,2*S*)-**3**, 24.2 min; t_R (1*R*,2*R*)-**3**, 30.0 min (Daicel CHIRALCEL OB-H (4.6 × 250 mm); *n*-heptane/2-propanol, 98:2; 0.5 mL/min; 220 nm). For (1*R*,2*R*)-**3,** the

110

enantiomeric excess measured was 99% and for (1S,2S)-**3**, the minor enantiomer was not detected (*ee*>99%).

19. The products have the following physicochemical characteristics (data reported for (1S,2S)-**3**): $[\alpha]_D^{25}$ +82.2 (*c* 1.05, MeOH) for (1S,2S)-**3**, $[\alpha]_D^{25}$ −79.8 (*c* 1.22, MeOH) for (1R,2R)-**3**; mp 91 °C; ^{1}H NMR (400 MHz, CDCl$_3$) δ: 0.92–1.02 (m, 1 H), 1.18–1.30 (m, 3 H), 1.68–1.78 (m, 2 H), 1.99–2.06 (m, 1 H), 2.10–2.22 (m, 1 H), 2.29 (ddd, J = 4.0, 9.4, 11.4 Hz, 1 H), 3.20 (dt, J = 4.5,10.1 Hz, 1 H), 3.70 (d, J = 12.9 Hz, 1 H), 3.96 (d, J = 12.9 Hz, 1 H),7.22–7.28 (m, 1 H), 7.30–7.35 (m, 4 H); ^{13}C NMR (100 MHz, CDCl$_3$) δ: 24.5, 25.3,30.8, 33.4, 50.9, 63.3, 74.1, 127.1, 128.2, 128.6, 140.8; IR (KBr) 3295, 3060, 2933, 2854, 1602, 1496, 1449, 1356, 1292, 1219, 1152, 1077 cm^{-1}; MS (EI, 70 eV) *m/z* : 205 (36), 146 (90), 114 (10), 91 (100); Anal. Calcd for C$_{13}$H$_{19}$NO (205.30): C, 76.06; H, 9.33; N, 6.82. Found: C, 75.87; H, 9.18; N, 6.72.

Safety and Waste Disposal Information

All hazardous materials should be handled and disposed of in accordance with "Prudent Practices in the Laboratory"; National Academy Press; Washington, DC, 1995.

3. Discussion

Enantiopure amino alcohols are versatile synthetic intermediates for the preparation of a wide variety of natural products and biologically active compounds.[2] Also their importance in asymmetric synthesis, where the need for chiral auxiliaries and ligands is continually increasing, has been well recognized in recent years.[3] Despite the great interest in this field and the impressive success that have been made in the development of new procedures for the preparation of optically pure vicinal amino alcohols, there are only few efficient procedures for the preparation of highly enantiomerically enriched aminocyclohexanols which are suitable for a broad variety of further derivatizations.[4] Among those few procedures, only Overman's aminolysis of cyclohexene oxide with an aluminum amide stemming from enantiomerically enriched methylbenzylamine and trimethylaluminum,[5] Jacobsen's Cr(III)/salen-catalyzed enantioselective ring-opening reaction of cyclohexene oxide by azidosilanes,[6] and the enzymatic resolution of racemic 2-azidocyclohexanol[7] have found occasional applications in synthesis.[8] The procedure described here is an improved version of the previously reported protocol for the resolution of racemic *trans*-2-(*N*-benzyl)amino-1-cyclohexanol (*rac*-**3**), which is easily available in 100 g scale by a solvent-free aminolysis of cyclohexene oxide at high temperature.[9] The advantages of this novel method are its preparative ease and its efficiency in large scale resolutions delivering both amino alcohol enantiomers with 99% ee by sequential use of inexpensive (*S*)- and (*R*)-mandelic acid. A simple aqueous workup procedure permits the isolation of the amino alcohol in analytically pure form and the recovery of mandelic acid in high yield. The synthetic usefulness of the method was demonstrated by debenzylation of enantiopure *trans*-2-(*N*-benzyl)amino-1-cyclohexanol (**3**) by hydrogenolysis, leading to the easily modifiable deprotected amino alcohol **5** (Scheme 1), which gave access to a broad variety of diversely substituted derivatives and their corresponding cis isomers **6-8**.[9] These enantiomerically pure aminocyclohexanols have been

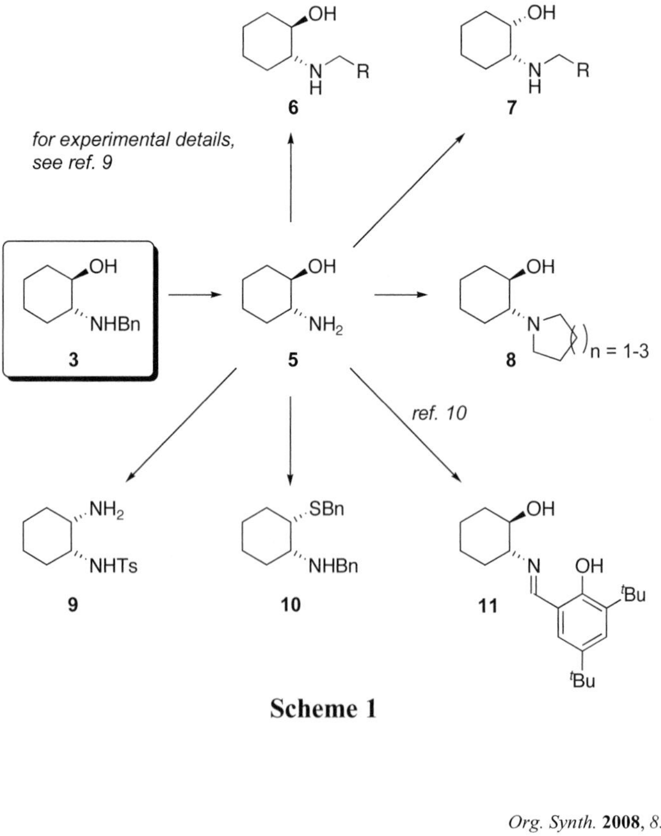

Scheme 1

applied as ligands in catalyzed, asymmetric phenyl transfer reactions to benzaldehyde and transfer hydrogenations of aryl ketones, leading to enantioselectivities of up to 96% ee.

In addition, it was demonstrated that enantiopure *trans*-2-aminocyclohexanol **5** is a valuable building block for the preparation of enantiomerically pure *cis*-diaminocyclohexane **9** and chiral *cis*-amino thiol derivative **10**. The latter compound showed an impressive activity in the iridium-catalyzed transfer hydrogenation reaction. Syntheses of new analogues of well-known classes of amino alcohol derived ligands, like oxazolines, oxazaborolidines and Schiff bases **11**, starting from **5** are currently in progress.[10]

Figure 1

Figure 2

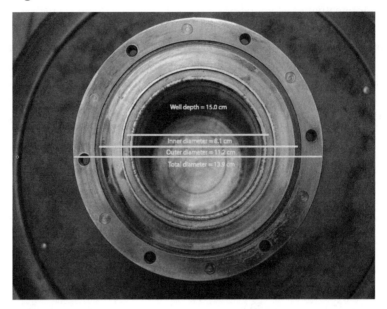

1. Institut für Organische Chemie der RWTH Aachen, Landoltweg 1, D-52056 Aachen, Germany.

2. For reviews, see: (a) Bergmeier, S. C. *Tetrahedron* **2000**, *56*, 2561–2576; (b) *Carbohydrate Mimics*; Chapleur, Y., Ed.; Wiley-VCH: Weinheim, 1998; (c) *The Merck Index*, 12th Edition.; Chapman & Hall: New York, 1996; (d) Shaw, G. In *Comprehensive Heterocyclic Chemistry II*; 5th Edition.; Katrizky, A. R.; Rees, C. W.; Scriven, E. F. V., Eds.; Pergamon: New York, 1996; Vol. 7, p 397; (e) Coppola, G. M.; Schuster, H. F. *Asymmetric Synthesis. Construction of Chiral Molecules Using Amino Acids*; Wiley: New York, 1987.

3. For reviews, see: (a) Fache, F.; Schulz, E.; Tommasino, M. L.; Lemaire, M. *Chem. Rev.* **2000**, *100*, 2159–2232; (b) Ager, D. J.; Prakash, I.; Schaad, D. R. *Chem. Rev.* **1996**, *96*, 835–876.

4. (a) González-Sabín, J.; Gotor, V.; Rebolledo, F. *Tetrahedron: Asymmetry* **2004**, *15*, 1335–1341; (b) Guangyou, Z.; Yuquing, L.; Zhaohui, W.; Nohira, H.; Hirose, T. *Tetrahedron: Asymmetry* **2003**, *14*, 3297–3300; (c) Bertau, M.; Bürli, M.; Hungerbühler, E.; Wagner, P.

Tetrahedron: Asymmetry **2001**, *12*, 2103–2107; (d) Periasamy, M.; Kumar, N. S.; Sivakumar, S.; Rao, V. D.; Ramanathan, C. R.; Venkatraman, L. *J. Org. Chem.* **2001**, *66*, 3828–3833; (e) Lu, X.; Xu, Z.; Tang, G. *Org. Process Res. Dev. J.* **2001**, *5*, 184–185; (f) Forró, E.; Szakonyi, Z.; Fülöp, F. *Tetrahedron: Asymmetry* **1999**, *10*, 4619–4626; (g) Nugent, W. A. *J. Am. Chem. Soc.* **1992**, *114*, 2768–2769; (h) Li, G.; Chang, H.-T.; Sharpless, K. B. *Angew. Chem.* **1996**, *108*, 449–452; *Angew. Chem., Int. Ed. Engl.* **1996**, *35*, 451–454; (i) Rudolph, J.; Sennhenn, P. C.; Vlaar, C. P.; Sharpless, K. B. *Angew. Chem.* **1996**, *108*, 2991–2995; *Angew. Chem., Int. Ed. Engl.* **1996**, *35*, 2810–2813; (j) Li, G.; Angert, H. H.; Sharpless, K. B. *Angew. Chem.* **1996**, *108*, 2995–2999; *Angew. Chem., Int. Ed. Engl.* **1996**, *35*, 2813–2817.

5. Overman, L. E.; Sugai, S. *J. Org. Chem.* **1985**, *50*, 4154–4155.

6. Martínez, L. E.; Leighton, J. L.; Carsten, D. H.; Jacobsen, E. N. *J. Am. Chem. Soc.* **1995**, *117*, 5897–5898.

7. (a) Ami, E.; Ohrui, H. *Biosci. Biotechnol. Biochem.* **1999**, *63*, 2150–2156; (b) Honig, H.; Seufer-Wasserthal, P. *J. Chem. Soc., Perkin Trans. 1* **1989**, 2341–2345; (c) Faber, K.; Hönig, H.; Seufer-Wasserthal, P. *Tetrahedron Lett.* **1988**, *29*, 1903–1904.

8. (a) Govindaraju, T.; Kumar, V. A.; Ganesh, K. N. *J. Org. Chem.* **2004**, *69*, 1858–1865; (b) Arndt, H.-D.; Knoll, A.; Koert, U. *Angew. Chem.* **2001**, *113*, 2137–2140; *Angew. Chem., Int. Ed.* **2001**, *40*, 2076–2078; (c) Nishida, A.; Shirato, F.; Nakagawa, M. *Tetrahedron: Asymmetry* **2000**, *11*, 3789–3805; (d) Fukazawa, T.; Shimoji, Y.; Hashimoto, T. *Tetrahedron: Asymmetry* **1996**, *7*, 1649–1658.

9. Schiffers, I.; Rantanen, T.; Schmidt, F.; Bergmans, W.; Zani, L.; Bolm, C. *J. Org. Chem.* **2006**, *71*, 2320–2331.

10. Schiffers, I.; Bolm, C.; *unpublished.*

Appendix
Chemical Abstracts Nomenclature; (Registry Number)

Cyclohexene oxide; (286-20-4)

Benzylamine; (100-46-9)

(*S*)-(+)-Mandelic acid; (17199-29-0)

(*R*)-(-)-Mandelic acid; (611-71-2)

(S)-Mandelic acid salt of (1R,2R)-*trans*-2-(N-benzyl)amino-1-cyclohexanol; (882409-01-0)

(R)-Mandelic acid salt of (1S,2S)-*trans*-2-(N-benzyl)amino-1-cyclohexanol; (882409-00-9)

(1R,2R)-*trans*-2-(N-Benzyl)amino-1-cyclohexanol; (141553-09-5)

(1S,2S)-*trans*-2-(N-Benzyl)amino-1-cyclohexanol; (322407-34-1)

Carsten Bolm was born in Braunschweig in 1960. He studied chemistry at the TU Braunschweig (Germany) and at the University of Wisconsin, Madison (USA). In 1987 he obtained his doctorate with Professor Reetz in Marburg (Germany). After postdoctoral training with Professor Sharpless at MIT, Cambridge (USA), Carsten Bolm worked in Basel (Switzerland) with Professor Giese to obtain his habilitation. In 1993 he became Professor of Organic Chemistry at the University of Marburg (Germany), and since 1996 he has a chair of Organic Chemistry at the RWTH Aachen (Germany).

Ingo Schiffers was born in Mönchengladbach (Germany) and studied chemistry at the RWTH Aachen University (Germany). In 2002 he completed his doctoral studies under the supervision of Professor Bolm in which he investigated the asymmetric methanolysis of meso-anhydrides. Since 2003 he is holding a position as a permanent researcher in the group of Professor Bolm.

Eric Woerly was born in Illinois and raised in Indiana. He received his B.S. degree in Chemistry from Indiana University Purdue University Indianapolis (IUPUI), where he performed undergraduate research with Professors Martin J. O'Donnell and William L. Scott. His research focused on the solid-phase synthesis of oligopeptides. In 2007, he began his graduate studies at the University of Illinois at Urbana Champaign. His work on this Organic Syntheses check was completed under the guidance of Professor Scott Denmark.

Aurélie Toussaint was born in Besançon (France). In 2004, she received her diploma in engineering at the Ecole Supérieure de Physique et de Chimie Industrielles (Paris, France) and her master diploma in organic chemistry (Université Pierre et Marie Curie, Paris, France). Since October 2004, she is a PhD student in the research group of Prof. Andreas Pfaltz at the University of Basel.

CATALYTIC ENANTIOSELECTIVE ADDITION OF TERMINAL ALKYNES TO ALDEHYDES: PREPARATION OF (S)-(-)-1,3-DIPHENYL-2-PROPYN-1-OL AND (S)-(-)-4-METHYL-1-PHENYL-2-PENTYN-1,4-DIOL

Submitted by Ryo Takita, Shinji Harada, Takashi Ohshima, Shigeki Matsunaga, Masakatsu Shibasaki.[1]

Checked by Shaun Fontaine, Julia Robinson, and Rick L. Danheiser.

1. Procedure

A. (S)-(–)-1,3-Diphenyl-2-propyn-1-ol (3). A flame-dried, 50-mL, two-necked, round-bottomed flask equipped with a glass stopper, argon inlet adapter, and teflon-coated magnetic stir bar is charged with (S)-BINOL (0.230 g, 0.80 mmol) (Note 1) and then is transferred to a glove box and charged with $InBr_3$ (0.291 g, 0.80 mmol) (Note 2). The flask is removed from the glove box and equipped with a rubber septum and a flame-dried reflux condenser fitted with an argon inlet adapter. Dichloromethane (4.0 mL) (Note 3) and then benzaldehyde (4.1 mL, 40 mmol) (Note 4) are added via syringe at room temperature. The resulting solution is stirred at 20-25 °C for 15 min, after which dicyclohexylmethylamine (0.86 mL, 4.0 mmol) (Note 5) is added by syringe. After 10 min, phenylacetylene (8.8 mL, 80 mmol) (Note 6) is added in one portion by syringe and the resulting yellow solution is stirred at 40 °C (oil bath temperature) for 48 h (Note 7). The reaction mixture is then allowed to cool to room temperature, then is poured into a 200-mL separatory funnel (Note 8), and is washed with 50 mL

118

of aqueous 1 M HCl solution. The aqueous phase is separated and extracted with two 50-mL portions of dichloromethane, and the combined organic layers are dried over Na_2SO_4 (2 g) (Note 9), filtered, and concentrated by rotary evaporation (room temperature, 20 mmHg) to afford 13.1 g of a yellow oil. This material is purified by silica gel column chromatography (elution with ethyl acetate/hexanes) (Note 10) to furnish 6.10-6.45 g (73-77%) of **3** as a pale yellow solid (Note 11). The enantiomeric excess is determined to be 98% by CSP-HPLC (Note 12).

  B.   (S)-(–)-4-Methyl-1-phenyl-2-pentyn-1,4-diol (**5**). A flame-dried, 50-mL, two-necked, round-bottomed flask equipped with a glass stopper, argon inlet adapter, and Teflon-coated magnetic stir bar is charged with (*S*)-BINOL (0.573 g, 2.0 mmol) (Note 1) and then transferred to a glove box and charged with $InBr_3$ (0.713 g, 2.0 mmol) (Note 2). The flask is removed from the glove box and equipped with a rubber septum and a flame-dried reflux condenser fitted with an argon inlet adapter. Dichloromethane (10 mL) (Note 3) and then benzaldehyde (4.1 mL, 40 mmol) (Note 4) are added via syringe at room temperature. The resulting solution is stirred at 20-25 °C for 15 min, after which dicyclohexylmethylamine (2.1 mL, 10 mmol) (Note 5) is added by syringe. After 10 min, 2-methyl-3-butyn-2-ol (**4**) (7.8 mL, 80 mmol) (Note 13) is added in one portion by syringe and the resulting yellow solution is stirred at 40 °C (oil bath temperature) for 48 h (Note 14). The reaction mixture is then allowed to cool to room temperature, poured into a 200-mL separatory funnel (Note 8), and washed with 50 mL of aqueous 1 M HCl solution. The aqueous phase is separated and extracted with two 50-mL portions of dichloromethane, and the combined organic layers are dried over Na_2SO_4 (Note 9), filtered, and concentrated to afford 13.6 g of a yellow oil. This material is purified by silica gel column chromatography (elution with ethyl acetate/hexanes) (Note 15) to furnish 6.37-6.87 g (83-89%) of **5** as a colorless solid (Note 16). The enantiomeric excess is determined to be 99% by CSP-HPLC (Notes 17 and 18).

2. Notes

  1.   (*S*)-BINOL (chemical purity 99.8%, enantiomeric purity 99.8%) was purchased by the submitters from Fujimoto Bunshi Kagaku and by the checkers from Aldrich Chemical Company, Inc. (chemical purity 99%, enantiomeric purity 99%).   The reagent was further purified by

recrystallization from diethyl ether/hexane according to the following procedure. (S)-BINOL (5.0 g) was dissolved in hot diethyl ether (60 mL) and hexane (100 mL, room temperature) was added. The solution was allowed to stand at room temperature for 2 d and the resulting colorless crystals were collected by filtration to afford 3.9 g of (S)-BINOL which was used after drying at 60 °C (0.3 mmHg) for 24 h. The material purchased by the submitters contained a pale brown impurity which was removed prior to recrystallization by dissolving 10 g of the (S)-BINOL in 100 mL of ether, adding 3 g of activated charcoal, stirring for 1 h, and then filtering the mixture through Celite® and concentrating by rotary evaporation (room temperature, 20 mmHg).

2.   InBr$_3$ (99.999%) was purchased from Aldrich Chemical Company, Inc. and used without further purification. InBr$_3$ must be stored and handled in a glove box as the quality of the reagent was found to be important for the success of the reaction.

3.   Anhydrous dichloromethane was purchased by the submitters from Kanto Chemical and used without further purification. The checkers obtained dichloromethane from Mallinckrodt Baker (>99.8%) and purified it by pressure filtration through activated alumina. In general, the solvent concentration for this reaction is set to 0.2 M based on InBr$_3$ (5 mL of solvent per 1 mmol of InBr$_3$).

4.   Benzaldehyde (98+%) was purchased by the submitters from Wako Pure Chemical Industries and distilled from CaH$_2$ prior to use. The checkers obtained benzaldehyde (99.5%) from Aldrich Chemical Company, Inc. and distilled it at 57-58 °C (8.2 mmHg) from CaH$_2$ the same day it was used.

5.   Dicyclohexylmethylamine (>97.0%) was purchased by the submitters from Tokyo Chemical Industry Co., Ltd. and distilled from CaH$_2$ prior to use. The checkers obtained dicyclohexylmethylamine (97%) from Aldrich Chemical Company, Inc. and distilled it at 82-85 °C (110 mmHg) from CaH$_2$ on the same day it was used.

6.   Phenylacetylene (2) (>97%) was purchased by the submitters from Tokyo Chemical Industry Co., Ltd. and by the checkers from Fluka. The alkyne (50 mL) was passed through a pad of activated alumina (diameter = 2.5 cm, height = 3.3 cm) and then distilled at 73 °C (77 mmHg). The purity of the alkyne was found to have a large effect on its reactivity in this reaction, and the alkyne is best distilled immediately prior to use. The

order of addition (aldehyde first, then amine, and finally alkyne) was also found to be important to obtain the best results.

7. The reaction was monitored by thin layer chromatography (TLC) on Merck or EMD precoated 0.25 mm silica gel 60 F254 plates, visualization with 254 nm UV light followed by a p-anisaldehyde stain, elution with hexane/ethyl acetate/4:1 , R_f (3) = 0.32, R_f (benzaldehyde) = 0.52.

8. The flask was washed with small portions of dichloromethane (50 mL total) which were added to the separatory funnel.

9. Anhydrous sodium sulfate was purchased from Nacalai Tesque, Inc. or Mallinckrodt.

10. Column chromatography was performed by the checkers using 200 g of Sorbent Technologies 230-400 mesh silica gel 60 (6 x 14 cm column) and 100-mL fractions were collected (1500 mL of hexane/ethyl acetate, 95:5 and then 3500 mL of hexane/ethyl acetate, 92:8). The desired product was obtained in fractions 21-47, which were combined and concentrated by rotary evaporation at room temperature (20 mmHg) and then dried at 4 mmHg overnight to provide a pale yellow solid. The submitters obtained the product as a pale-yellow oil.

11. The product exhibits the following physicochemical properties: pale yellow solid; mp 40-42 °C; $[\alpha]^{21}_D$ –1.85 (c 1.25, CHCl$_3$); IR (neat): 3345, 2229, 1598, 1489, 1454, 1030, 756, 690 cm^{-1}; ^{1}H NMR (400 MHz, CDCl$_3$) δ: 2.38 (d, J = 6.0 Hz, 1 H), 5.71 (d, J = 6.4 Hz, 1 H), 7.30-7.51 (m, 8 H), 7.62-7.65 (m, 2 H); ^{13}C NMR (400 MHz, CDCl$_3$) δ: 65.3, 86.9, 88.8, 122.6, 126.9, 128.5, 128.6, 128.8, 128.9, 131.9, 140.8; HRMS (ESI (+)) m/z 231.0770 [M+Na]$^+$; Anal. Calcd. for C$_{15}$H$_{12}$O: C, 86.51; H, 5.81. Found: C, 86.47; H, 5.73.

12. Enantiomeric excess was determined to be 98% by HPLC. t_R (minor) 11.0 min, t_R (major) 18.3 min (DAICEL CHIRALCEL OD-H,; hexane/2-propanol, 9:1; 1.0 mL/min; 254 nm, 22 °C).

13. 2-Methyl-3-butyn-2-ol (4) (>98%) was purchased by the submitters from Tokyo Chemical Industry Co., Ltd. and by the checkers from City Chemical Company and was distilled from CaSO$_4$ (DRIERITE®) at 100 °C (760 mmHg) prior to use.

14. The reaction was monitored by thin layer chromatography (TLC) on Merck or EMD precoated 0.25 mm silica gel 60 F254 plates, visualization with 254 nm UV light followed by a p-anisaldehyde stain,

elution with hexane/ethyl acetate, 2:1, R_f (5) = 0.29, R_f (benzaldehyde) = 0.71.

15. Column chromatography was performed by the checkers using 100 g of Sorbent Technologies 230-400 mesh silica gel 60 (5.5 x 11 cm column) and 100-mL fractions were collected (1000 mL of hexane/ethyl acetate, 85:15, then 1000 mL each of 75:25, 65:45, and finally 60:40 hexane/ethyl acetate). The desired product was obtained in fractions 15-39, which were combined and concentrated by rotary evaporation at room temperature (20 mmHg). The residue was then dissolved in 20 mL of dichloromethane and concentrated by rotary evaporation to remove residual ethyl acetate, and the resulting solid was dried at 4 mmHg overnight.

16. The product exhibits the following physicochemical properties: colorless solid; mp 69-70 °C (lit.[7] 66 °C); $[\alpha]^{20}_D$ –15.4 (c 3.31, CHCl₃); IR (neat) 3335, 2982, 1455, 1165, 950, 699 cm⁻¹; ¹H NMR (400 MHz, CDCl₃) δ: 1.52 (s, 6 H), 3.15 (br s, 1 H), 3.47 (br s, 1 H), 5.44 (s, 1 H), 7.29-7.38 (m, 3 H), 7.49-7.51 (m, 2 H); ¹³C NMR (400 MHz, CDCl₃) δ: 31.3, 31.4, 64.4, 65.4, 82.0, 91.5, 126.9, 128.4, 128.7, 140.7 ; HRMS (ESI (+) m/z 213.0878 [M+Na]⁺; Anal. Calcd. for $C_{12}H_{14}O_2$: C, 75.76; H, 7.42. Found: C, 75.71; H, 7.44.

17. Enantiomeric excess was determined to be 99% by HPLC; t_R (major) 12.9 min, t_R (minor) 16.4 min (DAICEL CHIRALCEL OD-H, eluent: hexane/2-propanol, 9:1; 0.5 mL/min; 254 nm, 25 °C). The checkers did not observe the presence of this minor isomer).

18. The submitters reported the further purification of the product by recrystallization which was performed as follows: 6.71 g of 5 was dissolved in warm dichloromethane (12 mL at 35 °C) and to this solution was slowly added 15 mL of hexane over 10 min. The solution was left standing overnight and the resulting solid was collected on a Buchner funnel under vacuum, then was washed with 30 mL of hexane, and dried at 4 mmHg (room temperature) overnight to afford 6.47 g of a white crystalline solid. The enantiomeric excess of this material was determined to be >99% ee by HPLC (Note 17).

Safety and Waste Disposal Information

All hazardous materials should be handled and disposed of in accordance with "Prudent Practices in the Laboratory"; National Academy Press; Washington, DC, 1995.

3. Discussion

The addition of terminal alkynes to aldehydes, especially in an enantioselective manner, is of great interest because of the versatility of the corresponding propargylic alcohols.[2] Stoichiometric amounts of strong bases such as organolithium, organomagnesium, or dialkylzinc reagents are widely used for this type of reaction with chiral ligands or chiral Lewis acid complexes. However, intrinsic drawbacks, such as the use of excess amounts of metal reagents and a separate step for metal acetylide preparation, make it difficult to achieve an efficient process. Thus, the use of only catalytic amounts of chiral metal salts and the use of terminal alkynes directly as a substrate under a simple operation is desirable. The first example of a catalytic system ($Zn(OTf)_2$, N-methylephedrine, and Et_3N) was reported by Carreira and co-workers.[3] Their system provides the corresponding propargylic alcohols in a highly enantioselective manner. Aromatic aldehydes, however, cannot be used in the catalytic system due to the Cannizzaro reaction.[3a]

The procedure herein describes the practical preparation of chiral propargylic alcohols under simple and mild conditions catalyzed by an indium(III)/BINOL complex.[4] The catalytic system was developed based on our concept of bifunctional catalysis.[5] The dual activation of soft pro-nuclophiles (alkynes) and hard electrophiles (aldehydes) by indium(III)[6] enabled a wide substrate generality, especially in terms of aliphatic and aromatic aldehydes. In addition, the simple operation in this procedure with the readily available chiral ligand, BINOL, provides an attractive method for the preparation of optically active propargylic alcohols.

The examples of substrate scope are summarized in Tables 1 and 2.[4a] The catalytic system (10 mol %) is applicable for a wide variety of both aliphatic (Table 1) and aromatic aldehydes (Table 2). Phenylacetylene shows high reactivity while the use of alkyl- and alkenylacetylene exhibits slower reaction rates. The limitations of this system should be also noted: the use of acetylene gas, silylacetylenes (e.g. (trimethylsilyl)acetylene), or alkynes having ester functionality (e.g. propargyl propionate) yields no or little amount of the corresponding products under the reaction conditions.

It is noteworthy that the reaction with 2-methyl-3-butyn-2-ol (**4**) proceeds smoothly (Procedure B), affording the corresponding products in good chemical yield and high enantioselectivity. In terms of synthetic utility, the use of an acetylene equivalent is fascinating. 2-Methyl-3-butyn-2-ol is the protected form of acetylene, similar to (trimethylsilyl)acetylene, and it is easy to convert the corresponding products to versatile terminal alkynes (i.e., chiral 3-hydroxy-1-alkyne derivatives),[7] which can be further transformed into more complex and useful molecules via Sonogashira coupling reactions, alkylations, and so on. The examples of substrate scope with 2-methyl-3-butyn-2-ol are summarized in Table 3.[4b]

Table 1

$$R^1\text{CHO} + H\text{---}\equiv\text{---}R^2 \xrightarrow[\substack{\text{Cy}_2\text{NMe (50 mol \%)}\\ \text{CH}_2\text{Cl}_2\text{ (2.0 M), 40 °C}}]{\substack{\text{InBr}_3\text{ (10 mol \%)}\\ \text{(S)-BINOL (10 mol \%)}}} R^1\text{---}\overset{\text{OH}}{\underset{}{}}\text{---}R^2$$

(2.0 equiv)

entry	aldehyde	alkyne	time (h)	yield (%)	ee (%)
1	⬡-CHO	H—≡—Ph	9	95	98
2	⬡-CHO	H—≡—(CH$_2$)$_2$Ph	36	77	>99
3	⬡-CHO	H—≡—(CH$_2$)$_3$CN	42	93	99
4	⬡-CHO	H—≡—(CH$_2$)$_3$Cl	42	91	85
5	⅄CHO	H—≡—Ph	25	85	96
6	⅄CHO	H—≡—(CH$_2$)$_2$Ph	48	46	98
7[a]	⌬∼CHO	H—≡—Ph	24	85	98

[a] The aldehyde was slowly added over 22 h.

Table 2

$$R^1CHO \ + \ H{\equiv}R^2 \ (2.0 \ equiv) \xrightarrow[\substack{Cy_2NMe \ (50 \ mol \ \%) \\ CH_2Cl_2 \ (2.0 \ M), \ 40 \ ^\circ C}]{\substack{InBr_3 \ (10 \ mol \ \%) \\ (S)\text{-BINOL} \ (10 \ mol \ \%)}} \ R^1\text{-CH(OH)-C}{\equiv}\text{C-}R^2$$

entry	aldehyde	alkyne	time (h)	yield (%)	ee (%)
1	Ph–CHO	H≡–Ph	24	84	95
2	Ph–CHO	H≡–(CH₂)₂Ph	48	70	98
3	Ph–CHO	H≡–(CH₂)₃CN	48	81	98
4	Ph–CHO	H≡–cyclohexyl	48	77	89
5	Ph–CHO	H≡–cyclopropyl	48	74	83
6	MeO–C₆H₄–CHO	H≡–Ph	48	77	97
7	F–C₆H₄–CHO	H≡–Ph	24	75	95
8	F–C₆H₄–CHO	H≡–(CH₂)₂Ph	45	61	99
9	thiophene-3-CHO	H≡–Ph	29	80	97
10	furan-3-CHO	H≡–Ph	20	84	98

Table 3

entry	aldehyde	catalyst (x mol %)	time (h)	yield (%)	ee (%)
1	Ph—CHO	10	39	88	97
2 [a]	Ph—CHO	5	24	88	99
3 [b]	Ph—CHO	2	48	82	98
4	Cl—C6H4—CHO	10	48	87	99
5	MeO—C6H4—CHO	10	48	62	99
6	furyl-CHO	10	40	89	93
7	thienyl-CHO	10	41	87	99
8	(CH3)2CHCH2-CHO	10	19	81	99
9	(CH3)2CH-CHO	10	22	91	98
10	cyclohexyl-CHO	10	19	97	99
11	(CH3)3C-CHO	10	22	94	98
12	Ph—CH=CH—CHO	10	39	40	99

[a] The reaction was performed in CH$_2$Cl$_2$ (4.0 M).

[b] The reaction was performed in ClCH$_2$CH$_2$Cl (10 M) at 80 °C.

1. Graduate School of Pharmaceutical Sciences, The University of Tokyo, Hongo 7-3-1, Bunkyo-ku, Tokyo 113-0033, Japan.

2. For reviews, see (a) Cozzi, P. G.; Hilgraf, R.; Zimmermann, N. *Eur. J. Org. Chem.* **2004**, 4095-4105 (b) Pu, L. *Tetrahedron* **2003**, *59*, 9873-9886.

3. (a) Anand, N. K.; Carreira, E. M. *J. Am. Chem. Soc.* **2001**, *123*, 9687-9688. (b) Fässler, R.; Tomooka, C. S.; Frantz, D. E.; Carreira, E. M. *Proc. Natl. Acad. Sci. U.S.A.* **2004**, *101*, 5843-5845. See also: (c) Frantz, D. E.; Fässler, R.; Carreira, E. M. *J. Am. Chem. Soc.* **1999**, *121*, 11245-11246. (d) Frantz, D. E.; Fässler, R.; Carreira, E. M. *J. Am. Chem. Soc.* **2000**, *122*, 1806-1807. (e) Frantz, D. E.; Fässler, R.; Tomooka, C. S.; Carreira, E. M. *Acc. Chem. Res.* **2000**, *33*, 373-381.

4. (a) Takita, R.; Yakura, K.; Ohshima, T.; Shibasaki, M. *J. Am. Chem. Soc.* **2005**, *127*, 13760-13761. (b) Harada, S.: Takita, R.; Ohshima, T.; Matsunaga, S.; Shibasaki, M. *Chem. Commun.* **2007**, *9*, 948-950.

5. (a) Kanai M.; Kato, N.; Ichikawa, E.; Shibasaki, M. *Synlett* **2005**, 1491-1508. (b) Shibasaki, M.; Yoshikawa, N. *Chem. Rev.* **2002**, *102*, 2187-2210. (c) Shibasaki, M.; Sasai, H.; Arai, T. *Angew. Chem., Int. Ed. Engl.* **1997**, *36*, 1236-1256. (d) Ma, J.-A.; Cahard, D. *Angew. Chem., Int. Ed.* **2004**, *43*, 4566-4583. (e) *Multimetallic Catalysts in Organic Synthesis*; Shibasaki, M., Yamamoto, Y., Eds.; Wiley-VCH: Weinheim, 2004.

6. (a) Takita, R.; Fukuta, Y.; Tsuji, R.; Ohshima, T.; Shibasaki, M. *Org. Lett.* **2005**, *7*, 1363-1366. (b) For the example using stoichiometric amounts of InBr$_3$, see: Sakai, N.; Hirasawa, M.; Konakahara, T. *Tetrahedron Lett.* **2003**, *44*, 4171-4182.

7. Boyall, D.; López, F.; Sasaki, H.; Frantz, D.; Carreira, E. M. *Org. Lett.* **2000**, *2*, 4233-4236. See, also ref 4(b).

Appendix
Chemical Abstracts Nomenclature; (Registry Number)

(*S*)-BINOL: [1,1'-Binaphthalene]-2,2'-diol, (1*S*)-: (18531-99-2)

Indium bromide: (13465-09-3)

Benzaldehyde: (100-52-7)

(*S*)-(–)-1,3-Diphenyl-2-propyn-1-ol: Benzenemethanol, α-(2-phenylethynyl)-, (α*S*)-; (132350-96-0)

Dicyclohexylmethylamine: Cyclohexanamine, *N*-cyclohexyl-*N*-methyl-; (7560-83-0)

Phenylacetylene: Ethynylbenzene; (536-74-3)

2-Methyl-3-butyn-2-ol; (115-19-5)

(S)-(−)-4-Methyl-1-phenyl-2-pentyn-1,4-diol: (321855-44-1)

Masakatsu Shibasaki received his PhD. from the University of Tokyo in 1974 under the direction of the late Professor Shun-ichi Yamada before doing postdoctoral studies with Professor E. J. Corey at Harvard University. He joined Teikyo University in 1977, moved to Sagami Chemical Research Center, and took up a professorship at Hokkaido University, before returning to the University of Tokyo as a professor in 1991. He has received many awards, including the Pharmaceutical Society of Japan Award (1999), ACS Award (Arthur C. Cope Senior Scholar Award) (2002), Japan Academy Prize (2005), and ACS Award (Creative Work in Organic Synthesis) (2008).

Ryo Takita was born in 1978 in Sapporo, Japan. He obtained his PhD from The University of Tokyo in 2006 under the supervision of Professor Masakatsu Shibasaki, working on the development of an indium(III) catalyst system for asymmetric alkynylation of carbonyl compounds. From 2006 to 2007, he worked as a postdoctoral fellow at Massachusetts Institute of Technology with Professor Timothy M. Swager on the synthesis of electroactive hinge molecules. He is currently an assistant professor at Institute for Chemical Research, Kyoto University.

Shinji Harada was born in 1980 in Tokyo, Japan. After obtaining his B.S. degree from the University of Tokyo in 2002, he received Ph.D. from the University of Tokyo in 2007 under the supervision of Prof. Masakatsu Shibasaki. Then, he joined the faculty at Chiba University, where he is currently an assistant professor in Prof. Atsushi Nishida's group. His research interests are catalytic asymmetric reaction and natural product synthesis.

128

Takashi Ohshima received his PhD. from The University of Tokyo in 1996 under the direction of Professor Masakatsu Shibasaki. On the following year, he joined Otsuka Pharmaceutical Co., Ltd. for one year. After two years as a postdoctoral fellow at The Scripps Research Institute with Professor K. C. Nicolaou (1997-1999), he returned to Professor Shibasaki's group as an assistant professor. In 2005, he moved to Osaka University, where he is currently associate professor of chemistry. He has received the Fujisawa Award in Synthetic Organic Chemistry (2001) and Pharmaceutical Society of Japan Award for Young scientists (2004).

Shigeki Matsunaga was born in 1975 in Kyoto and received his Ph. D. from the University of Tokyo under the direction of Prof. M. Shibasaki. He started his academic career in 2001 as an assistant professor in Prof. Shibasaki's group at the University of Tokyo. He is the recipient of the Yamanouchi Award of Synthetic Organic Chemistry, Japan (2001), and The Chemical Society of Japan Award for Young Chemists (2007).

Shaun Fontaine (born 1984) graduated with a B.S in Biochemistry and Chemistry from the University of California, San Diego where he carried out undergraduate research in the laboratory of Professor Michael D. Burkart. In 2006 he joined the graduate program at the Massachusetts Institute of Technology and is currently studying the total synthesis of biologically active alkaloids in the laboratory of Professor Rick Danheiser.

Julia Robinson (born 1984) graduated with a B.A. in Chemistry from Reed College in Portland, Oregon, where she worked in the laboratory of Patrick G. McDougal. She then joined the doctoral program at the Massachusetts Institute of Technology where she is currently an NSF Graduate Fellow pursuing Ph.D. research in the laboratory of Professor Rick Danheiser.

COPPER-CATALYZED THREE-COMPONENT REACTION OF 1-ALKYNES, SULFONYL AZIDES, AND WATER: N-(4-ACETAMIDOPHENYLSULFONYL)-2-PHENYLACETAMIDE

Submitted by Seung Hwan Cho, Seung Jun Hwang, and Sukbok Chang.[1]
Checked by Melissa J. Leyva and Jonathan A. Ellman.[2]

1. Procedure

A single-necked, 250-mL round-bottomed flask equipped with a magnetic stir bar was charged with copper(I) cyanide (45.0 mg, 0.5 mmol) (Note 1) and 4-acetamidobenzenesulfonyl azide (12.01 g, 50.0 mmol) (Note 2). Distilled water (100 mL) was added to the flask, and then phenylacetylene (6.04 mL, 55.0 mmol) (Note 3) was added via syringe at room temperature. To this stirred mixture, triethylamine (7.67 mL, 55.0 mmol) (Note 4) was slowly added via syringe over one min at room temperature (Note 5). The reaction mixture was then vigorously stirred while open to air for 4 h at room temperature (Note 6). The reaction was quenched by the addition of a saturated aqueous ammonium chloride solution (20 mL) via syringe at room temperature. Methanol (50 mL) (Note 7) was added, and then the mixture was stirred for an additional 20 min at the same temperature. The resulting mixture was filtered through a Celite pad (7.0 g) (Note 8), and the pad was washed with methanol (5 x 50 mL). The filtrate was concentrated to half volume using a rotary evaporator (20 mmHg, water bath temperature 28 °C), and then the remaining solution was treated with 1N HCl (50 mL) at room temperature to make it slightly acidic (Note 9). The solution was concentrated using a rotary evaporator (20 mmHg, water bath temperature 28 °C) (Note 10), and then the resulting solid was collected by suction filtration with the aid of distilled water (Note 11). The solid was washed with a pre-cooled (0 °C) mixture of diethyl ether and isopropyl alcohol (20:1, 50 mL) followed by diethyl ether (50 mL) (Note 12). The remaining solid was transferred to a pre-weighed 250-mL round-bottomed flask (Note 13) and then was dried under vacuum at room temperature for 12 h to yield a light yellowish solid (14.4 g, 86%) (Note 14). To obtain analytically pure product, the crude material was dissolved in about 800 mL of boiling methanol in a 2-L Erlenmeyer flask (Note 15). The flask was cooled to room temperature and let stand for 30 h. Upon storage in a – 4 °C refrigerator for 3 days, the crystalline product was collected by suction filtration and was washed with 30 mL of pre-cooled (0 °C) methanol (Note 16). The solid was transferred to a pre-weighed 250-mL round-bottomed flask and then was dried under vacuum at room temperature for 10 h to afford N-(4-acetamidophenylsulfonyl)-2-phenylacetamide as a white solid (9.91 g, 60%) (Notes 17, 18).

2. Notes

1. Copper(I) cyanide (99.98%) was purchased from Aldrich Chemical Co., Inc and was used as received.
2. 4-Acetamidobenzenesulfonylazide (97%) was purchased from Aldrich Chemical Co., Inc and was used as received.
3. Phenylacetylene (98%) was purchased from Aldrich Chemical Co., Inc and was used as received.
4. The submitters purchased triethylamine (99%) from Junsei. Chemical Co., Ltd., and the reagent was distilled from sodium before use. The checkers purchased triethylamine (99%) from Fisher Scientific Inc., and the reagent was distilled from calcium hydride before use.
5. During the addition of triethylamine, the original heterogeneous reaction mixture turned to a yellowish solution.
6. The internal temperature of the reaction mixture increases to 36 °C over 2 h, and the color becomes dark red. The progress of the reaction was monitored by TLC until complete conversion was observed, and during this time the internal temperature returned to room temperature. For TLC analysis, EMD Chemicals, Inc. silica gel 60 F_{254} TLC plates were used. 4-Acetamidobenzenesulfonyl azide and N-(4-acetamidophenyl sulfonyl)-2-phenylacetamide have R_f values of 0.52 and 0.25, respectively (dichloromethane/methanol, 15:1).
7. The submitters purchased methanol (99.9%) from Merck & Co., Inc., and it was used as received. The checkers purchased methanol (HPLC Grade) from Fisher Scientific Inc. and it was used as received.

8. Celite 545® was purchased by the submitters from Daejung Co., Ltd., and by the checkers from Fisher Scientific Inc., and the reagent was used as received. A glass filter (60 mL) with a medium porosity fritted disc was used. A round-bottomed flask (1 L) was used to collect the filtrate.

9. When the 1 N HCl solution was poured into the mixture, a slight exotherm was observed (internal temperature: 29.3 °C) and a white slurry formed. The solution was measured to have a pH of 3.

10. As the solution was concentrated, solid began to precipitate in the bottom of the round-bottomed flask.

11. A 60-mL glass filter with a medium porosity fritted disc was used. Solid that adhered to the flask was transferred using distilled water (two 35-mL portions).

12. The submitters report 2% of product was lost during this process of filtration and washing, which was measured using an internal standard (1,1,2,2,-tetrachloroethane).

13. A round-bottomed flask of joint size 24/40 was used to minimize the loss of product during the transfer.

14. The submitters report a yield of 12.0-12.3 g, 72.2-74.0% without further purification. However, the elemental analysis that was reported was 0.85% low on carbon; Anal. Calcd. for $C_{16}H_{16}N_2O_4S$: C, 57.82; H, 4.85; S, 9.65; N, 8.43; Found: C, 56.97; H, 4.93; S, 10.10; N, 8.59.

15. The Erlenmeyer flask was heated on a hot plate while stirring with a magnetic stir bar and boiling methanol was added until the solid dissolved.

16. A porcelain Büchner filter funnel fitted with 90-mm Whatman 1 qualitative filter paper was used.

17. A half-scale run afforded 5.03 g (61%) of analytically pure product.

18. The product exhibits the following physicochemical properties: R_f = 0.25 (6.25% methanol in dichloromethane); mp 225–226 °C (decomp); ^{1}H NMR (400 MHz, DMSO-d_6) δ: 2.09 (s, 3 H), 3.53 (s, 2 H), 7.14-7.15 (d, J = 7.2 Hz, 2 H), 7.22-7.29 (m, 3 H), 7.76-7.84 (m, 4 H), 10.40 (s, 1 H), 12.26 (s, 1 H); ^{13}C NMR (100 MHz, DMSO-d_6) δ: 24.2, 42.0, 118.4, 126.9, 128.4, 129.0, 129.3, 132.5, 134.0, 143.9, 169.2, 169.4; IR (neat): 3371, 3241, 3175, 3132, 2970, 1738, 1653, 1590, 1455, 1366, 1167 cm^{-1}. Elemental analysis; Calcd. for $C_{16}H_{16}N_2O_4S$: C, 57.82; H, 4.85; S, 9.65; N, 8.43; Found: C, 57.77; H, 4.75; S, 9.62; N, 8.43.

Safety and Waste Disposal Information

All hazardous materials were disposed in accordance with "Prudent Practices in the Laboratory"; National Academic Press; Washington, DC, 1998.

3. Discussion

The amide group is a key motif in chemistry and biology.[3] Although traditional chemical methods for amide synthesis rely heavily on an interconversion strategy between carbonyl groups or their equivalent reactive compounds,[4] the lability of those functional groups often restricts their ubiquitous applications. Hence, the development of alternative routes has been challenging to synthetic chemists.[5]

Recently, we have reported the highly efficient Cu-catalyzed three-component reactions of terminal alkynes, sulfonyl azides, and amines or alcohols to afford amidines or imidates.[6] It is believed that the reaction proceeds via a common ketenimine intermediate, which is generated from the copper-mediated intermolecular cycloaddition of azides and alkynes followed by the release of N_2.[7] On the basis of this postulate, a novel non-conventional amide synthesis could be realized by allowing the plausible ketenimine intermediates, generated in situ under the reaction conditions, to react with water.[8]

This present procedure describes a convenient approach for the preparation of N-sulfonylamides from the reaction of terminal alkynes, sulfonyl azides, and water in the presence of a Cu(I) catalyst. It should be mentioned that water is employed in this case as both a solvent and reagent. The reaction proceeds smoothly at room temperature within a few hours and generates molecular nitrogen as a sole byproduct. A broad range of both terminal alkynes and sulfonyl azides are readily employed under the reaction conditions as demonstrated in Table 1. In addition, a range of copper(I) reagents can be readily used as the catalyst with a similar efficiency. For example, product yields were not significantly changed upon using either CuI or CuCN catalyst, although the latter displayed a slightly faster reaction rate.

Table 1. Preparation of Various *N*-Sulfonylamides.[a]

Entry	Alkyne	Time (h)	Isolated Yield (%)[b]
1		4	84
2		5	88
3		2	82
4		3	86
5		2	84
6[c]		4	85

[a] Reaction conditions: 4-acetamidobenzenesulfonyl azide (5.0 mmol), alkyne (5.5 mmol), CuCN (2.0 mol %), Et$_3$N (5.5 mmol) in water (15 mL) at room temperature. [b] The average yield of three runs. [c] CuCN (4.0 mol %) was used relative to azide, and the yield is of bisamide product.

1. Department of Chemistry, Korea Advanced Institute of Science and Technology (KAIST), Daejeon, 305-701, Republic of Korea, E-mail: sbchang@kaist.ac.kr

2. Department of Chemistry, University of California, Berkeley, CA 94720

3. Humphrey, J. M.; Chamberlin, A. R. *Chem. Rev.* **1997**, *97*, 2243-2266.

4. Bailey, P. D.; Collier, I. D.; Morgan, K. M. In *Comprehensive Organic Functional Group Transformations*; Katritzky, A. R.; Meth-Cohn, O.;

Rees, C. W. Eds.; Pergamon: Cambridge, 1995; Vol. 5, Chapter 6.

5. (a) Bray, B. L. *Nat. Rev. Drug Discovery* **2003**, *2*, 587-593. (b) Albericio, F. *Curr. Opin. Chem. Biol.* **2004**, *8*, 211-221. (c) Shangguan, N.; Katukojvala, S.; Greenberg, R.; Williams, L. J. *J. Am. Chem. Soc.* **2003**, *125*, 7754-7755. (d) Beller, M.; Seayad, J.; Tillack, A.; Jiao. H. *Angew. Chem., Int. Ed.* **2004**, *43*, 3368-3398. (e) Uenoyama, Y.; Fukuyama, T.; Nobuta, O.; Matsubara, H.; Ryu, I. *Angew. Chem., Int. Ed.* **2005**, *44*, 1075-1078.

6. (a) Bae, I.; Han, H.; Chang, S. *J. Am. Chem. Soc.*, **2005**, *127*, 2038-2039. (b) Yoo, E. J.; Bae, I.; Cho, S. H.; Han, H.; Chang, S. *Org. Lett.* **2006**, *8*, 1347-1350.

7. Yoo, E. J.; Ahlquist, M.; Kim, S. H.; Bae, I.; Fokin, V. V.; Sharpless, K. B.; Chang, S. *Angew. Chem., Int. Ed.* **2007**, *46*, 1730-1733.

8. (a) Cho, S. H.; Yoo, E. J.; Bae, I.; Chang, S. *J. Am. Chem. Soc.* **2005**, *127*, 16046-16047. (b) Cassidy, M. P.; Raushel, J.; Fokin, V. V. *Angew. Chem., Int. Ed.* **2006**, *45*, 3154-3157. (c) Cho, S. H.; Chang, S. *Angew. Chem., Int. Ed.* **2007**, *46*, 1897-1900.

Appendix
Chemical Abstracts Nomenclature; (Registry Number)

Copper Cyanide; (544-92-3)
Phenylacetylene; (536-74-3)
4-Acetamidobenzenesulfonyl azide; (2158-14-7)
Triethylamine; (121-44-8)

Sukbok Chang received his B.S. degree in chemistry from Korea University in 1985, and M.S. degree from KAIST under the guidance of Professor Sunggak Kim in 1987. In 1996, he earned his Ph.D. degree in organic chemistry at Harvard University working with Professor Eric N. Jacobsen. After postdoctoral experience with Professor Robert H. Grubbs at Caltech, he joined the faculty at Ewha Womans University in Seoul in 1998, and then moved to KAIST in 2002, where he is focusing on the development, understanding, and synthetic applications of transition metal catalysis.

Seung Hwan Cho was born in Gwang-Ju, Republic of Korea, in 1983. He received his B.S. degree in chemistry from Korea Advanced Institute of Science of Technology (KAIST) in 2005. He is currently in a M.S.-Ph.D. joint program at KAIST under the guidance of Professor Sukbok Chang. His graduate research has focused on development of new organometallic chemistry and studies on the biological activities of new types of heterocyclic compounds.

Seung Jun Hwang was born in Seoul, Republic of Korea, in 1984. He graduated from KAIST with a B.S. degree in chemistry in 2007. He is currently in a master course in the group of Professor Sukbok Chang at KAIST. He is interested in finding new methodologies using organometallic chemistry for synthetic applications.

Melissa J. Leyva was born in El Paso, Texas in 1982. She received her B.S. degree in Chemistry at the University of Texas, El Paso in 2005. She then began her doctoral studies at the University of California, Berkeley under the direction of Professor Jonathan A. Ellman. Her graduate research has focused on the identification of novel inhibitors for therapeutically important proteases.

PREPARATION OF CYCLOHEPTANE-1,3-DIONE VIA REDUCTIVE RING EXPANSION OF 1-TRIMETHYLSILYLOXY-7,7-DICHLOROBICYCLO[3.2.0]HEPTAN-6-ONE
(Cycloheptane-1,3-dione)

Submitted by Nga Do, Ruth E. McDermott, and John A. Ragan.[1a]
Checked by Andrew Martins and Mark Lautens.

1. Procedure

A. 1-Trimethylsilyloxy-7,7-dichlorobicyclo[3.2.0]heptan-6-one (Note 1). A 500-mL, three-necked, round-bottomed flask equipped with a nitrogen outlet, internal temperature probe, 125-mL addition funnel (capped with a septum and nitrogen inlet), and a 65 x 20 mm egg-shaped magnetic stir bar is purged with nitrogen, then charged with 1-trimethylsilyloxycyclopentene (20.8 g, 133 mmol) (Note 2), hexanes (208 mL), and triethylamine (22.3 mL, 16.2 g, 160 mmol, 1.2 equiv) (Note 3). The addition funnel is charged with hexanes (100 mL) and dichloroacetyl chloride (12.8 mL, 19.6 g, 133 mmol, 1.0 equiv), and this solution is added dropwise to the vigorously stirred reaction mixture at a rate that maintains effective stirring and an internal temperature below 30 °C (the addition requires 30-40 min). A white precipitate (Et₃N•HCl) forms during the addition, turning to a brown slurry

138

upon complete addition. The slurry is stirred for 12 h, at which time GC analysis shows approximately 10% starting enol ether (Note 4). An additional portion of dichloroacetyl chloride (1.5 mL, 2.3 g, 16 mmol, 0.12 equiv) in 3 mL of hexanes is added to the reaction mixture via the addition funnel. After 1 h (Note 5), GC/MS analysis shows no remaining starting material. A sintered-glass Büchner funnel (8 cm diameter, 350 mL capacity) is charged with 40 g of silica gel (Note 6) slurried in hexanes, and the reaction mixture is filtered with suction through this pad into a 2-L filter flask rinsing with an additional 600 mL of hexanes in three portions. The clear, pale-yellow filtrate is concentrated on the rotary evaporator (25 mmHg, room temperature water bath) to provide the product as a pale-yellow liquid which crystallizes to a yellow solid when stored in the freezer (7 °C) (29.9 g, 112 mmol, 84% yield) (Note 7). This material is used without further purification in the next step.

B. *1-Trimethylsilyloxybicyclo[3.2.0]heptan-6-one.* A 1-L, four-necked, round-bottomed flask equipped with an overhead stirrer, adapter with internal temperature probe, reflux condenser with nitrogen inlet, and a septum with nitrogen outlet is purged with nitrogen, then charged with 2-propanol (350 mL, Note 3) and nitrogen is bubbled through the solvent for 15 min via a 30-cm, 18 gauge needle. The mechanical stirrer, fitted with a 60 X 20 mm Teflon paddle, is started at 120-150 rpm. 1-Trimethylsilyloxy-7,7-dichlorobicyclo[3.2.0]heptan-6-one (29.9 g, 112 mmol) is dissolved in 50 mL of 2-propanol (gentle heating is required) and added to the reaction flask, rinsing with an additional 50 mL, then 20 mL of 2-propanol. Nitrogen is then bubbled through the solution for 15 min. The septum is removed and replaced with a glass funnel, through which 10% Pd/C (5.98 g, 50% w/w in water) is added, the septum is replaced, and nitrogen is bubbled through the solution for 5 min. The septum is removed again, replaced with a glass funnel, and sodium formate is added in a single portion (38.1 g, 559 mmol, 5.0 equiv, Note 3), rinsing with 25 mL of 2-propanol The septum is replaced and nitrogen is bubbled through the solution for 5 min. The flask is heated in an oil bath to an internal temperature of 80 °C, and is maintained at this temperature with stirring for 18 h. The reaction mixture is cooled to room temperature and a 60-µL aliquot is removed and diluted with 600 µL of ether. A 4-µL sample of this solution is analyzed by GC (Note 4), and shows complete conversion. The slurry is filtered with suction through a pad of Celite (8 cm diameter sintered glass funnel, 350 mL volume, with a 4 cm pad of Celite), into a 1-L filter flask rinsing with four 50-mL portions of

2-propanol. The filtrate is concentrated on the rotary evaporator (25 mmHg, room temperature water bath) to a cloudy, yellow oil, which is then diluted with 150 mL of methyl *t*-butyl ether (MTBE, Note 3) and then is transferred to a 250-mL separatory funnel. The organic phase is washed with three portions of half-saturated brine (75 mL each), once with saturated brine, then is dried over sodium sulfate, filtered, and concentrated on the rotary evaporator (25 mmHg, room temperature water bath) to provide the product as a clear, orange oil (17.2 g, 86.7 mmol, 77% yield) (Note 8) with traces of MTBE and 2-propanol. This material is used without further purification in the next reaction.

C. Cycloheptane-1,3-dione. A three-necked, 250-mL, round-bottomed flask equipped with an internal temperature probe, 50-mL addition funnel with nitrogen inlet, a septum with nitrogen outlet and magnetic stir bar is purged with nitrogen and charged with 1-trimethylsilyloxybicyclo[3.2.0]-heptan-6-one (17.1 g, 81.0 mmol) and 33 mL each of 2-propanol and water. The solution is cooled to an internal temperature of 0-5 °C with an ice-water bath, and 26 mL of 2:1 water-acetic acid (v/v) is added dropwise via the addition funnel over 30 min. At this rate of addition, the internal temperature remains below 5 °C throughout the addition. The solution is then allowed to stir for 16 h and gradually come to room temperature. A 60-μL aliquot is removed and diluted with 600 μL of ether from which a 4-μL sample is used for GC analysis (Note 4), which shows complete conversion to cycloheptane-1,3-dione. The reaction mixture is poured into 200 mL of MTBE in a 500-mL separatory funnel. The layers are separated, and the aqueous phase is extracted with an additional 100 mL of MTBE. The organic solutions are combined, washed with 50 mL of brine, and dried over sodium sulfate. Filtration through cotton, rinsing with 50 mL MTBE and concentration on the rotary evaporator (25 mmHg, 30 °C water bath) provides the crude product as a clear, amber oil. ^{1}H NMR analysis shows the diketone contaminated with residual MTBE and acetic acid. The oil is dissolved in 50 mL of dichloromethane, and treated with 22.5 g of silica gel (Note 6). Concentration on the rotary evaporator provides a tan, free-flowing solid, which is placed on the top of a 7.5 x 23 cm column of silica gel packed as a slurry in hexane/Et$_2$O, 2:1. The column is eluted with hexane/Et$_2$O, 2:1 (500 mL) and 1:1 (3.8 L), followed by combination of the product-containing fractions (TLC in hexane/Et$_2$O, 1:1, visualization with *p*-anisaldehyde stain). The combined fractions are concentrated via rotary evaporator, under vacuum (25 mmHg, room temperature water bath) to

Org. Synth. **2008**, *85*, 138-146

provide cycloheptane-1,3-dione as a clear, pale-yellow oil (6.69 g, 53.1 mmol, 61% yield). This material is approximately 95% pure as determined by ^{1}H NMR analysis and is suitable for most subsequent uses, although the chromatographed material fails combustion analysis. If necessary, further purification can be achieved by short path vacuum distillation (Notes 9, 10, and 11).

2. Notes

1. This procedure is a modification of that reported previously by the submitters.[3]

2. 1-Trimethylsilyloxycyclopentene (97%) was purchased from Aldrich Chemical Company, Inc., and used as received. It can also be prepared from cyclopentanone.[4] The reported yields were obtained using a freshly-opened bottle; older bottles became colored and cloudy, and afforded lower yields.

3. Dichloroacetyl chloride (98%, freshly opened bottle), methyl t-butyl ether (99.8%, HPLC grade) and sodium formate (99+%) were purchased from Aldrich Chemical Company. Glacial acetic acid (99%), hexane (99%) and isopropyl alcohol (99.5%) were purchased from Fisher Scientific. Triethylamine (98%) was purchased from ACP Chemicals. Pd/C (10 wt%) was purchased from Alfa Aesar. All reagents and solvents were used as received.

4. A Hewlett-Packard GC with an HP-5 0.25 mm x 60 m, 0.25 μm column was used. Oven temperature program: 1 min at 40 °C, 20 °C/min ramp to 300 °C, 4 min hold at 300 °C. The carrier gas (He) was held at a constant flow rate of 1.0 mL/min. Detection was by flame ionization detector.

Compound	Retention time (min)
1-Trimethylsilyloxycyclopentene	9.1
1-Trimethylsilyloxy-7,7-dichlorobicyclo[3.2.0]heptan-6-one (**1**)	12.9
1-Trimethylsilyloxy-bicyclo[3.2.0]heptan-6-one (**2**)	11.4
Cycloheptane-1,3-dione (**3**)	10.9

5. GC analysis after 30 minutes still showed a peak for the silyl enol ether (~7% remaining); however, the addition of extra Et$_3$N and/or dichloroacetyl chloride did not change the product/silyl enol ether ratio.

6. Silica gel (40-63 μm) was purchased from Silicycle Inc.

7. ^{1}H NMR (CDCl$_3$, 400 MHz) δ: 0.25 (s, 9 H), 1.50-1.63 (m, 1 H), 1.86-1.98 (m, 2 H), 2.00-2.11 (m, 2 H), 2.51-2.58 (m, 1 H), 3.66 (br d, $J =$ 8.6 Hz, 1 H); ^{13}C NMR (100 MHz, CDCl$_3$) δ 1.6, 25.2, 29.1, 38.1, 67.9, 87.8, 92.2, 199.3; IR thin film (cm^{-1}) 844, 1106, 1253, 1324, 1804, 2961; MS (EI) m/z (relative intensity): 266 (1), 149 (20), 93 (12), 84 (12), 79 (17), 75 (12), 73 (100), 55 (12). This data is in agreement with that reported by Krepski and Hassner.[5]

8. ^{1}H NMR (CDCl$_3$, 400 MHz) δ: 0.17 (s, 9 H), 1.47-1.65 (m, 1 H), 1.76-2.00 (m, 4 H), 2.11-2.18 (m, 1 H), 2.92-3.01 (m, 1 H), 3.24-3.39 (m, 2 H); ^{13}C NMR (100 MHz, CDCl$_3$) δ 1.5, 25.8, 28.7, 40.5, 59.7, 70.3, 77.4, 212.4; IR neat (cm^{-1}) 844, 1087, 1224, 1253, 1782, 2956MS (EI) m/z (relative intensity): 170 (18), 169 (100), 157 (75), 156 (95), 155 (25), 75 (90), 73 (68). This data is in agreement with that reported by Pak.[6]

9.   The chromatographically-purified diketone (3.00 g) is placed in a 25-mL round bottomed flask equipped with a short-path distillation apparatus (8-cm stillhead length with a 7-cm jacketed water-cooled condenser and 4 pronged distribution adapter) and heated in an oil bath (bath temperature reached ~105 °C) under reduced pressure (0.20 – 0.22 mmHg). The product distilled over at a head temperature of 81-82 °C, to afford 2.34 g (78 % recovery) of analytically pure diketone (Note 10).

10.   The product exhibits the following physicochemical properties: ^{1}H NMR (CDCl$_3$, 400 MHz) δ: 1.96-2.03 (m, 4 H), 2.56-2.63 (m, 4 H), 3.61 (s, 2 H); ^{13}C NMR (CDCl$_3$, 100 MHz) δ: 25.1, 44.2, 59.9, 205.1; MS (EI) m/z (relative intensity): 126 (68), 98 (100), 83 (37), 70 (43), 55 (49); IR (film): 3608 , 2943, 2867, 1714, 1694, 1455, 1207, 1134, 924 cm^{-1}. HRMS (EI): calcd for C$_7$H$_{10}$O$_2$: 126.0681 (M$^+$), found: 126.0679. Anal. Calcd for C$_7$H$_{10}$O$_2$: C, 66.65; H, 7.99; Found: C, 66.34; H, 8.14. This data is in agreement with that reported previously.[2f]

11.   The diketone should be stored cold under nitrogen. The diketone is unstable to base, readily undergoing a retro-Dieckmann cyclization to form 6-oxoheptanoic acid.[7] The checkers report that storing under nitrogen is not necessary, and that the product did not noticeably degrade after several months in the freezer at -10 °C.

Safety and Waste Disposal Information

All hazardous materials should be handled and disposed of in accordance with "Prudent Practices in the Laboratory"; National Academy Press; Washington, DC, 1995.

3. Discussion

Several syntheses of cycloheptane-1,3-dione have been reported in the literature.[2] At the time this work was initiated, we required a practical synthesis capable of providing multi-kilogram quantities of this intermediate. None of the existing literature syntheses were deemed adequate for this purpose due to a combination of practicality considerations and the use of heavy metal or potentially explosive reagents (e.g. ethyl diazoacetate,[2a] PhHg(CBr₃),[2c] ClCH₂OCH₃,[2e] or Hg(OAc)₂).[2f] We were aware of Hassner's preparation of the bicyclic adduct of dichloroketene and 1-trimethylsilyloxycyclopentene,[5] and Pak's observation that the dechlorinated derivative of this diketone was converted to cycloheptane-1,3-dione upon treatment with fluoride ion.[6] However, the conditions described for the reduction (stoichiometric amount of Bu₃SnH) were not practical for the large scale preparation of an intermediate destined to be converted into clinical supplies. We found and reported that a Zn/AcOH/aq 2-propanol system effected the desired reduction, desilylation and ring expansion in a single-pot sequence.[3] Unfortunately, upon scale-up this reaction suffered a dramatic decrease in yield, due to formation of a mixture of the desired diketone and 2-acetylcyclopentanone,[7] an observation also made by others.[8]

We believe the reason for this formation of consitutuional isomers to be competition between two reaction pathways, as shown below.[7] In path A, reduction of both chlorine atoms prior to desilylation leads to clean formation of the desired diketone (3). However, if desilylation precedes chlorine atom reduction (path B), then the retro-aldol reaction generates a 2-acetyl-substituted cyclopentanone (4) by rupture of the C_1-C_7 bond. Chlorine atom reduction then generates 2-acetylcyclopentanone (5). These results are consistent with observations made by Hassner and Krepski on related dichlorocyclobutanones.[5]

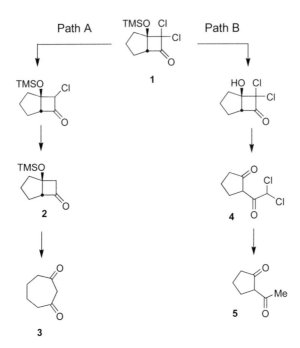

In this improved procedure we report that transfer hydrogenolysis with sodium formate efficiently reduces the chlorine atoms without any desilylation. Acid-mediated desilylation then cleanly forms the desired diketone, with none of the isomeric product being formed. Although this improved process involves three separate operations, there is no need to purify either intermediate, leading to an efficient preparation of the title diketone.

1. Pfizer Global Research & Development, Chemical Research & Development, Groton, CT, 06340 (john.a.ragan@pfizer.com).
2. (a) Eistert, B.; Haupter, F.; Schank, K. *Liebigs Ann. Chem.* **1963**, *665*, 55-67; (b) Maclean, I.; Sneeden, R. P. A. *Tetrahedron* **1965**, *21*, 31-34. (c) Hutmacher, H.-M.; Krüger, H.; Musso, H. *Chem. Ber.* **1977**, *110*, 3118-3125; (d) Suzuki, M.; Watanabe, A.; Noyori, R. *J. Am. Chem. Soc.* **1980**, *102*, 2095-2096; (e) Nishiguchi, I.; Hirashima, T.; Shono, T.; Sasaki, M. *Chem. Lett.* **1981**, 551-554; (f) Bhushan, V.; Chandrasekaran, S. *Synth. Commun.* **1984**, *14*, 339-345.

3. Ragan, J. A.; Makowski, T. W.; am Ende, D. J.; Clifford, P. J.; Young, G. R.; Conrad, A. K.; Eisenbeis, S. A. *Org. Process Res. & Dev.* **1998**, *2*, 379-381.
4. House, H.; Czuba, L. J.; Gall, M.; Olmstead, H. D. *J. Org. Chem.* **1969**, *34*, 2324-2336.
5. Krepski, L.; Hassner, A. *J. Org. Chem.* **1978**, *43*, 3173-3179.
6. Pak, C. S.; Kim, S. K. *J. Org. Chem.* **1990**, *55*, 1954-1957.
7. Ragan, J. A.; Murry, J. A.; Castaldi, M. J.; Conrad, A. K.; Jones, B. P.; Li, B.; Makowski, T. W.; McDermott, R. E.; Sitter, B. J.; White, T. D.; Young, G. R. *Org. Process Res. & Dev.* **2001**, *5*, 498-507.
8. Professor Leo Paquette, Ohio State University, *private communication.*

Appendix
Chemical Abstract Nomenclature (Registry Number)

Cycloheptane-1,3-dione (1194-18-9)
1-Trimethylsilyloxybicyclo[3.2.0]heptan-6-one (125302-44-5)
1-Trimethylsilyloxy-7,7-dichlorobicyclo[3.2.0]heptan-6-one (66324-01-4)
Dichloroacetyl chloride (79-36-7)
1-Trimethylsilyloxycyclopentene (19980-43-9)
Acetic acid (64-19-7)

John Ragan was born and raised in Kansas City, Missouri. He received his B.S. degree from MIT in 1985, where he performed undergraduate research with Rick Danheiser supporting the total synthesis of anatoxin a. He then joined Stuart Schreiber's group at Yale University, moving with Stuart to Harvard in 1988. He completed his Ph.D. in 1990 on the total synthesis of FK506. He then joined Clayton Heathcock's group at the University of California-Berkeley as an American Cancer Society podstodoctoral fellow, where he worked on the biomimetic synthesis of daphniphyllum alkaloids. He joined Pfizer Global Research & Development in 1992, where he has worked in Discovery Chemistry and Chemical Research & Development. He is currently an Associate Research Fellow. Outside of work he enjoys jazz trumpet and Texas Hold 'Em poker.

Nga Do was born in 1973 and grew up in Indiana. She attended Purdue University where she earned a B.S. in Chemistry and conducted undergraduate research in the labs of Professor Mark Lipton. She then joined Professor Scott Rychnovsky's lab at the University of California, Irvine, receiving her M.S. in 1998. She then joined the Chemical Research and Development group at Pfizer in Groton, Connecticut.

Ruth McDermott was born in Boston, MA, and received a B.S. in Chemistry from the University of Massachusetts at Boston. She completed her undergraduate research under the direction of J.-P. Anselme. She is currently part of the Chemical Research and Development group at Pfizer in Groton, CT, where she has been employed for 20 years. She enjoys photography and participating in short-term mission trips in the US and abroad.

Andrew Martins was born in Toronto, Ontario, Canada, and raised in nearby Brampton, Ontario. His post-secondary education began at the University of Waterloo where he completed the Honours Co-operative Chemistry program and performed his honours B.Sc. thesis work in the laboratories of Professor Éric Fillion. He then joined the group of Professor Mark Lautens at the University of Toronto where he is currently a graduate student.

SYNTHESIS OF ENANTIOPURE DI(*TERT*-BUTYL) (2*S*,4*S*)-4-HYDROXY-6-OXO-1,2-PIPERIDINEDICARBOXYLATE. A USEFUL BUILDING BLOCK FOR THE PREPARATION OF 4-HYDROXYPIPECOLATE DERIVATIVES

Submitted by Olivier Chaloin,[1] Frédéric Cabart,[2] Julien Marin,[1,3] Haixiang Zhang,[2] and Gilles Guichard.[1]
Checked by Masakatsu Shibasaki and Hisashi Mihara.

1. Procedure

A.   2-tert-Butoxycarbonylamino-4-(2,2-dimethyl-4,6-dioxo-[1,3]dioxan-5-yl)-4-oxo-butyric acid tert-butyl ester. A one-necked, 500-mL, round-bottomed flask containing a magnetic stir bar and fitted with a three-way stopcock fitted with a rubber septum and an argon inlet is charged with 15.0 g (51.85 mmol) of *N*-α-*tert*-butoxycarbonyl-L-aspartic acid α-*tert*-butyl ester (Boc-L-Asp-O*t*-Bu) (Note 1) and 150 mL of dichloromethane (Note 2). The resulting solution is cooled in an ice bath at 0 °C (ice bath temperature) whereupon 7.85 g (54.44 mmol, 1.05 equiv) of 2,2-dimethyl-1,3-dioxane-4,6-dione (Meldrum's acid) (Note 3) and 9.50 g (77.78 mmol, 1.5 equiv) of 4-dimethylaminopyridine (DMAP) (Note 4) are added. *N*-(3-Dimethylaminopropyl)-*N*'-ethylcarbodiimide hydrochoride (EDC·HCl) (14.91 g, 77.78 mmol, 1.5 equiv) (Note 5) is then added portion-wise to the reaction mixture at 0 °C over 5 min. The three-way adapter is returned to the neck and reaction mixture is allowed to warm to room temperature and is stirred under argon for 3 h (Note 6). The solution is diluted with 50 mL of dichloromethane before being transferred to a 500-mL separatory funnel and is then washed successively with two, 150-mL portions of 1 M aq KHSO₄ solution and 150 mL of brine. The organic layer is dried over Na₂SO₄ (25-30 g), then is filtered through a cotton plug in a funnel, collected in a 500-mL flask and concentrated by rotary evaporation (40 °C, 6 mm Hg) to provide 21.5 g of crude **2** as a clear, pale-yellow oil (Note 7).

  B. Di(tert-butyl) (2S)-4,6-dioxo-1,2-piperidinedicarboxylate. Ethyl acetate (330 mL) (Note 8) is added to the 500-mL flask containing **2**, which is then equipped with a magnetic stir bar and a reflux condenser. The solution is heated in an oil bath at 80 °C (bath temperature) for 3 h (Note 9). The solution is allowed to cool before being transferred to a 1-L separatory funnel and is washed successively with 200 mL of 1 N aq KHSO₄ solution and 200 mL of brine. The organic layer is dried over Na₂SO₄ (25-30 g), filtered through a cotton plug in a funnel, and concentrated by rotary evaporation (34-40 ˚C, 6 mm Hg). The residual crystalline material is slurried in 100 mL of cyclohexane, then is collected by filtration in a Hirsch funnel and is washed with two, 50-mL portions of cyclohexane and then dried under reduced pressure (40 °C, 6 mm Hg) to afford 12.61 g (78%, 2 steps from **1**) of dioxopiperidine **3** as a white solid (Notes 10, 11).

  C. Di(tert-butyl) (2S,4S)-4-hydroxy-6-oxo-1,2-piperidinedicarbox-ylate. A two-necked, 500-mL, round-bottomed flask fitted with a rubber septum, argon needle inlet and thermometer is charged with dioxopiperidine **3** (12.61 g, 40.25 mmol) and 150 mL of dichloromethane (Note 12). The resulting solution is cooled in an ice bath at 0 °C (bath temperature) while 25 mL of AcOH (Note 13) is added. Sodium borohydride (NaBH₄) (4.57 g, 120.75 mmol, 3.0 equiv) (Note 14) is then added portion-wise to the reaction mixture at such a rate as to keep the internal temperature below 10 °C. The reaction mixture is allowed to warm to room temperature and is stirred for 12 h (Note 15). The reaction mixture is neutralized with 150 mL of 1 M aq KHSO₄ solution and then is transferred to a 500-mL separatory funnel. The aqueous phase is separated and the organic layer is dried over Na₂SO₄ (25-30 g), filtered through a cotton plug in a funnel, and concentrated by rotary evaporation (40 °C, 6 mm Hg). Dichloromethane (250 mL) (Note 12) is added to the 500-mL flask, which is then equipped with a magnetic stir bar. The resulting solution is cooled in an ice bath at 0 °C while 25 mL AcOH (Note 13) is added. Sodium borohydride (NaBH₄) (1.52 g, 40.25 mmol, 1.0 equiv) (Note 14) is then added portion-wise to the reaction mixture over 5 min. The reaction mixture is allowed to warm to room temperature and is stirred for 4 h. The reaction mixture is neutralized with 100 mL of a 1 M aq KHSO₄ solution and is transferred to a 500-mL separatory funnel. The aqueous phase is separated and the organic layer is dried over Na₂SO₄ (25-30 g), filtered through a cotton plug in a funnel, and concentrated by rotary evaporation (40 °C, 6 mm Hg) (Notes 16, 17). The residue is dissolved in 200 mL of ethyl acetate, and the solution transferred to a 500-mL separatory funnel where it is washed successively with two, 150-mL portions of 1 M aq KHSO₄ solution and two, 150-mL portions of sat. aq sodium bicarbonate solution and 150 mL of brine. The organic layer is dried over Na₂SO₄ (25-30 g), filtered through a cotton plug in a funnel, and concentrated by rotary evaporation (40 °C, 6 mm Hg). The residual crystalline solid is dissolved in 23 mL of dichloromethane. Diisopropyl ether (150 mL) is slowly added and the resulting slurry is triturated for 1 h. The white solid that precipitates is collected by filtration, then is washed with 50 mL of diisopropyl ether, 50 mL of cyclohexane and then is dried under reduced pressure (25 °C, 6 mm Hg and then 0.04 mm Hg) to afford 6.25 g (49.3%) of **4**. The filtrate is concentrated and the residue is dissolved in 9 mL dichloromethane. After addition of diisopropyl ether (100 mL) and a magnetic stir bar, the mixture is stirred for 10 h at room temperature to afford a second crop (1.60 g), which is collected by filtration and is combined with the first crop of crystals to give a final yield of 7.85 g (62 %) (Note 18).

2. Notes

1. *N*-α-*tert*-butoxycarbonyl-L-aspartic acid α-*tert*-butyl ester (98+%), was purchased from Novabiochem by the submitters and from Waterstone Technology Product List by the checkers, was used as received.

2. Dichloromethane was purchased from WAKO Pure Chemicals and was used as received.

3. 2,2,5-Trimethyl-1,3-dioxane-4,6-dione (Meldrum's acid) (98%) was purchased from Aldrich Chemical Co., Inc and was used as received.

4. 4-DMAP (99%) was purchased from Aldrich Chemical Co., Inc and was used as received.

5. *N*-(3-Dimethylaminopropyl)-*N'*-ethylcarbodiimide hydrochoride (EDC·HCl) was purchased from Senn Chemicals by the submitters and from Watanabe Chemical Industries, LTD by the checkers, was used as received.

6. The progress of the reaction was followed by TLC analysis on silica gel with cyclohexane/ ethyl acetate/acetic acid, 7:3:0.5) as eluent. Visualization of the TLC plates was performed with ninhydrin. Boc-Asp-O*t*-Bu has R_f = 0.31 (red) and product **2** has R_f = 0.20 (dark red).

7. Compound **2** is used in the next step without further purification. The submitters observed partial spontaneous rearrangement to dioxopiperidine **3** upon drying compound **2** under reduced pressure for 5 days. Compound **2** has the following spectroscopic properties: ^{1}H NMR (500 MHz, CDCl$_3$,) δ: 1.42 (s, 9 H), 1.44 (s, 9 H), 1.74 (s, 3 H), 3.51 (m, 1 H), 3.66 (m, 1 H), 4.67 (m, 1 H), 5.27 (br, 1 H).

8. Ethyl acetate (puriss. p.a. ACS reagent grade) was purchased from Aldrich Chemical Co., Inc and used as received.

9. The progress of the reaction was followed by TLC analysis on silica gel with cyclohexane/ethyl acetate/acetic acid, 7:3:0.5 as eluent. Visualization of the TLC plates was performed with ninhydrin; dioxopiperidine **3** has R_f = 0.29 (light orange).

10. Dioxopiperidine **3** exhibits the following physicochemical properties: C$_{18}$ RP-HPLC t_R 23.4 (HPLC analysis was performed on a CAPCELL PAC C$_{18}$ column (5 μm, 4.6 mm × 250 mm) (*SHISEIDO*) by using a linear gradient of A (0.1% TFA in H$_2$O) and B (0.08% TFA in CH$_3$CN) at a flow rate of 1.2 mL/min with UV detection at 214 nm; linear gradient, 20–80% B, 20 min); colorless crystals; $[\alpha]_D^{23}$ = +107.2 (c = 1.02, CHCl$_3$); mp 114–116 °C; ^{1}H NMR (500 MHz, CDCl$_3$) δ: 1.46 (s, 9 H), 1.55

(s, 9 H), 2.83 (dd, J = 17.7, 7.0 Hz, 1 H), 3.03 (dd, J = 17.7, 2.1 Hz, 1 H), 3.37 (d, J = 19.5 Hz, 1 H), 3.53 (d, J = 19.6 Hz, 1 H), 5.07 (dd, J = 7.1, 2.2 Hz, 1 H); ^{13}C NMR (125 MHz, CDCl$_3$) δ: 27.8 (3 CH$_3$), 27.9 (3 CH$_3$), 40.9 (CH$_2$), 50.4 (CH$_2$), 54.6 (CH), 84.0 (C), 84.7 (C), 151.2 (C), 165.3 (C), 168.9 (C), 200.2 (C); IR (KBr) 2981, 1766, 1725, 1644, 1600, 1278, 1147 cm^{-1}; ESI-MS [M$^+$+Na] 336. Anal. Calcd for C$_{15}$H$_{23}$NO$_6$: C, 57.50; H, 7.40; N, 4.47. Found: C, 57.22; H, 7.32; N, 4.38.

11.   Dioxopiperidine **3** exists in equilibrium with di(*tert*-butyl) (2*S*)-4-hydroxy-6-oxo-3,6-dihydro-1,2(2*H*)-pyridinedicarboxylate, the thermodynamically stable enol form. The product **3** exists exclusively in the keto form in CDCl$_3$, whereas in DMSO-d_6, the ^{1}H and ^{13}C NMR spectra show the enol form only: ^{1}H NMR (500 MHz, (CD$_3$)$_2$SO) δ: 11.2 (s, 1H), 4.95 (d, J = 1.6 Hz, 1H), 4.85 (dd, J = 6.9, 1.5 Hz, 1H), 3.04 (ddd, J = 17.7, 6.9, 1.6 Hz, 1H), 2.59 (dd, J = 17.7, 1.5 Hz, 1H), 1.46 (s, 9H), 1.41 (s, 9H); ^{13}C NMR (125 MHz, (CD$_3$)$_2$SO) δ: 169.6, 168.2, 164.1, 152.0, 96.2, 81.6, 81.4, 55.0, 30.3, 27.7, 27.5.

12.   Dichloromethane was freshly distilled over from calcium hydride under argon.

13.   AcOH (99.5+%), purchased from Prolabo by the submitters and from Wako Pure Chemical Industries, Ltd. by the checkers), was used as received.

14.   Sodium borohydride (NaBH$_4$) (98+%) was purchased from Acros Organics by the submitters (KANTO Chemical CO., Inc.) and was used as received.

15.   At that stage, C$_{18}$ RP-HPLC analysis of the reaction mixture reveals that 11% of dioxopiperidine **3** is still present (see Note 16 for analysis conditions). On this scale, as well as on larger scale, the submitters found that a prolonged reaction time was not sufficient to bring the reaction to completion and that satisfactory results were obtained by performing the reduction a second time after work-up.

16.   Analytical C$_{18}$ RP-HPLC: CAPCELL PAC C$_{18}$ column (5 μm, 4.6 mm × 250 mm) (*SHISEIDO*) (linear gradient, 20–80% B, 20 min) of the crude product reveals that the reduction is complete (< 2 % of **3**) and that the diastereomeric ratio between di(*tert*-butyl) (2*S*,4*S*)-4-hydroxy-6-oxo-1,2-piperidinedicarboxylate (**4**) and di(*tert*-butyl) (2*S*,4*R*)-4-hydroxy-6-oxo-1,2-piperidinedicarboxylate (***epi*-4**) is 91:9. The elution times (t_R) for **3**, **4** and ***epi*-4** are 23.4, 20.6, and 21.4 min, respectively.

17. After this workup, the product still contained acetic acid which was detrimental to the recrystallization. A second extractive workup as described was needed to remove all of the acetic acid.

18. (2S,4S)-4-Hydroxy-6-oxo-1,2-piperidinedicarboxylate **4** was obtained in high diastereomeric purity (dr > 99:1 for the combined crops as determined by analytical C_{18} RP-HPLC (linear gradient, 20–80% B, 20 min)) (Note 10). The enantiomeric ratio was > 99:1 as determined by HPLC analysis on an OJ-H column (5 μm, 3.9 m × 150 mm) by elution with 5% isopropanol in n-hexane at a flow rate of 1 mL/min with UV detection at 214 nm. The elution times (t_R) for **4** and ***ent*-4** are 11.8 and 8.4 min, respectively. Lactam **4** exhibited the following physicochemical properties: C_{18} RP-HPLC t_R 20.6 min (linear gradient, 20–80% B, 20 min); $[\alpha]_D^{21}$ = –23.7 (c = 1.02, CHCl₃); colorless crystals; mp 128–129 °C; ^{1}H NMR (500 MHz, CDCl₃) δ: 1.46 (s, 9 H), 1.51 (s, 9 H), 2.11 (d, J = 4.0 Hz, 1 H), 2.22 (ddd, J = 14.1, 6.4, 2.8 Hz, 1 H), 2.38-2.43 (m, 1 H), 2.60 (ddd, J = 17.4, 4.3, 1.9 Hz, 1 H), 2.77 (dd, J = 17.4, 4.9 Hz, 1 H), 4.26 (m, 1 H), 4.65 (dd, J = 6.8, 3.7 Hz, 1 H); ^{13}C NMR (125 MHz, CDCl₃) δ: 27.8 (3 CH₃), 27.9 (3 CH₃), 32.9 (CH₂), 43.2 (CH₂), 56.1 (CH), 64.2 (CH), 82.5 (C), 83.5 (C), 151.8 (C), 168.6 (C), 171.2 (C); IR(KBr) 3504, 2975, 1766, 1706, 1284, 1139; ESI-MS [M⁺+Na] 338; Anal. Calcd for $C_{15}H_{25}NO_6$: C, 57.13; H, 7.99; N, 4.44. Found: C, 56.88; H, 7.84; N, 4.33.

Safety and Waste Disposal Information

All hazardous materials should be handled and disposed of in accordance with "Prudent Practices in the Laboratory"; National Academy Press; Washington, DC, 1995.

3. Discussion

Five-membered ring 2,4-dioxo-1-pyrrolidinecarboxylates (chiral tetramic acids) obtained from Meldrum's acid and N-protected α-amino acids have received considerable attention as precursors of statine [(3S,4S)-4-amino-3-hydroxy-6-methylheptanoic acid] derivatives.[4-8] Their stereocontrolled reduction leads exclusively to the *cis*-4-hydroxy derivative. However, there have been relatively few reports on the synthesis and reactivity of N-acylated 4,6-dioxopiperidines, their six-membered counterparts.[9]

By analogy with the synthesis of chiral tetramic acids,[4] condensation of *N*-protected 3-aminopropanoic acids with Meldrum's acid in the presence of EDC·HCl and DMAP, followed by cyclization at 80 °C in ethyl acetate, provides a convenient and short entry to enantiopure *N*-acylated 4,6-dioxopiperidines (e.g. **3**).[10] *N*-Acylated 4,6-dioxopiperidines exist in equilibrium with the thermodynamically stable enol form.

We found that treatment of various *N*-acylated 4,6-dioxopiperidines with $NaBH_4$ in CH_2Cl_2/AcOH (10:1) for 16–72 h resulted in a quantitative and stereoselective reduction of the keto functionality.[10] The selectivity of the reaction was significantly influenced by the bulk of the side chain at C2, the lowest and highest selectivities being observed for the methyl (dr 84:16) and isopropyl groups (dr >99:1), respectively. In the case of a carboxylate side chain at C2, the *tert*-butyl ester group (in **3**) exerts a stronger stereodirecting effect than the corresponding benzyl ester. The stereoselective reduction in step (C) afforded **4** in diastereomerically pure form (dr >99:1) following a single recrystallizaton. The absolute configuration at C4 was confirmed by X-ray crystal structure determination.[10]

This three-step procedure allows for the preparation of (2*S*,4*S*)-4-hydroxy-6-oxo-1,2-piperidinedicarboxylate **4** in 46% overall yield from Boc-L-Asp-OtBu and using common and cheap reagents. Hydroxylated lactam **4** is a versatile intermediate that can be readily converted to several useful amino acid derivatives.[10,11] Reduction of the lactam moiety by treatment with BH_3·SMe, and subsequent deprotection/protection afforded the enantiopure protected *cis*-4-hydroxypipecolic acid **5,** an amino acid constituent of natural antibiotics and several drug candidates, in 83% yield. The corresponding *N*-Fmoc protected *trans*-(2*S*,4*S*)-4-hydroxypipecolate **6** was obtained by further inversion of the C4 stereocenter under Mitsunobu conditions (57% from **4**). Alternatively, ring opening of **4** under reductive conditions ($NaBH_4$, EtOH) following protection of the hydroxyl group gave the corresponding 1,3-diol (85% (TBDPS protecting group); 83% (TBDMS protecting group)), which can be further transformed to the 4-hydroxylysine derivative **7** or to enantiomerically pure lactone **8**.

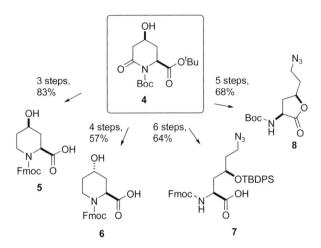

1. CNRS, Institut de Biologie Moléculaire et Cellulaire, Immunologie et Chimie Thérapeutiques, 15 rue René Descartes, F-67084 Strasbourg Cedex, France,

2. NeoMPS, 7 rue de Boulogne, F-67100 Strasbourg, France.

3. Present Address : Novalyst, 23 rue du Loess, BP 20, 67037 Strasbourg Cedex 02, France.

4. Jouin, P.; Castro, B. *J. Chem. Soc., Perkin Trans. 1* **1987**, 1177-1182.

5. Fehrentz, J.-A.; Bourdel, E.; Califano, J.-C.; Chaloin, O.; Devin, C.; Garrouste, P.; Lima-Leite, A.-C.; Llinares, M.; Rieunier, F.; Vizavonna, J.; Winternitz, F.; Loffet, A.; Martinez, J. *Tetrahedron Lett.* **1994**, *35*, 1557-1560.

6. Galeotti, N.; Poncet, J.; Chiche, L.; Jouin, P. *J. Org. Chem.* **1993**, *58*, 5370-5376.

7. Ma, D.; Ma, J.; Ding, W.; Dai, L. *Tetrahedron: Asymmetry* **1996**, *7*, 2365-2370.

8. Decicco, C. P.; Grover, P. *J. Org. Chem.* **1996**, *61*, 3534-3541.

9. Murray, P. J.; Starkey, I. D. *Tetrahedron Lett.* **1996**, *37*, 1875-1878.

10. Marin, J.; Didierjean, C.; Aubry, A.; Casimir, J.-R.; Briand, J.-P.; Guichard, G. *J. Org. Chem.* **2004**, *69*, 130-141.

11. Marin, J.; Briand, J. P.; Guichard, G. *Eur. J. Org. Chem.* **2008**, 1005-1012.

Appendix
Chemical Abstracts Nomenclature; (Registry Number)

N-α-*tert*-Butoxycarbonyl-L-aspartic acid α-*tert*-butyl ester (Boc-L-Asp-
O*t*-Bu); (34582-32-6)

2,2-Dimethyl-1,3-dioxane-4,6-dione (Meldrum's acid); (2033-24-1)

N-(3-Dimethylaminopropyl)-*N'*-ethylcarbodiimide hydrochoride
(EDC·HCl); (25952-53-8)

2-*tert*-Butoxycarbonylamino-4-(2,2-dimethyl-4,6-dioxo-[1,3]dioxan-5-yl)-4-
oxo-butyric acid *tert*-butyl ester; (10950-77-9)

Di(*tert*-butyl) (2*S*)-4,6-dioxo-1,2-piperidinedicarboxylate; (653589-10-7)

Sodium borohydride; (16940-66-2)

Di(*tert*-butyl) (2*S*,4*S*)-4-hydroxy-6-oxo-1,2-piperidinedicarboxylate;
(653589-16-3)

Gilles Guichard was born in Orléans (France) in 1969. He studied chemistry at the Ecole Nationale Supérieure de Chimie in Toulouse and received his Ph.D. in 1996 from the University Louis Pasteur (Strasbourg) under the guidance of Dr. Jean-Paul Briand working on the synthesis and immune recognition of pseudopeptides. After post-doctoral research with Prof. Dieter Seebach at the ETH Zürich (Switzerland) working on beta-peptides, he returned to Strasbourg as a CNRS Chargé de Recherche in the Institute of Molecular and Cellular Biology and was promoted CNRS Directeur de Recherche in 2006. His research is mainly focused on biomimetic chemistry of peptides, folding and biomolecular recognition.

Olivier Chaloin received his Ph.D. in chemistry from the University of Montpellier II in 1996. After postdoctoral studies with Prof. G. Flouret at Northwestern University (Chicago, USA) and with Prof. R. Offord in Geneva (Switzerland), he joined the CNRS and moved to the Institute of Molecular and Cellular Biology in Strasbourg (France). His research interests focus on peptide, pseudopeptide and protein synthesis.

Frédéric Cabart was born in 1972 in France. He received his B.S. and M.S. degrees in Chemistry from the Université de Nancy I, France in 1996 and 1997 respectively. He has been employed at NeoMPS S.A. France since 1998, where he currently works as Research & Development Chemist involved in the syntheses of pharmaceutical building blocks from laboratory scale to kilogram scale.

Julien Marin was born in Metz (France) in 1976. He studied chemistry at the Ecole Nationale Supérieure de Chimie in Mulhouse and received his Ph.D. in 2003 from the University Louis Pasteur (Strasbourg) under the guidance of Dr. Gilles Guichard working on the synthesis and auto-immune recognition of glycopeptides. After post-doctoral research with Prof. Stephen Hanessian at the University of Montreal (Canada) working on the design and synthesis of new inhibitors of endothelin converting enzymes, he joined the contract research organization Novalyst Discovery in 2006. His current position is project manager in the organic custom-synthesis department.

Haixiang Zhang was born in 1963 in China. He received his Ph.D. in Organic Chemistry under the supervision of Prof. G. Balavoine and Dr. F. Guibé from the Université de Paris-Sud, France in 1990. He then joined the Propeptide/Isochem S.A. France as Research & Development Chemist. In 2000, he moved to current position as Research & Development Manager at NeoMPS S.A., France. His current interests focus on the organic synthetic chemistry in pharmaceutical industry.

 Hisashi Mihara was born in 1981 in Shiga, Japan. He received his M.S. degree in 2006 from the University of Tokyo. He is pursuing a Ph.D. degree at the Graduate School of Pharmaceutical Sciences, The University of Tokyo, under the guidance of Professor Masakatsu Shibasaki. His current interest is catalytic enantioselective total synthesis of biologically active compounds.

SYNTHESIS OF (2R,3R)-2,3-DIMETHYL-1,4-BUTANEDIOL BY OXIDATIVE HOMOCOUPLING OF (4S)-ISOPROPYL-3-PROPIONYL-2-OXAZOLIDINONE
[2,3-dimethylbutane-1,4-diol, (2R, 3R)-]

Submitted by Chong-Dao Lu and Armen Zakarian.[1]
Checked by Katrien Brak and Jonathan A. Ellman.

1. Procedure

A. *(4S)-Isopropyl-3-propionyl-2-oxazolidinone (2)*. An oven-dried, 250-mL, three-necked, round-bottomed flask is equipped with a mechanical stirrer, an internal thermometer, and an adapter equipped with an argon inlet and a rubber septum. The flask is flushed with argon and is charged with 333 mL of dry tetrahydrofuran (THF) (Note 1) and (4S)-isopropyl-2-oxazolidinone (8.61 g, 66.7 mmol) (Note 2). The mixture is cooled to –75 °C in a dry ice-acetone bath. At this temperature, a 2.57 M solution of *n*-butyllithium in hexane (27.2 mL, 70.0 mmol) (Note 3) is added dropwise

Org. Synth. **2008**, *85*, 158-171
Published on the Web 4/17/2008

over 20 min using a syringe at such a rate that the reaction temperature remains below –72 °C. The reaction mixture is stirred at this temperature for 25 min and propionyl chloride (6.29 g, 5.94 mL, 68.0 mmol) is added dropwise over 10 min using a syringe while maintaining an internal temperature below –73 °C. The reaction mixture is stirred at this temperature for 30 min, and the reaction is then quenched with sat. aq. ammonium chloride solution (100 mL). The reaction mixture is transferred to a 1-L separatory funnel and the organic phase is collected. The aqueous phase is extracted with ethyl acetate (3 × 100 mL). The combined organic phases are washed with brine (100 mL). Drying with anhydrous sodium sulfate (40 g), filtration, and removal of the solvent under reduced pressure on a rotary evaporator (30 °C, 40 mmHg) afford the crude product as a light yellow oil. The oil is purified by silica gel column chromatography (Note 4). Fractions containing the product (Note 5) are concentrated by rotary evaporation (30 °C, 40 mmHg) and are dried (Note 6) to yield 11.0 g (89%) of (4S)-isopropyl-3-propionyl-2-oxazolidinone (**2**) as a colorless oil (Note 7).

B.   *(2R,3R)-1,4-Bis[(4S)-4-isopropyl-2-oxo-(1,3-oxazolidine-3-yl)]-2,3-dimethylbutane-1,4-dione (3)*. An oven-dried, 500-mL, three-necked, round-bottomed flask is equipped with an internal thermometer, a 125-mL pressure-equalizing addition funnel fitted with a rubber septum, an adapter equipped with an argon inlet and a rubber septum, and a Teflon-coated magnetic stirring bar (Note 8). The flask is flushed with argon and charged with dry THF (200 mL) (Note 1) and diisopropylamine (6.37 g, 8.83 mL, 63.0 mmol) (Note 9). The mixture is cooled to –75 °C in a dry ice-acetone bath. At this temperature, a 2.52 M solution of butyllithium in hexane (23.8 mL, 60.0 mol) (Note 3) is added dropwise over 10 min using a syringe while maintaining an internal temperature below –70 °C. The reaction mixture is stirred at this temperature for 60 min and a solution of (4S)-isopropyl-3-propionyl-2-oxazolidinone (9.26 g, 50.0 mmol) in THF (50 mL) is added dropwise over 20 min from the dropping funnel, which is rinsed with THF (5 mL), all while maintaining an internal temperature below –71 °C. The reaction mixture is stirred at –75 °C for 45 min. Then, titanium (IV) chloride (23.7 g, 13.7 mL, 125 mol) (Note 10) is added dropwise over 15 min using a syringe while maintaining an internal temperature below –70 °C (Note 11). After being stirred for another 60 min at –78 °C, the mixture is allowed to warm to ambient temperature. After being stirred overnight (17 h), the reaction is quenched at ambient temperature with 1 N aq. hydrochloric acid (4 × 25 mL), and ethyl acetate (100 mL) is added (Note 12). The reaction mixture is transferred to a 1-L separatory funnel and the organic phase is collected. The aqueous phase is extracted with ethyl acetate (2 × 100 mL). The combined organic phases are washed with brine (100 mL). Drying the extracts with anhydrous sodium sulfate (35 g), filtration, and removal of the solvent under reduced pressure on a rotary evaporator (35 °C, 40 mmHg) afford the crude product as a light yellow solid (Note 13). The solid is transferred to a 500-mL, round-bottomed flask equipped with a reflux condenser. Ethyl acetate (60 mL) is added, and the solution is heated to reflux in a mineral oil bath (Note 14). Hexanes (150 mL) are added slowly through the condenser while maintaining the solution at reflux (Note 15). The resulting suspension is allowed to cool to room temperature and then placed in a freezer at –20 °C. After standing at –20 °C overnight, the crystals are collected by filtration and are rinsed with 40 mL of hexane/EtOAc, 4:1. The product is dried in a 100-mL round-bottomed flask fitted with a vacuum adapter and evacuating the flask (0.02 mmHg) for 24 h. Thus, 3.65 g (40%) of (2R,3R)-1,4-bis[(4S)-4-isopropyl-2-oxo-(1,3-oxazolidine-3-yl)]-2,3-dimethylbutane-1,4-dione (**3**) is obtained as white, cotton-like crystals (Notes 16 and 17).

C. *(2R,3R)-2,3-Dimethylsuccinic acid (4)*. A 250-mL, single-necked, round-bottomed flask under an argon atmosphere equipped with a Teflon-coated magnetic stirring bar is charged with 3.68 g (10.0 mmol) of **3**, THF (50 mL), and water (40 mL) (Note 18). After purging the reaction flask with argon, the resulting mixture is cooled to 0 °C (bath temperature) in an ice-water bath. To this solution is added via syringe 30% aq.hydrogen peroxide (11 mL, ~102 mmol) over 5 min (Note 19), followed by addition of lithium hydroxide monohydrate (1.68 g, 40 mmol) in one portion. The resulting solution is allowed to warm to ambient temperature. After being stirred for 18 h (Note 20), the reaction mixture is cooled to 0 °C (bath temperature) in an ice-water bath, and the reaction is quenched with 1.5 M aqueous Na_2SO_3 solution (40 mL). The resulting mixture is transferred to a 500-mL separatory funnel and is extracted with dichloromethane (3 × 80 mL) (Note 21). The aqueous layer is transferred to a 250-mL single-neck round-bottomed flask equipped with a Teflon-coated magnetic stirring bar and is cooled to 0 °C (bath temperature) in an ice-water bath. Concentrated aq. hydrochloric acid (12 M, 20 mL) is added over 5 min. The resulting solution is transferred to a 250-mL separatory funnel and is extracted with ethyl acetate (3 × 60 mL). The combined organic phases are washed with brine (100 mL). Drying the organic extracts with anhydrous sodium sulfate (15 g), filtration, and removal of the solvent under reduced pressure on a rotary evaporator (30 °C, 40 mm Hg) afford the crude product as a white solid. The product is dried in a 100-mL, round-bottomed flask fitted with a vacuum adapter and evacuating the flask (0.02 mmHg) for 24 h to afford 1.34 g, (92%) of 2R,3R)-2,3-dimethylsuccinic acid as a colorless solid (Note 22).

D. *(2R,3R)-2,3-Dimethylbutane-1,4-diol (5)*. An oven-dried, 250-mL, three-necked, round-bottomed flask is equipped with a rubber septum, a 50-mL pressure-equalizing addition funnel fitted with a rubber septum, a reflux condenser fitted with an argon inlet adapter, and a Teflon-coated magnetic stirring bar. The flask is charged with 0.857 g (22.6 mmol) of lithium aluminum hydride and THF (56 mL) (Notes 1 and 23). (*CAUTION*: lithium aluminum hydride is very reactive and must be handled carefully to avoid contact with water). The resulting suspension is cooled to 0 °C (bath temperature) in an ice-water bath, and a solution of (2R,3R)-2,3-dimethylsuccinic acid (1.10 g, 7.53 mmol) in THF (24 mL) is added dropwise over 10 min from the dropping funnel (Note 24), which is rinsed with THF (4 mL). The resulting suspension is heated at reflux in a mineral oil bath for 12 h (Note 25). The reaction mixture is cooled to 0 °C (bath temperature) in an ice-water bath and is diluted with ethyl ether (80 mL). Water (0.9 mL) is added dropwise slowly (Note 26), followed after 5 min by 3 M aq. sodium hydroxide solution (0.9 mL) and, after 5 min, more water (2.6 mL). The resulting mixture is allowed to warm to ambient temperature and is stirred at this temperature for 2 h. The white precipitate is removed by filtration of the mixture through a ceramic Büchner funnel lined with filter paper (Whatman 90 mm thick qualitative circles) fitted onto a 100-mL filter flask. The white residue is transferred to a 50-mL, round-bottomed flask equipped with a Teflon-coated magnetic stirring bar. After breaking the precipitate into small pieces with a metal spatula, THF (20 mL) is added and then the suspension is heated at reflux in a mineral oil bath with vigorous stirring for 60 min (Note 27). The precipitate is removed by filtration through a medium-porosity, glass-fritted Büchner funnel into a 100-mL filter flask and is washed with ethyl ether (3 × 20 mL). The combined filtrates are dried with anhydrous sodium sulfate (6 g), filtered, and concentrated under reduced pressure on a rotary evaporator (25 °C, 40 mmHg) to afford the crude product as a light yellow oil (Note 28). The product is purified by Kugelrohr distillation (Note 29) to afford 0.80 g (90%) of (2R,3R)-2,3-dimethylbutane-1,4-diol as a viscous, clear oil (Notes 30, 31, and 32).

2. Notes

1.   The checkers obtained tetrahydrofuran by filtration through an alumina column under a N_2 atmosphere on a GlassContour system. The submitters distilled tetrahydrofuran from sodium benzophenone ketyl under an atmosphere of argon.

2.   The checkers purchased (4*S*)-isopropyl-2-oxazolidinone from Acros, Inc. The submitters purchased (4*S*)-isopropyl-2-oxazolidinone from Aldrich Chemical Company, Inc. or prepared it from L-valine according to the following procedure (the chemicals were purchased from Aldrich Chemical Company, Inc. and were used as received): **A**) Reduction of L-valine to (*S*)-valinol with LiAlH₄ according to the procedure reported in reference 2: L-valine (40.0 g, 340 mmol) was added slowly, in small portions, to a two-neck 2-L flask equipped with a Teflon-coated magnetic stirring bar, a reflux condenser, and an an argon inlet adapter on top of the condenser, containing a mixture of lithium aluminum hydride (25.9 g, 680 mol) in 800 mL of dry THF (Note 1) cooled to 0 °C with an ice-water bath. After all of the L-valine had been added (about 1 h), the reaction mixture was heated at reflux for 10 hours. The solution was cooled to 0 °C, and excess LiAlH₄ was quenched with 100 mL of aq. sodium hydroxide (2.0 M). The precipitate was filtered off and extracted with boiling THF (100 mL) for an hour and filtered off again. The combined organic filtrates were dried with sodium sulfate, filtered, and the solvent was removed under reduced pressure on a rotary evaporator (20 mmHg) to give 28.5 g of (*S*)-valinol (276 mmol, 81% yield) as a colorless oil, which was used in the next step without further purification. **B**) Preparation of (4*S*)-isopropyl-2-oxazolidinone from (*S*)-valinol and diethyl carbonate according to the procedure in reference 3: A 100-mL flask equipped with a 10-cm Vigreux column and a Teflon-coated magnetic stirring bar was charged with (*S*)-valinol (28.5 g, 276 mmol), diethyl carbonate (36.8 mL, 304 mmol), and anhydrous potassium carbonate (3.87 g, 28 mmol). The mixture was heated at 135 °C (temperature of the oil bath) until no more ethanol distilled (32 mL EtOH was distilled off after 5 h of heating). The resultant mixture was cooled to room temperature and dissolved in diethyl ether (800 mL), and the solution was filtered through a 2-cm pad of Celite to remove the potassium carbonate. The filtrate was concentrated under reduced pressure on a rotary evaporator (20 mmHg) to 250 mL and then was placed into a 0 °C refrigerator. After 12 h, the crystals were collected by filtration. The filtrate was concentrated to 60 mL and

162   *Org. Synth.* **2008**, *85*, 158-171

placed in a 0 °C refrigerator overnight followed by filtration to provide additional product. The combined crystals were dried under vacuum to afford (4S)-isopropyl-2-oxazolidinone 29.9 g (232 mmol, 84% yield) as white needles.

3.   *n*-Butyllithium in hexane was purchased from Acros Chemical Company, Inc. and was titrated according to the method of Gilman.[4]

4.   Flash chromatography was performed by using an 8-cm diameter column packed with 250 g of silica gel (MP Silitech 32-63 D 60 Å). The product was eluted with 2 L of hexane/EtOAc, 17:3, followed by 1 L of hexane/EtOAc, 4:1, and 0.5 L of hexane/EtOAc, 3:1. After collection of 500 mL of eluent, 50 mL-fractions were collected. The submitters report that the product can be purified by distillation at 102–106 °C/0.75 mmHg. The checkers found that distillation did not successfully separate remaining starting material from the product.

5.   Thin layer chromatography (TLC) analysis is used to identify product fractions. TLCs are performed by using Merck silica gel 60 F_{254} analytical plates; detection with UV or by dipping in a $KMnO_4$ solution which was prepared from $KMnO_4$ (3 g), K_2CO_3 (20 g), NaOH 5% (5 mL) in H_2O (300 mL) followed by heating. Product R_f = 0.38 and starting material R_f = 0.06 in hexane/EtOAc, 7:3.

6.   To ensure complete removal of ethyl acetate and hexanes, anhydrous toluene (2 x 20 mL; obtained by filtration through a drying column on a GlassContour system) is added to the purified product and concentrated by rotary evaporation followed by drying under high vacuum (0.02 mm Hg) for 12 h.

7.   The product, (4S)-isopropyl-3-propionyl-2-oxazolidinone, is commercially available from Aldrich Chemical Company. The procedure provided here is based on the procedure in reference 5a. The product displayed the following physicochemical properties: $[\alpha]_D^{20}$ +96.3 (CH_2Cl_2, c 1.0), lit.[5b] $[\alpha]_D^{20}$ +96.8 ($CHCl_3$, c 8.7); ^{1}H NMR (400 MHz, $CDCl_3$) δ: 0.87 (d, J = 7.0 Hz, 3 H), 0.91 (d, J = 7.0 Hz, 3 H), 1.17 (t, J = 7.3 Hz, 3 H), 2.31–2.44 (m, 1 H), 2.84–3.04 (m, 2 H), 4.20 (dd, J = 3.0 Hz, 9.0 Hz, 1 H), 4.24 (at, J = 8.4 Hz, 1 H), 4.43 (dt, J = 8.4 Hz, 3.3 Hz, 1 H). ^{13}C NMR (100 MHz, $CDCl_3$) δ: 8.5, 14.8, 18.1, 28.5, 29.3, 58.5, 63.5, 154.3, 174.2. HRMS (FAB) *m/z*: calcd. for $C_9H_{16}NO_3$ [M+H] 186.1130, found 186.1128.

8.   The checkers used an egg-shaped Fisherbrand 19 mm x 41 mm stirring bar.

9. Diisopropylamine was purchased from Aldrich Chemical Company, Inc. and was distilled under nitrogen (atmospheric pressure) from calcium hydride prior to use.

10. Titanium (IV) chloride was purchased from Aldrich Chemical Company, Inc. and was used as received.

11. The reaction mixture becomes viscous as titanium chloride is added at low temperature, therefore it is important to ensure that efficient stirring is maintained during the addition of the titanium reagent. The reaction mixture turns dark purple and yellow smoke forms upon addition of the titanium chloride. The smoke thickens during the progress of the titanium chloride addition, and disappears 5-10 min after the addition is complete.

12. Aqueous hydrochloric acid (37%, ~12 N) certified A.C.S. PLUS was purchased from Fisher Chemical Fisher Scientific and diluted to the concentration of 1 N with water. The 1 N HCl is added in four portions with 1 min intervals while maintaining the reaction temperature below 40 °C.

13. The submitters noted that ^{1}H NMR analysis of the crude product mixture indicated 65-70% yield of the homocoupling product, the remainder of the material being mostly the starting propionyl oxazolidinone. Complete conversion was never achieved under a variety of reaction conditions.

14. The solid fully dissolves upon bringing the solution to reflux.

15. The product precipitates during the addition of hexanes.

16. The product has the following physicochemical properties: mp 249–250 °C (lit.[6] mp 252–253 °C); $[\alpha]_D^{20}$ +40.3 (CH$_2$Cl$_2$, c 1.0), lit.[6] $[\alpha]_D^{20}$ +40.5 (CHCl$_3$, c 1.0); ^{1}H NMR (400 MHz, CDCl$_3$) δ: 0.84–0.92 (m, 12 H), 1.22 (d, J = 6.0 Hz, 6 H), 2.19–2.32 (m, 2 H), 4.00–4.12 (m, 2 H), 4.20 (dd, J = 3.0, 9.0 Hz, 2 H), 4.26 (at, J = 8.6 Hz, 2 H), 4.36 (dt, J = 8.4, 3.0 Hz, 2 H). ^{13}C NMR (100 MHz, CDCl$_3$) δ: 14.5, 15.2, 18.1, 28.1, 41.4, 58.8, 63.1, 153.6, 177.2; IR (film) cm^{-1}: 2970, 1768, 1691, 1384, 1232, 1194, 1099; HRMS (FAB) m/z: calcd for C$_{18}$H$_{28}$N$_2$O$_6$Na [M+Na] 391.1845, found 391.1854. Anal. Calcd. for C$_{18}$H$_{28}$N$_2$O$_6$: C, 58.68; H, 7.66; N, 7.60. Found: C, 58.73; H, 7.84; N, 7.60.

17. On half scale (25 mmol), the checkers obtained 2.57 g (56% yield) of (2R,3R)-1,4-bis[[(4S)-4-isopropyl-2-oxo-(1,3-oxazolidine-3-yl)]-2,3-dimethylbutane-1,4-dione.

18. The (4S)-isopropyl-3-propionyl-2-oxazolidone does not fully dissolve in THF and a thick slurry forms upon addition of the water.

19. 30% Hydrogen peroxide was purchased from Aldrich Chemical Company, Inc. and was used as received.

20. Reaction progress was monitored by the disappearance of starting material by thin layer chromatography (TLC). TLC analysis was performed using Merck silica gel 60 F_{254} analytical plates; detection with UV or by dipping in a $KMnO_4$ solution, which was prepared from $KMnO_4$ (3 g), K_2CO_3 (20 g), NaOH 5% (5 mL) in H_2O (300 mL) followed by heating. Starting material R_f = 0.6 with CH_2Cl_2/MeOH, 20:1 as the eluent (the product remained at the baseline under these elution conditions).

21. The combined organic phases are washed with brine (100 mL). Drying the organic extracts over anhydrous sodium sulfate, filtration, and removal of the solvent under reduced pressure on a rotary evaporator (40 mmHg) affords the recovered (4S)-(–)-isopropyloxazolidin-2-one as a crystalline solid (2.30 g, 89%).

22. The product has the following physicochemical properties: mp 130–131 °C (lit.[7] mp 134–135 °C); $[\alpha]_D{}^{20}$ +7.4 (H_2O, c 1.0), (lit.[7] $[\alpha]_D{}^{20}$ +8.4 (H_2O, c 4.0); ^{1}H NMR (400 MHz, CD_3OD) δ: 1.16 (d, J = 6.8 Hz, 6 H), 2.68–2.80 (m, 2 H), 4.92 (br s, 2 H); ^{13}C NMR (100 MHz, CD_3OD) δ: 14.0, 42.8, 179.1; IR (film) cm^{-1}: 2982, 2909, 1701, 1461, 1419, 1283, 1213, 932; HRMS (FAB) m/z: calcd. for $C_6H_{11}O_4$ [M+H] 147.0657, found 147.0658. Anal. Calcd. for $C_6H_{10}O_4$: C, 49.31; H, 6.90. Found: C, 49.46; H, 7.01.

23. Lithium aluminum hydride was purchased from Aldrich Chemical Company, Inc. and was used as received.

24. The checkers observed vigorous bubbling upon addition of (2R,3R)-2,3- dimethylsuccinic acid.

25. Reaction progress was monitored by thin layer chromatography (TLC). TLC analysis was performed using Merck silica gel 60 F_{254} analytical plates; detection by dipping in a solution of bromocresol green [prepared by dissolving bromocresol green (0.04 g) in EtOH (100 mL) and slowly dripping in a 0.1 M solution of NaOH until the solution turns pale blue]. Starting material R_f = 0.1 (yellow) and product R_f = 0.2 (white) with CH_2Cl_2/MeOH, 20:1 as the eluent.

26. Water reacts violently with lithium aluminum hydride (LAH) necessitating slow addition of the water and cooling with a 0 °C ice-water bath.

27. This procedure extracts any product trapped in the white precipitate. The majority of the product (~ 80%) is in the filtrate.

28. The product has moderate volatility due to its low molecular weight. Care must therefore be taken to minimize loss during concentration. No loss was observed during concentration on a rotary evaporator (40 mmHg), but the product should not be placed under high vacuum for extended time periods.

29. The distillation was performed at 110 °C, 10 mmHg. A 10-mL collection flask was cooled with an ice-water cooling bath.

30. The product has the following physicochemical properties: bp 105–110 °C, 10 mmHg. $[\alpha]_D^{20}$ +10.4 (CH_2Cl_2, c 1.0), (lit.[8] $[\alpha]_D^{20}$ +8.0 (CH_2Cl_2, c 1.0)). ^1H NMR (500 MHz, $CDCl_3$) δ: 0.91 (d, J = 6.8 Hz, 6 H), 1.66–1.75 (m, 2 H), 2.40 (br s, 2 H), 3.52 (dd, J = 10.8, 5.8 Hz, 2 H), 3.64 (dd, J = 10.8, 3.8 Hz, 2 H). ^{13}C NMR (125 MHz, $CDCl_3$) δ: 13.8, 37.8, 65.9; IR (film) cm^{-1}: 3274, 2959, 2920, 2871, 1452, 1384, 1040; HRMS (FAB) m/z: calcd for $C_6H_{15}O_2$ [M+H] 119.1072, found 119.1072. Anal. Calcd. for $C_6H_{14}O_2$, C, 60.98; H, 11.94. Found: C, 60.76; H, 11.79.

31. The diastereomeric purity was determined by the checkers to be >98% by 125 MHz ^{13}C NMR (S/N = 46:1) and comparison to literature values for the meso diol.[9] ^{13}C NMR (125 MHz, $CDCl_3$) δ: 13.6, 38.9, 65.5 for the meso diol.

32. The checkers established enantiomeric purity by HPLC analysis of the (R)-MTPA and (S)-MTPA diesters, [prepared according to the procedure reported in reference 10 using (2R,3R)-2,3-dimethylbutane-1,4-diol (3.0 mg, 0.025 mmol), MTPA-Cl (0.038 mL, 0.20 mmol), Et_3N (0.035 mL, 0.25 mmol), and DMAP (25 mg, 0.20 mmol), in CH_2Cl_2 (0.12 mL, 0.2M) at rt for 12 h]. The enantiomeric ratio was determined to be 99.2:0.8 by HPLC analysis using an Agilent 1100 series LC equipped with a silica normal phase column (Microsorb Si 100 Å packing) with a multiwavelength detector (hexanes/tert-butyl methyl ether, 97:3, 1 mL/min, λ = 210 nm, t_R (R, R, R, R) = 37.7 min, t_R (S, R, R, S) = 38.9 min). The submitters established enantiomeric purity by comparison of the 500 MHz ^1H NMR spectrum of the (R)-MTPA diester derivative with the ^1H NMR spectrum of a sample prepared from racemic 2,3-dimethylsuccinic acid (Aldrich). A comparison of the vicinal methyl groups was the most informative and indicated that none of the minor enantiomer could be detected.

Safety and Waste Disposal Information

All hazardous materials should be handled and disposed of in accordance with "Prudent Practices in the Laboratory"; National Academy Press; Washington, DC, 1995.

3. Discussion

(2*R*,3*R*)-2,3-Dimethyl-1,4-butanediol is a valuable C_2-symmetric building block that has potential utility for the preparation of novel asymmetric catalysts and has been employed in studies related to natural product synthesis. (2*R*,3*R*)-2,3-Dimethylsuccinic acid, which serves as an intermediate in its preparation, is also of value. Several enantioselective syntheses of the diol and the diacid have been documented in the literature. One of the earliest is a report by McCasland and Proskow that is based on resolution of racemic dimethylsuccinic acid by fractional crystallization of its dibasic salt with brucine.[7] The diacid was subsequently reduced to the diol via its dimethyl diester (Scheme 1). Recently, this procedure was adopted by Widenhoefer and co-workers to access (2*R*,3*R*)-2,3-dimethyl-1,4-butanediol via (2*R*,3*R*)-2,3-dimethylsuccinic acid in 62% ee.[11] The somewhat tedious fractional recrystallization and poor availability of racemic 2,3-dimethylsuccinic acid[12] make this method less appealing.[13]

Scheme 1

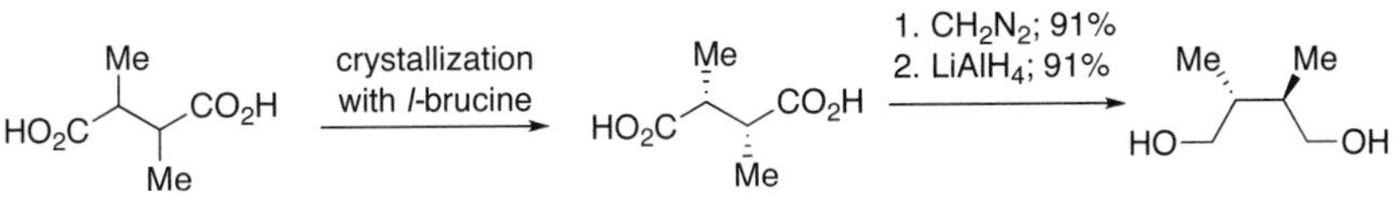

Chan and co-workers developed a method capitalizing on diastereoselective alkylation of chiral *N*-acylnorbornene sultams with alpha-haloesters that is suitable for the synthesis of (2*R*,3*R*)-2,3-dimethyl-1,4-butanediol (Scheme 2).[14] Although this approach leads to highly enantiopure (2*R*,3*R*)-2,3-dimethyl-1,4-butanediol, multistep preparation of the requisite chiral auxiliary, which requires chromatographic separations, significantly detracts from the practicality of this method.

Scheme 2

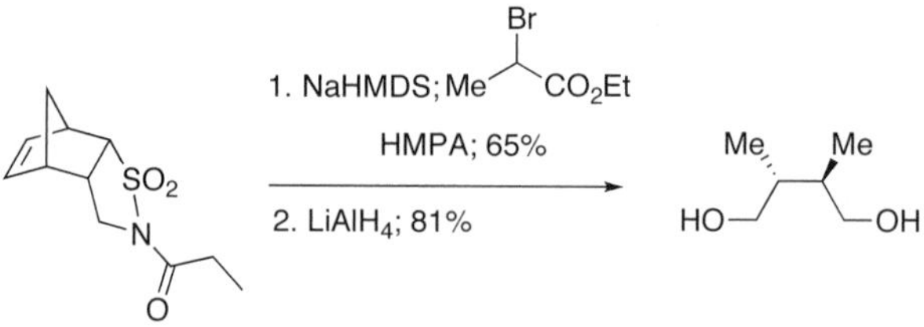

Another chiral auxiliary method suitable for the preparation of the diol has been reported by Feringa and co-workers (Scheme 3).[15] Sequential treatment of 5-(l-Menthyloxy)-2(5H)-furanone with tri(methylthio)methyl-lithium and iodomethane at –90 °C affords butenolide **1**, which is desulfurized with Raney Ni and then reduced with lithium aluminum hydride to provide enantiopure (2R,3R)-2,3-dimethyl-1,4-butanediol.

Scheme 3

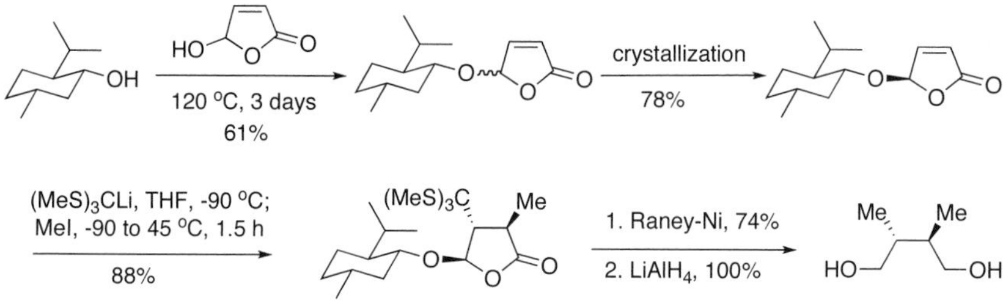

Oxidative homocoupling of lithiated chiral N-acyloxazolidinones described by Kise and co-workers provided the basis for the procedure described here. It appears to be the most direct method for the synthesis of both enantioenriched (2R,3R)-2,3-dimethyl-1,4-butanediol and (2R,3R)-dimethylsuccinic acid (Scheme 4). Some of the advantages of this approach are the wide availability of the Evans chiral auxiliary, which serves as the source of chirality in this case, and the experimental simplicity of the reactions. Titanium tetrachloride, CuCl$_2$, PhI(OAc)$_2$, and iodine have been employed as oxidants in the homocoupling reaction. After careful screening, we found that TiCl$_4$ gives the most consistent results and yields compared to those obtained with the other reagents. Although Kise and co-workers reported yields in the range of 65–68%, we reproducibly obtained the homocoupling product in 40–45% yield after recrystallization.[16] The

168

procedure employing TiCl$_4$ has also proven to be the most convenient experimentally.

Scheme 4

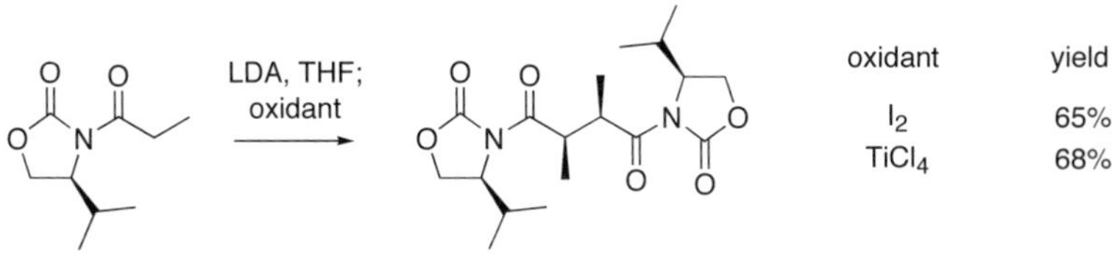

oxidant	yield
I$_2$	65%
TiCl$_4$	68%

Subsequent removal of the chiral auxiliary with lithium hydroperoxide, according to the protocol developed by Evans and co-workers,[17] and direct reduction of the diacid with lithium aluminum hydride delivered (2R,3R)-2,3-dimethyl-1,4-butanediol of high enantiomeric purity (>99:1 er, 30% overall yield) as determined by 500 MHz ^{1}H NMR spectroscopy of the derived (R)-Mosher diester and analytical HPLC. The overall preparation requires only four steps from commercially available starting materials and a single chromatographic purification.

1. Department of Chemistry and Biochemistry, Florida State University, Tallahassee, FL 32306-4390
2. Granander, J.; Sott, R.; Hilmersson, G. *Tetrahedron* **2002**, *58*, 4717-47-25.
3. Evans, D. A.; Mathre, D. J.; Scott, W. L. *J. Org. Chem.* **1985**, *50*, 1830-1835.
4. Gilman, H.; Cartledge, F. K.; *J. Organomet. Chem.* **1964**, *2*, 447-454.
5. (a) Entwistle, D. A.; Jordan, S. I.; Montgomery, J.; Pattenden, G. *Synthesis* **1998**, 603-612. (b) Evans, D. A.; Bartroli, J.; Shih, T. L. *J. Am. Chem. Soc.* **1981**, *103*, 2127-2129.
6. Kise, N.; Ueda, T.; Kumada, K.; Terao, Y.; Ueda, N. *J. Org. Chem.* **2000**, *65*, 464-468.
7. McCasland, G. E.; Proskow, S. *J. Am. Chem. Soc.* **1956**, *78*, 5646-5652.
8. Pelc, M. J.; Zakarian, A. *Org. Lett.* **2005**, *7*, 1629-1631.
9. Kpegba, K.; Metzner, P.; Rakotonirina, R. *Tetrahedron Lett.* **1986**, *27*, 1505-1508.
10. Braddock, D. C.; Bhuva, R.; Millan, D. S.; Perez-Fuertes, Y.; Roberts, C. A.; Sheppard, R. N.; Solanki, S.; Stokes, E. S. E.; White, A. J. P. *Org. Lett.* **2007**, *9*, 445-448.

11. Perch, N. S.; Pei, T.; Widenhoefer, R. A. *J. Org. Chem.* **2000**, *65*, 3836-3845.

12. Racemic 2,3-dimethylsuccinic acid is available from the Sigma-Aldrich Co. as a ~2:1 mixture with the *meso*-isomer.

13. Sutton, S. C.; Nantz, M. H.; Hitchcock, S. R. *Org. Prep. Proc. Int.* **1992**, *24*, 39-43.

14. (a) Lin, J.; Chan, W. H.; Lee, A. W. M.; Wong, W. Y.; Huang, P. Q. *Tetrahedron Lett.* **2000**, *41*, 2949-2951. (b) Chan, W. H.; Lee, A. W. M.; Jiang, L. S.; Mak, T. C. W. *Tetrahedron: Asymmetry.* **1997**, *8*, 2501-2504.

15. (a) Jansen, J. F. G. A.; Feringa, B. L. *Tetrahedron Lett.* **1989**, *30*, 5481-5484. (b) Feringa, B. L.; De Lange, B.; DeJong, J. C. *J. Org. Chem.* **1989**, *54*, 2471-2475.

16. Kise, N.; Ueda, T.; Kumada, K.; Terao, Y.; Ueda, N. *J. Org. Chem.* **2000**, *65*, 464–468.

17. Evans, D. A.; Britton, T. C.; Ellman, J. A. *Tetrahedron Lett.* **1987**, *28*, 6141-6144.

Appendix
Chemical Abstracts Nomenclature; (Registry Number)

(4*S*)-Isopropyl-2-oxazolidinone: (4*S*)-4-(1-Methylethyl)-2-oxazolidinone; (17016-83-0)

Propionyl chloride: Propanoyl chloride; (79-03-8)

n-Butyllithium; (109-72-8)

(4*S*)-Isopropyl-3-propionyl-2-oxazolidinone: (4*S*)-4-(1-Methylethyl)-3-(1-oxopropyl)-2-oxazolidinone; (77877-19-1)

(2*R*,3*R*)-1,4-Bis[(4*S*)-4-isopropyl-2-oxo-(1,3-oxazolidine-3-yl)]-2,3-dimethylbutane-1,4-dione: (4*S*,4'*S*)-3,3'-[(2*R*,3*R*)-2,3-Dimethyl-1,4-dioxo-1,4-butanediyl]bis[4-(1-methylethyl)-2-oxazolidinone; (259540-48-2)

Diisopropylamine: 2-Propanamine, *N*-(1-methylethyl)-; (108-18-9)

Titanium (IV) chloride (7550-45-0)

(2*R*,3*R*)-2,3-Dimethylsuccinic acid; (5866-39-7)

Hydrogen peroxide; (7722-84-1)

(2*R*,3*R*)-2,3-Dimethylbutane-1,4-diol: (2*R*,3*R*) 2,3-Dimethyl-1,4-butanediol; (127253-15-0)

Lithium aluminum hydride; (16853-85-3)

Armen Zakarian was born in Moscow in 1973 and completed his undergraduate studies at Moscow State University in 1994. His Diploma research was completed at the Zelinsky Institute of Organic Chemistry with Dr. Vladimir Borodkin. He received his Ph.D. under the direction of Professor Robert A. Holton at Florida State University in 2001, and then spent two years (2002-2004) in the laboratories of Professor Larry E. Overman (University of California, Irvine) as a postdoctoral research associate. In August 2004, he joined the faculty at the Department of Chemistry and Biochemistry, Florida State University, and will be moving to the University of California, Santa Barbara in 2008. His research interests include the total synthesis of natural products, bioorganic chemistry, and the development of synthetic methodology.

Chongdao Lu was born in 1979 and carried out his undergraduate studies at Yunnan University (1996-2000). He received his doctorate degree in 2005 from Chengdu Institute of Organic Chemistry, working under the direction of Professors Wen-Hao Hu and Ai-Qiao Mi on novel methodology involving ylide chemistry. In 2005, he joined the group of Armen Zakarian, where he is a postdoctoral research associate. His current interests are the total synthesis of natural products and the development of new methodology for organic synthesis. He is a recipient of the MDS Postdoctoral Fellowship (2006-present).

Katrien Brak was born in 1983 in Leuven, Belgium. She earned a B.S. degree in chemistry from the Massachusetts Institute of Technology in 2005. That year, she began her doctoral studies at the University of California, Berkeley in the laboratories of Prof. Jonathan A. Ellman. Her graduate work has focused on the development of nonpeptidic protease inhibitors as well as the development of methods for the asymmetric synthesis of amines.

2-METHYLENECYCLOPROPANECARBOXYLIC ACID
ETHYL ESTER

Submitted by Mark E. Scott, Nai-Wen Tseng, and Mark Lautens.[1a]
Checked by Pascal Dubé and John A. Ragan.[1b]

1. Procedure

A. 2-Bromo-2-methyl-cyclopropanecarboxylic acid ethyl ester (**1**). A flame-dried, two-necked, 50-mL round-bottomed flask containing a magnetic stirring bar is fitted with septa to prevent exposure to air. Needles are then used for both an argon inlet and gas outlet to a bubbler. To this flask is added $Rh_2(OAc)_4$ (47.4 mg, 0.214 mmol Rh, 0.2 mol%) (Note 1) followed by 2-bromopropene (18.5 mL, 25.2 g, 0.208 mol, 2.18 equiv) (Note 2) by syringe. The argon inlet is removed and ethyl diazoacetate (11.0 mL, 11.1 g, 95.2 mmol, 1.00 equiv) (Note 3) is added by syringe pump over a period of two days (Note 4). Once the addition is complete, the reaction mixture is stirred for an additional 8 h (Note 5). The gas outlet is then removed and one septum is replaced with a one-piece, vacuum-jacketed short path distillation head containing an 18 mm (length) × 13 mm (OD) Vigreaux column. The excess 2-bromopropene is removed by distillation (760 mmHg, bp 42–45 °C) (Note 6). The reaction flask is cooled to room temperature, vacuum is applied (18–22 mmHg), and the flask is gradually heated in an oil bath to a bath temperature of 80 °C. This distillation

Org. Synth. **2008**, *85*, 172-178
Published on the Web 5/13/2008

provides 15.3–18.0 g (78–91% yield) of the bromo ester **1** (bp 64–73 °C, 18–22 mmHg) as a clear, colorless liquid (Notes 7–9).

B. 2-Methylene-cyclopropanecarboxylic acid, ethyl ester (**2**). A flame-dried, two-necked, 250-mL round-bottomed flask containing a magnetic stirring bar is fitted with a septum and a condenser containing a septum. An argon inlet and gas outlet is introduced through the condenser septa using needles. Under argon atmosphere, sodium hydride (6.10 g, 0.153 mol, 1.77 equiv) (Note 10) and diethyl ether (110 mL) (Note 11) are added. Bromo ester **1** (17.9 g, 0.0864 mol, 1.00 equiv) is then added *via* cannula using 10 mL of diethyl ether to complete the transfer. The argon inlet needle is removed, leaving the reaction solution vented to a bubbler. The reaction mixture is brought to reflux (oil bath set to 45 °C) before ethanol (1.1 mL, 0.019 mol, 0.22 equiv) (Note 12) is added (CAUTION: addition of the ethanol should be done slowly over the course of several minutes to prevent excessive hydrogen gas evolution). The resulting, light-brown solution is stirred at reflux for 15 h (reaction progress can be monitored by GC-MS, Note 13). The reaction mixture is then cooled and the precipitated tan solid is filtered through a 1 cm pad of Celite® in a sintered glass Büchner funnel (8.5 cm ID). The residue is washed with diethyl ether (2 × 50 mL) and the resulting yellow filtrate is distilled at 760 mmHg using a 12 cm (length) × 17 mm OD Vigreaux column and short path distillation apparatus (18 mm (length) × 13 mm (OD)) to remove the diethyl ether. Vacuum is then applied (13–14 mmHg) and the resulting yellow oil residue is then distilled to afford 7.75 g (71% yield) of 2-methylenecyclopropane carboxylic acid ethyl ester (**2**) (40–44 °C, 13–14 mmHg) as a clear, colorless liquid. A second distillation (Note 14), necessary to remove a trace impurity, is performed to provide 7.28 g (66% yield) of analytically pure product (Note 15).

2. Notes

1. Rhodium acetate dimer was purchased from Strem Chemicals, Inc. and was used as received.

2. 2-Bromopropene (99%, stabilized with copper) was purchased from Alfa Aesar and was stored at 5 °C until it was used as received.

3. Ethyl diazoacetate (≤ 15% dichloromethane) was purchased from Aldrich and was stored at 5 °C until it was used as received. ^{1}H NMR analysis found that the dichloromethane content was 15 mol% = 9 wt%.

This value was then used to accurately determine the number of moles of reagent added to the reaction [(Molar ratio)×(MW ratio) = (0.85/0.15)×(156.14/84.93) = 10.4 = 0.91/0.09]. The density for ethyl diazoacetate (d = 1.085) was used for determining the mass of the starting material.

4.   A KD Scientific (model 100) syringe pump was used to deliver the ethyl diazoacetate over a period of two days at approximately 0.20 mL/hour.

5.   The submitters stirred the reaction mixture for 3 hours after complete addition. In principle this additional time is unnecessary.

6.   Approximately 8.8 g of 2-bromopropene containing approximately 14% by mass of dichloromethane (by ^{1}H NMR) could be recycled from this reaction. Periodic cooling using a dry ice/acetone bath should also be used in order to prevent evaporation of the 2-bromopropene mixture in the receiving flask.

7.   GC analysis (see Note 13 for method) revealed that the product mixture was >98% pure (trans:cis ratios ranged from 1.5–2.0:1. t_R(trans), 4.04 min; t_R(cis), 4.56 min). The product exhibited the following physicochemical properties: IR (film) cm^{-1}: 2980, 1725, 1378, 1267, 1184, 1147, 1088, 1056, 854; ^{1}H NMR (400 MHz, CDCl$_3$) δ: 1.11 (dd, 1 H cis, J = 8.7, 6.0 Hz), 1.17 (t, 3 H trans, J = 7.5 Hz), 1.19 (t, 3 H cis, J = 7.5 Hz), 1.31 (t, 1 H trans, J = 6.6 Hz), 1.47 (dd, 1 H trans, J = 9.6, 6.7 Hz), 1.59–1.70 (m, 2 H cis), 1.72 (s, 3 H cis), 1.74 (s, 3 H trans), 2.18 (dd, 1 H trans, J = 9.5, 7.0 Hz), 4.04 (q, 2 H trans, J = 7.1 Hz), 4.09 (dq, 2 H cis, J = 7.1, 1.3 Hz); ^{13}C NMR (100 MHz, CDCl$_3$) δ: 14.0, 14.1, 22.6, 23.6, 24.0, 28.4, 29.3, 30.5, 32.7, 32.9, 60.7, 60.8, 168.8, 169.8. Anal. Calcd for C$_7$H$_{11}$O$_2$Br: C, 40.60; H, 5.35. Found: C, 40.39; H, 5.44.

8.   In this case, the isomers refer to the relationship between the bromine and ester moieties. ROESY analysis of the mixture in benzene–d_6 was used to determine the configuration for both isomers.

9.   The checkers found that it was important to maintain sufficient vacuum during the distillation (ca. 15 mmHg) to keep the temperature near the 40–44 °C cited in the procedure in order to avoid decomposition of the product.

10. Sodium hydride (60% in mineral oil) was purchased from Aldrich and used as received.

11. The checkers used BHT-stabilized, anhydrous diethyl ether purchased from JT Baker, which was used as received. The submitters used diethyl ether distilled from Na/benzophenone immediately prior to use.

12. The checkers used ethanol purchased from Pharmco-AAPER (200 proof, ACS grade), which was used as received. The submitters used ethanol that was distilled from magnesium/iodine and was kept over 4Å molecular sieves under argon until used.

13. GC/MS analysis was performed on an HP 6890 Series GC coupled with an Agilent 5973 mass spectrometer with an HP-1 column (0.2 mm x 12 m, 0.33 μm film thickness). A mass spec (EI) detector was utilized with a 1 min solvent delay. Injection size was 1 μL. Inlet was in split mode, with an injection temperature of 250 °C, flow rate (He) of 1.0 mL/min, and a column gradient from 50 °C for 30 seconds, then ramp at a rate of 10 °C/min to 275 °C and hold for 1 min. Samples for GC analysis were prepared by removing a 10 μL aliquot from the reaction and diluting to 1 mL with acetonitrile.

14. The second distillation may not be necessary depending on the intended use for the product.

15. GC analysis (see Note 13 for method) revealed that the product was >98% pure (t_R, 2.19 min) and exhibited the following physicochemical properties: IR (film) cm^{-1}: 2982, 1732, 1446, 1370, 1333, 1296, 1262, 1177, 1110, 1050, 906, 720; ^{1}H NMR (400 MHz, CDCl$_3$) δ: 1.16 (t, 3 H, J = 7.0 Hz), 1.50–1.55 (m, 1 H), 1.69–1.72 (m, 1 H), 2.13–2.16 (m, 1 H), 4.04 (q, 2 H, J = 7.0 Hz), 5.40–5.42 (m, 2 H); ^{13}C NMR (100 MHz, CDCl$_3$) δ: 11.1, 13.9, 17.8, 60.4, 104.2, 130.1, 171.7; Anal. Calcd. for C$_7$H$_{10}$O$_2$: C, 66.65; H, 7.99. Found: C, 66.49; H, 7.69.

Safety and Waste Disposal Information

All hazardous materials should be handled and disposed of in accordance with "Prudent Practices in the Laboratory"; National Academy Press; Washington, DC, 1995.

3. Discussion

Previous methods to prepare this versatile methylenecyclopropane building block have been reported in low yields via the cyclopropanation of either 2-bromopropene (copper-catalyzed)[2] or ethyl buta-2,3-dienoate (Simmons-Smith cyclopropanation).[3] A subsequent report for the preparation of this methylenecyclopropylcarboxylic acid ethyl ester was reported in much higher yields using a rhodium diacetate catalyst.[4] Although this latter report was higher yielding, yields for this process were sometimes variable due to the volatility of the product and intermediates. This approach removes any unnecessary manipulation steps, thereby preventing possible loss due to evaporation while still obtaining high chemical yields of the product. In addition, this modified method allows for the reuse of approximately one third of the unused starting material, making this method highly cost efficient.

The resulting 2-methylenecyclopropanecarboxylic acid ethyl ester has been extensively used for the preparation of model compounds for the study of the hypoglycemic activity of methylenecyclopropylglycine,[4,5] and for the synthesis of a variety of antiviral nucleoside analogues.[6] In addition, the methylenecyclopropanecarboxylic acid ethyl ester has also been shown to be a useful reagent for the synthesis of functionalized polymers containing a highly strained cyclopropane moiety in the polymer backbone.[7] A summary of these applications, as well as additional synthetic transformations[8] from either the ester or its derivatives, are detailed below.

1. (a) Davenport Research Laboratories, Department of Chemistry, University of Toronto, Toronto, Ontario, Canada, M5S 3H6. (b) Chemical Research & Development, Pfizer Global Research & Development, Groton, Connecticut, 06340.

2. (a) Carbon, J. A.; Martin, W. B.; Swett, L. R. U.S. Patent 2956077, **1960**. (b) Carbon, J. A.; Martin, W. B.; Swett, L. R. *J. Am. Chem. Soc.* **1958**, *80*, 1002.

3. Black, D. K.; Landor, S. R. *J. Chem. Soc. C* **1968**, 288-290.

4. Lai, M.-T.; Liu, H.-W. *J. Am. Chem. Soc.* **1990**, *112*, 4034-4035.

5. (a) Baldwin, J. E.; Widdison, W. C. *J. Am. Chem. Soc.* **1992**, *114*, 2245-2551. (b) Lai, M.-T.; Liu, L.-D.; Liu, H.-W. *J. Am. Chem. Soc.* **1991**, *113*, 7388-7397. (c) Li, D.; Agnihotri, G.; Dakoji, S.; Oh, E.; Lantz, M.; Liu, H.-W. *J. Am. Chem. Soc.* **1999**, *121*, 9034-9042.

6. (a) Zhou, S.; Zemlicka, J. *Tetrahedron* **2005**, *61*, 7112-7116. (b) Zemlicka, J.; Qui, Y.-L.; Drach, J. C.; Ptak, R. G. U.S. Patent 6352991, **2002**. (c) Guan, H.-P.; Ksebati, M. B.; Cheng, Y.-C.; Drach, J. C.; Kern, E. R.; Zemlicka, J. *J. Org. Chem.* **2000**, *65*, 1280-1290. (d) Cheng, C.; Shimo, T.; Somekawa, K.; Baba, M. *Tetrahedron* **1998**, *54*, 2031-2040. (e) Qiu, Y.-L.; Hempel, A.; Camerman, N.; Camerman, A.; Geiser, F.; Ptak, R. G.; Breitenbach, J. M.; Kira, T.; Li, L.; Gullen, E.; Cheng, Y.-C.; Drach, J. C.; Zemlicka, J. *J. Med. Chem.* **1998**, *41*, 5257-5264. (f) Zhou, S.; Kern, E. R.; Gullen, E.; Cheng, Y.-C.; Drach, J. C.; Matsumi, S.; Mitsuya, H.; Zemlicka, J. *J. Med. Chem.* **2004**, *47*, 6964-6972.

7. (a) Takeuchi, D.; Anada, K.; Osakada, K. *Macromolecules* **2002**, *35*, 9628-9633. (b) Osakada, K.; Takeuchi, D. Japanese Patent JP2003026783, **2003**.

8. (a) Crandall, J. K.; Conover, W. W.; Komin, J. B. *J. Org. Chem.* **1975**, *40*, 2042-2044. (b) Scott, M. E.; Schwarz, C. A.; Lautens, M. *Org. Lett.* **2006**, *8*, 5521-5524. (c) Lautens, M.; Han, W.; Liu, J. H.-C. *J. Am. Chem. Soc.* **2003**, *125*. 4028-4029. (d) Lautens, M.; Han, W. *J. Am. Chem. Soc.* **2002**, *124*, 6312-6316. (e) Cheng, C.; Shimo, T.; Somekawa, K.; Kawaminami, M. *Tetrahedron Lett.* **1997**, *38*, 9005-9008. (f) Corlay, H.; James, I. W.; Fouquet, E.; Schmidt, J.; Motherwell, W. B. *Synlett* **1996**, 990-992. (g) Corlay, H.; Fouquet, E.; Motherwell, W. B. *Tetrahedron Lett.* **1996**, *37*, 5983-5986.

Appendix
Chemical Abstracts Nomenclature; (Registry Number)

Rhodium acetate dimer: Rhodium, tetrakis[μ-(acetato-κO:κO')]di-, (Rh-Rh); (15956-28-2)

2-Bromopropene: 1-Propene, 2-bromo-; (557-93-7)

Ethyl diazoacetate: Acetic acid, 2-diazo-, ethyl ester; (623-73-4)

Sodium hydride; (7646-69-7)

Cyclopropanecarboxylic acid, 2-bromo-2-methyl-, ethyl ester; (89892-99-9)

Cyclopropanecarboxylic acid, 2-methylene-, ethyl ester; (18941-94-1)

Mark Scott was born in Simcoe, Ontario, Canada. He received his Bachelor's degree (Honours) in Engineering Chemistry and later his M.Sc.Eng. in Chemical Engineering from Queen's University under the supervision of Profs. Scott Parent and Ralph Whitney. Following his Master's, he received his Ph.D. in Mark Lautens' research group at the University of Toronto where he studied Lewis acid-mediated ring expansions of methylenecyclopropanes. He has been the recipient of NSERC Postgraduate (M.Sc.Eng. and Ph.D.) Scholarships. He is currently an NSERC postdoctoral fellow at Princeton University under Prof. David MacMillan.

Nai-Wen Tseng was born in Taipei, Taiwan in 1979. He received a Bachelor of Science degree in Chemistry at the National Tsing Hua University in 2001 where he conducted his undergraduate research under Prof. Hsing-Jang Liu. He is currently pursuing his Ph.D. degree under the guidance of Prof. Mark Lautens at the University of Toronto. His research is focused on the development of new rhodium-catalyzed addition/cyclization reactions.

Pascal Dubé was born in 1978 in St-Pacôme, Québec, Canada. He received a B.Sc. degree in pharmaceutical chemistry in 2000 from Université de Sherbrooke. As part of a Co-Op program he did internships at Merck Frosst Inc. and Pfizer. He then received a Ph.D. from Université de Sherbrooke in 2005 under the guidance of Professor Claude Spino. His doctoral research involved the development of a synthetic route to quassinoids featuring the Diels-Alder cycloaddition of transmissible dienes. After a postdoctoral fellowship with Professor F. Dean Toste at UC Berkeley he joined the Pfizer Chemical Research and Development group in 2006.

REGIOSELECTIVE SYNTHESIS OF 1,3,5-TRISUBSTITUTED PYRAZOLES BY THE REACTION OF *N*-MONOSUBSTITUTED HYDRAZONES WITH NITROOLEFINS

Submitted by Xiaohu Deng and Neelakandha S. Mani.[1]
Checked by Kay M. Brummond, Daitao Chen and Josh Osbourn.[2]

1. Procedure

A.   *5-Benzo[1,3]dioxol-5-yl-3-(4-chlorophenyl)-1-methyl-1H-pyrazole (2).* A 500-mL, three-necked, round-bottomed flask equipped with a mechanical stirrer (fitted with a 7.5-cm glass paddle), a condenser (open to the atmosphere) and a rubber septum is charged with 4-chlorobenzaldehyde (5.00 g, 35.5 mmol, 1.25 equiv) and MeOH (200 mL) (Note 1). Methylhydrazine (1.9 mL, 1.66 g, 35.5 mmol, 1.25 equiv) is added to this solution dropwise via syringe over 3 min (Note 1). After 2 h at room temperature, the formation of hydrazone **1** is complete   (Note 2).   The septum is removed and 3,4-methylenedioxy-β-nitrostyrene (5.50 g, 28.5 mmol) is added as a solid in one portion (Note 3). The reaction mixture is stirred at room temperature open to air for 72 h (Notes 4-6). The condenser is replaced with a 125-mL, pressure-equalizing addition funnel and water

(60 mL) is added to the mixture over 20 min. The resulting suspension is stirred at room temperature for an additional 1 h. The yellow solid that is formed is collected on a 250-mL fritted glass Büchner funnel by vacuum filtration (washing with ca. 30 mL of 1:1 MeOH/H₂O) and is suction-dried overnight. The yellow solid is dissolved in a minimum amount of boiling MeOH, then after being cooled to room temperature, the product is collected by suction filtration in 125-mL fritted glass Büchner funnel as a fine brown powder (6.07 g, 19.3 mmol, 68%) (Note 7).

B. *1-Benzyl-3-(4-chlorophenyl)-5-p-tolyl-1H-pyrazole (4)*. In a 500-mL, one-necked, round-bottomed flask equipped with a 3.8-cm, egg-shaped Teflon-coated magnetic stirring bar and a condenser, 4-chlorobenzaldehyde (4.86 g, 34.5 mmol, 1.25 equiv) is dissolved in MeOH (150 mL) and then water (10 mL) is added (Note 8). Benzylhydrazine dihydrochloride (6.75 g, 34.5 mmol, 1.25 equiv) is added in one portion (Note 9). After stirring the mixture at room temperature for 3 h, hydrazone 3 is formed (Note 10). 4-Methyl-β-nitrostyrene (4.51 g, 27.7 mmol) (Note 11) is added in one portion and the reaction solution is stirred at room temperature open to air until complete (Notes 12, 13). The condenser is replaced with a 125-mL, pressure-equalizing addition funnel and water (50 mL) is slowly added into the mixture over 20 min and the resulting white suspension is stirred at room temperature for an additional 1 h. The resulting white solid is collected on a 250-mL fritted glass Büchner funnel by vacuum filtration (washing with ca. 30 mL of 1:1 MeOH/H₂O) and is suction-dried (ca. 10 min). The solid is transferred into a 50-mL, round-bottomed flask and further dried under vacuum at room temperature for 8 h. The title compound is obtained as a white solid (9.12 g, 25.3 mmol, 92%) (Note 14).

2. Notes

1. 4-Chlorobenzaldehyde (97% purity) and methylhydrazine (98% purity) were purchased from Aldrich. Methanol (HPLC grade) was purchased from EMD. All reagents were used as received without further purification. HPLC t_R = 8.68 min (Note 2) for 4-chlorobenzaldehyde.

2. The checkers monitored the progress of the reaction by ¹H NMR spectroscopy. (Typical procedure for ¹H NMR analysis: an aliquot of the reaction mixture was transferred into an NMR tube via pipette. The NMR tube was fitted with a rubber septum, connected to a vacuum with a needle, the solvent was evaporated and the residue was dissolved in CDCl₃). NMR

data for hydrazone **1**; ^{1}H NMR (500 MHz, CDCl$_3$) δ: 3.01 (s, 3 H), 7.34 (d, J = 8.5 Hz, 2 H), 7.49 (s, 1 H), 7.52 (d, J = 8.5 Hz, 2 H). The submitters monitored the formation of hydrazone **1** by HPLC analysis. All HPLC analyses were performed on a Hewlett Packard 1100 (Agilent ZORBAX® Eclipse XDB-C8, 5 μm, 4.6 x 150 mm, 1 mL/min, acetonitrile/water with 0.05% trifluoroacetic acid : 1% acetonitrile/99% water to 99% acetonitrile/1% water ramp over 8 min, then hold at 99% acetonitrile/1% water) HPLC t_R = 7.43 min for **1**.

3. 3,4-Methylenedioxy-β-nitrostyrene (98% purity) was purchased from Alfa Aesar and used as received. HPLC t_R = 8.83 min (Note 2).

4. Air is essential for the reaction to proceed.

5. The reaction time is dependent upon the scale and the stirring speed. ^{1}H NMR analysis of small aliquots is desirable to follow the reaction progress. The submitters reported a reaction time of 40 h for a 5 g reaction; the checkers observe a reaction time of 72 h for the same reaction scale.

6. As the reaction proceeds 3,4-methylenedioxy-β-nitrostyrene slowly dissolves and the product precipitates, evidenced by a brown heterogeneous mixture turning homogeneous and a yellow precipitate forming after 1-2 h.

7. The product displayed the following physicochemical properties: mp 117-120 °C; ^{1}H NMR (500 MHz, CDCl$_3$) δ: 3.90 (s, 3 H), 6.05 (s, 2 H), 6.52 (s, 1 H), 6.91-6.92 (m, 3 H), 7.37 (dt, J = 9.0, 2.0 Hz, 2 H), 7.75 (dt, J = 9.0, 2.0 Hz, 2 H); ^{13}C NMR (125.7 MHz, CDCl$_3$) δ: 37.7, 101.7, 103.3, 108.8, 109.4, 122.9, 124.3, 126.9, 129.0, 132.2, 133.5, 145.1, 148.1, 148.2, 149.5; IR (film) cm^{-1}: 2895 (w), 1554 (w), 1478 (s); MS m/z (relative intensity): 312 (M$^+$, 100%); HRMS-ESI m/z: [M]$^+$ calcd for C$_{17}$H$_{13}$ClN$_2$O$_2$ 312.0666; found, 312.0663. Anal. Calcd for C$_{17}$H$_{13}$ClN$_2$O$_2$: C, 65.29; H, 4.19; N, 8.96. Found: C, 65.09; H, 4.13; N, 9.00. The submitters reported HPLC t_R = 10.47 min. The crude pyrazole (prior to recrystallization) is contaminated with a small amount of the 1,4-addition product (E)-1-(1-(benzo[d][1,3]dioxol-5-yl)-2-nitroethyl)-2-(4-chlorobenzylidene)-1-methylhydrazine as evidenced by the three resonances: 5.41 (dd, J = 13.2, 9.5 Hz, 1 H), 4.98 (dd, J = 9.5, 4.8 Hz, 1 H), 4.60 (dd, J = 13.2, 4.8, 1 H). For complete characterization of the 1,4-addition product see the original disclosure of this procedure by the authors.[3]

8. Water (distilled) is necessary for hydrazone and pyrazole formation when the HCl salt of the hydrazine is used.

9. Benzylhydrazine dihydrochloride (97% purity) was purchased from Aldrich and used as received.

10. The checkers monitored the progress of the reaction by ^{1}H NMR spectroscopy. (Typical procedure for ^{1}H NMR analysis: an aliquot of the reaction mixture was transferred into an NMR tube via pipette. The NMR tube was fitted with a rubber septum, connected to a vacuum with a needle, the solvent was evaporated and the residue was dissolved in $CDCl_3$.) NMR data for hydrazone **3**; ^{1}H NMR (500 MHz, $CDCl_3$) δ: 4.44 (s, 2 H), 5.74 (br s, 1 H), 7.30 (d, J = 8.5 Hz, 2 H), 7.32-7.28 (m, 1 H), 7.36-7.37 (m, 4 H), 7.48 (d, J = 8.5 Hz, 2 H), 7.55 (s, 1 H). The submitters reported the HPLC t_R = 10.05 min for **3** (Note 2).

11. 4-Methyl-β-nitrostyrene (98% purity) was purchased from Alfa Aesar and used as received. HPLC t_R = 9.42 min (Note 2).

12. Whereas the submitters report a 48 h reaction time, the checkers observed a reaction time of 88 to 92 h. The progress of the reaction was monitored by ^{1}H NMR spectroscopy.

13. As the reaction proceeds the product precipitates as a white solid after ca. 1-2 h.

14. No purification is necessary for 1-benzyl-3-(4-chlorophenyl)-5-*p*-tolyl-1*H*-pyrazole (**4**). The product displayed the following physicochemical properties:: mp 132-133 °C; HPLC t_R =11.99 min (Note 2); R_f = 0.43 (hexanes/EtOAc, 4/1); ^{1}H NMR (500 MHz, $CDCl_3$) δ: 2.40 (s, 3 H), 5.38 (s, 2 H), 6.61 (s, 1 H), 7.11 (d, J = 7.0 Hz, 2 H), 7.21-7.32 (m, 7 H), 7.38 (dt, J = 9.0, 2.0 Hz, 2 H), 7.80 (dt, J = 9.0, 2.0 Hz, 2 H); ^{13}C NMR (125.7 MHz, CD_2Cl_2) δ: 21.5, 53.4, 103.7, 126.9, 127.1, 127.6, 127.7, 128.8, 129.0 (2C), 129.6, 132.3, 133.5, 137.9, 139.0, 146.0, 150.1; IR (film) cm^{-1}: 3028 (w), 2929 (w), 1490 (w), 1436 (s); MS m/z (relative intensity): 358 (M^+, 80%), 239 (26%), 91 (100%); HRMS-ESI m/z: $[M]^+$ calcd for $C_{23}H_{19}ClN_2$ 358.1237; found, 358.1225; Anal. Calcd for $C_{23}H_{19}ClN_2$, C, 76.98; H, 5.34; N, 7.81. Found: C, 76.85; H, 5.34; N, 7.84. The submitters reported greater than 98% purity based upon HPLC analysis and t_R = 11.99 min (Note 2).

Safety and Waste Disposal Information

All hazardous materials should be handled and disposed of in accordance with "Prudent Practices in the Laboratory"; National Academy Press; Washington, DC, 1995.

3. Discussion

Substituted pyrazoles are important synthetic targets in the pharmaceutical industry.[4] Regioselective synthesis of the pyrazole ring remains a significant challenge for organic chemists.[5] Inspired by the literature precedent reactions of hydrazones and nitroolefins[6] we developed a general, one-pot, regioselective synthesis of substituted pyrazoles from N-monosubstituted hydrazones and nitro-olefins.

The outcome of the reaction depended upon the nature of solvents used (Table 1). Non-polar solvents such as toluene are not a good choice for the reaction, with no reaction being observed. Aprotic polar solvents generally favored the formation of Michael addition product **3**, with the exception of CH_2Cl_2. Protic polar solvents favored the formation of pyrazole **2**. Although AcOH and CH_2Cl_2 also afforded high **2**:**3** ratios, the reaction in alcoholic solvents such as MeOH and EtOH was much cleaner and provided the best yields of **2**.

Table 1. Solvent Effect on the Pyrazole Formation Reaction.

solvents	Et₂O	DMF	DMF/H₂O	MeOH[b]	EtOH	IPA	AcOH
2:3 Ratio[a]	1:99	4:96	41:59	94:6	93:7	71:29	99:1

solvents	DCM	THF	EtOAc	CH₃CN[c]	Toluene	Pyridine	Et₃N
2:3 Ratio[a]	94:6	6:94	20:80	8:92	0:0	10:90	9:91

(a) Determined by HPLC analysis at 254 nm (UV), not corrected; (b) The isolated yield of compound 2 was 83%; (c) The isolated yield of compound 3 was 90%.

This pyrazole formation reaction is fairly general with nitroolefins bearing various substituents at the R^3 and R^4 positions (Table 2). Steric effects at R^2 position play a small role; for example, sterically congested 2,2-dimethylpropionaldehyde affords pyrazole **7** in 56% yield (Entry 5). In contrast, electronic effects are prominent. Whereas electron-donating or

Table 2. Three-component Reaction for Pyrazole Synthesis

entry	hydrazine	aldehyde	nitro-olefin	product	yield
1	MeNHNH$_2$			**2**	83%
2	MeNHNH$_2$			**4**	92%
3	MeNHNH$_2$			**5**	81%
4	MeNHNH$_2$			**6**	82%
5	MeNHNH$_2$			**7**	56%
6	MeNHNH$_2$			**8**	72%
7	MeNHNH$_2$			**9**	90%
8[a]	Ph–NHNH$_2$ · 2HCl			**10**	90%
9[a,b]	NHNH$_2$ · HCl			**11**	26%
10[a,c]	NHNH$_2$ · HCl			**12**	15%
11[c]	PhNHNH$_2$			**13**	42%
12	PhNHNH$_2$			**14**	76%

(a) 10:1 MeOH/H2O was used as solvent; (b) room temperature, 7 days; (c) reflux, 4 days

184

slightly electron-withdrawing aldehydes provide pyrazole products smoothly, electron-demanding 4-nitrobenzaldehyde gives the Michael addition product **9** in 90% yield (Entry 7), which does not cyclize even at reflux temperature. At the R^1 position, both electronic and steric effects are important. The pyrazole formation reaction proceeds smoothly with primary alkyl hydrazones such as methyl hydrazones (Entries 1-7) and benzyl hydrazone (Entry 8). However, as the bulk of the R^1 group increases, the reaction slows or requires higher temperature. For example, with isopropylhydrazine, the pyrazole formation reaction was not complete after 7 days at room temperature with only 26% isolated yield (Entry 9). In the case of *tert*-butylhydrazine, after refluxing for 4 days, a different 1-*t*-butyl-3,4-diarylpyrazole isomer **12** was isolated in 15% yield (Entry 10). An electronic effect related to the hydrazone substituents is apparent with electron-deficient phenyl hydrazines. In entry 11, reflux temperature was required for 4 days to give 42% yield of pyrazole **13**. Interestingly, this effect can be moderated through the use of an electron-donating aldehyde such as propionaldehyde, which gives a 76% yield of pyrazole **14** at room temperature (Entry 12).

A possible reaction pathway has been proposed, as shown in Scheme 1. The first step is a reversible cycloaddition to give a 4-nitro-pyrazolidine intermediate, probably proceeding through a 1,3-dipole intermediate generated *in situ*, as proposed in literature on the reaction of hydrazones with

Scheme 1. Proposed Mechanism

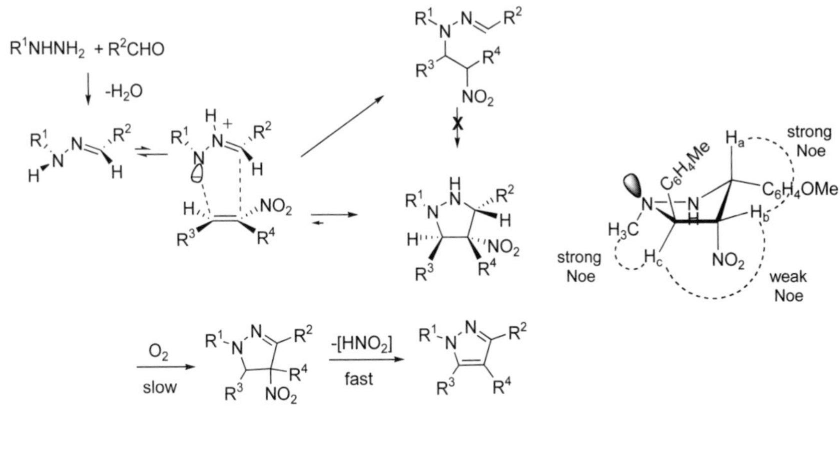

other electron-deficient olefins.[7] This process is in competition with an irreversible Michael addition process.[8] In one case, the key intermediate was directly observed on NMR experiments and was fully characterized. The key pyrazolidine intermediate then undergoes a slow oxidation by air, followed by a fast elimination of HNO_2[9] to afford the pyrazole product.

1. Johnson & Johnson Pharmaceutical R&D LLC, 3210 Merryfield Row, San Diego, CA 92121.
2. University of Pittsburgh, Department of Chemistry, Pittsburgh, PA 15260.
3. Deng, X.; Mani, N. S. *Org. Lett.* **2006**, *8*, 3505-3508.
4. Elguero, J.; Goya, P.; Jagerovic, N.; Silva, A. M. S. Pyrazoles as Drugs: Facts and Fantasies. In *Targets in Heterocyclic Systems*; Attanasi, O.A.; Spinelli, D., Eds.; Royal Society of Chemistry: Cambridge, **2002**, Vol. 6, 52-98.
5. (a) Elguero, J. *Comp. Heterocycl. Chem.* **1984**, *5*, 167. (b) Elguero, J. *Comp. Heterocycl. Chem. II* **1996**, *3*, 1-75, 817-932. (c) Makino, K.; Kim, H. S.; Kurasawa, Y. *J. Heterocycl. Chem.* **1998**, *35*, 489-497 and references cited therein.
6. (a) Snider, B. B.; Conn, R. S. E.; Sealfon, S. *J. Org. Chem.* **1979**, *44*, 218-221. (b) Gómez-Guillén, M.; Conde Jiménez, J. L. *Carbohydr. Res.* **1988**, *180*, 1-17. (c) Gómez-Guillén, M.; Hans-Hans, F.; Lassaletta Simon, J. M.; Martín-Zamora, M. E. *Carbohydr. Res.* **1989**, *189*, 349-358.
7. (a) Grigg, R.; Dowling, M.; Jordan, M. W.; Sridharan, V. *Tetrahedron* **1987***, 43*, 5873-5886. (b) Shimizu, T.; Hayashi, Y.; Miki, M.; Teramura, K. *J. Org. Chem.* **1987**, *52*, 2277-2285.
8. Huisgen, R. *J. Org. Chem.* **1968**, *33*, 2291-2297.
9. (a) Quiclet-Sire, B.; Zard, S. Z. *Synthesis* **2005**, 3319-3326. (b) Ballini, R.; Giantomassi, G. *Tetrahedron* **1995**, *51*, 4173-4182.

Appendix
Chemical Abstracts Nomenclature;
(Registry Number)

4-Chlorobenzaldehyde; (104-88-1)

Methylhydrazine; (60-34-4)

Benzylhydrazine dihydrochloride; (20570-96-1)

(*E*)-3,4-Methylenedioxy-β-nitrostyrene; (22568-48-5)

trans-p-Methyl-β-nitrostyrene: Benzene, 1-methyl-4-[(1*E*)-2-nitroethenyl]-; (5153-68-4)

5-Benzo[1,3]dioxol-5-yl-3-(4-chloro-phenyl)-1-methyl-1*H*-pyrazole: 1*H*-Pyrazole,   5-(1,3-benzodioxol-5-yl)-3-(4-chlorophenyl)-1-methyl-; (908329-89-5)

1-Benzyl-3-(4-chloro-phenyl)-5-*p*-tolyl-1*H*-pyrazole: 1*H*-Pyrazole,   3-(4-chlorophenyl)-5-(4-methylphenyl)-1-(phenylmethyl)-; (908329-95-3)

Xiaohu Deng was born in 1974 in Jingdezhen, China. He received a B.S. and a M.S. degree in chemistry under the supervision of Prof. Zi Gao from Fudan University in 1995 and 1997, respectively. After completing his doctoral dissertation in 2001 with Prof. Lanny S. Liebeskind at Emory University, he has been working in Johnson & Johnson Pharmaceutical R&D LLC in San Diego since. His research focuses on the design and development of practical synthetic methods to drug candidates.

Neelakandha Mani was born in 1957 in Kozhikode, India. He completed his Bachelors in Chemistry at Malabar Christian College and Ph.D. at Indian Institute of Science Bangalore with G.S.R. Subba Rao. He spent 3 years with Anthony G.M. Barrett (Northwestern and Colorado State University) and 4 years with Craig A. Townsend (Johns Hopkins University) as post-doctoral researcher. After 5 years as a scientist at Council of Scientific and Industrial Research (CSIR), India and 3 years as a senior scientist at Ligand Pharmaceuticals he moved to Johnson & Johnson in 2000. His research interests include synthesis of heterocycles, design of practical routes to drug molecules and development of environmentally friendly synthetic methods.

 Daitao Chen was born in Sichuan, P.R. China in 1974. He received his B.Sc. in 1996 and M.S. in 1999 at Peking University. He then moved to Emory University and obtained his Ph.D. in 2004 under the guidance of Dr. Debra Mohler. After working as a post-doctoral researcher in Dr. Kay Brummond's group at the University of Pittsburgh, he joined Scynexis, Inc (RTP, NC) in 2007. He is currently working on the DNDi team as a research scientist.

 Josh Osbourn was born in 1985 in Shinnston, West Virginia. In 2007, he graduated from West Virginia University with a B.S. degree in chemistry. During his time at WVU, he worked in the laboratory of Professor George O'Doherty and was supported through a Pfizer undergraduate research fellowship. In 2007, he entered the Ph.D. program at the University of Pittsburgh and is currently pursuing graduate studies in the laboratory of Professor Kay Brummond.

DIPHENYLDIAZOMETHANE

Submitted by Muhammad I. Javed and Matthias Brewer.[1]
Checked by Jonathan A. Ellman and Andy S. Tsai.

1. Procedure

Caution! Diazo compounds are toxic, irritating, and many compounds are explosive. Care should be taken when handling diazo compounds to limit personal exposure. It is prudent to handle diazo compounds behind a blast shield.

A 1-L, three-necked round-bottom flask is equipped with two rubber septa and an overhead mechanical stirrer fitted with a 9.5 x 2 cm rounded Teflon blade (Note 1). The flask is flame dried under a stream of nitrogen which is introduced and vented through the septa *via* 16 gauge needles. The exhaust septum is replaced with a thermometer adapter equipped with a low-temperature alcohol-based thermometer. The flask is charged with dimethyl sulfoxide (3.98 mL, 4.38 g, 56.0 mmol, 1.10 equiv) (Note 2) and anhydrous tetrahydrofuran (450 mL) *via* syringe (Note 3), and the solution is cooled to –55 °C (Notes 4 and 5) under a positive pressure of nitrogen with stirring. In a separate 100-mL, oven-dried, round-bottom flask is combined oxalyl chloride (4.67 mL, 6.79 g, 53.5 mmol. 1.05 equiv) (Note 6) and tetrahydrofuran (50 mL) *via* syringe. This solution is then added to the mixture over 10 min *via* cannula (Note 7), and the solution is maintained between –55 °C and –50 °C for 35 min. The reaction mixture is then cooled to –78 °C with the aid of a dry-ice acetone bath. To a separate 100-mL, oven-dried, round-bottomed flask is added benzophenone hydrazone (10.00 g, 51.0 mmol, 1.00 equiv) (Note 8) and triethylamine (15.05 mL, 10.84 g, 0.107 mol, 2.10 equiv) (Note 9) and tetrahydrofuran (50 mL). This solution is added to the reaction solution over 10 min *via* cannula (Note 10) to provide a deep-red solution containing a copious white precipitate. The reaction mixture is maintained at −78 °C for 30 min, and is then filtered while cold through a medium porosity sintered-glass funnel (Note 11) into a 2-L, round-bottom flask and the solid (Note 12) is rinsed with two 100-mL portions of tetrahydrofuran. The filtrate is concentrated at room temperature by rotary evaporation (15 mmHg) and then at 1.5 mmHg to provide diphenyldiazomethane (9.83 g, 99%) as a red oil that solidifies upon cooling (Note 13). The product is dissolved in pentane (120 mL) (Note 14), the solution is rapidly filtered through activated basic alumina (100 g) (Notes 15 and 16) supported in a medium porosity sintered glass funnel (Note 17) and the solids are rinsed with pentane (~300 mL) until the filtrate is colorless. The filtrate is concentrated at room temperature by rotary evaporation (15 mmHg) and then at 1.5 mmHg to provide diphenyldiazomethane (9.19 g, 93%) as an analytically-pure red crystalline solid (Notes 18 and 19).

2. Notes

1. When this procedure is run on smaller scale a Teflon-coated magnetic stirring bar can be used in place of the overhead mechanical stirrer.

2. Dimethyl sulfoxide was purchased from Acros and was dried over molecular sieves (<50 ppm H_2O).

3. Tetrahydrofuran was purified by passage through activated alumina using a GlassContour solvent purification system.[2]

4. The temperature was achieved and maintained by adding small pieces of dry-ice to an acetone bath.

5. The dimethyl sulfoxide must be dissolved prior to cooling; it freezes as a nearly insoluble mass on contact with precooled tetrahydrofuran.

6. Oxalyl chloride (98%) was purchased from Acros and was freshly distilled before use.

7. The internal temperature rose to −50 °C during this addition.

8. Benzophenone hydrazone was purchased from Acros and was purified by recrystallization from absolute ethanol before use. A 40 g sample of hydrazone was dissolved in 100 mL of boiling absolute ethanol. The solution was allowed to return to room temperature and then was cooled to −20 °C for 30 min. The precipitate was filtered through a 150-mL, medium-porosity, sintered-glass funnel and was washed with 30 mL of cold (0 °C) absolute ethanol. The recovered solid was dried under 1.5 mmHg vacuum for 12 h to return 34.7 g of pure, crystalline material.

9. Triethylamine was purchased from Acros and was freshly distilled from calcium hydride before use.

10. The internal temperature rose to –63 °C during this addition.

11. The funnel used was a 600-mL medium-porosity sintered-glass funnel purchased from Chemglass. The glass frit was 9.5 cm in diameter.

12. The white solid is triethylamine hydrochloride.

13. This material is approximately 95% pure based on ^{1}H NMR spectroscopy. Trace quantities of DMSO, unreacted hydrazone and benzophenone azine are observed.

14. Pentane of 98% or greater purity was purchased from Fisher Scientific and was used as received.

15. The submitters purchased activated Brockman Activity I basic alumina (50–200 micron) from Acros, which was used as received. The checkers purchased activated alumina, Brockman Activity I (60–325 mesh) from Fischer Scientific, which was used as received.

16. The filtration should be done such that the material is on the alumina column for no more than 5 min. When diphenyldiazomethane was allowed to remain in contact with the basic alumina, small quantities of an impurity were noted in the recovered product. GC and proton NMR analysis indicate that this impurity is tetraphenylethylene.

17. The funnel used was a 150-mL medium-porosity sintered-glass funnel purchased from Chemglass. The glass frit was 6 cm in diameter and the alumina bed had a height of 3 cm when contained in the funnel.

18. The product displays the following physicochemical properties: mp 29–31 °C; ^{1}H NMR (500 MHz, CDCl$_3$) δ: 7.42 (app t, J = 7.68 Hz, 4H), 7.38-7.31 (m, 4H), 7.22 (app t, J = 7.33 Hz, 2H); ^{13}C NMR (125 MHz, CDCl$_3$) δ 129.53, 129.12, 125.60, 125.16, 62.58;; IR (film) cm^{-1}: 2032, 1591, 1491, 1261, 745, 689. Anal. Calcd for $C_{13}H_{10}N_2$: C, 80.39; H, 5.19; N, 14.42. Found: C, 80.25; H, 5.21; N, 14.16.

19. Diphenyldiazomethane decomposes on standing.[6] To obtain acceptable elemental analysis results, freshly purified diphenyldiazomethane was shipped overnight at dry-ice temperature.

Safety and Waste Disposal Information

All hazardous materials should be handled and disposed of in accordance with "Prudent Practices in the Laboratory"; National Academy Press; Washington, DC, 1995.

3. Discussion

The preparation of diazo compounds is often achieved by the dehydrogenation of hydrazones.[3-5] Reagents that effect this transformation include mercury(II) oxide,[6] lead(IV) acetate,[7] manganese dioxide,[8] chromium(IV) oxide,[9] silver oxide,[8] nickel peroxide,[10] bis(acetylacetonato)-copper(II),[11] oxone,[12] metal hypochlorites,[13] barium manganate,[14] peracetic acid,[15] hypervalent organoiodo compounds,[16,17] and a cobalt Schiff base complex.[18] However, many of these reagents are hazardous and environmentally deleterious, and product yields often suffer due to reaction-byproduct induced decomposition of the diazo species. Additionally, the isolation and purification of the diazo product from the reaction mixture can be problematic and often involves potentially hazardous manipulation of the diazo solution, or substantial product loss during purification.

The procedure presented here[19] is a simple, convenient and high-yielding method to prepare and purify diphenyldiazomethane. This procedure, which uses chlorodimethylsulfonium chloride[20] to dehydrogenate benzophenone hydrazone, is advantageous to other routes because it does not rely on the use of environmentally-deleterious heavy-metal salts. The reaction between benzophenone hydrazone and chlorodimethylsulfonium chloride is highly efficient, and the crude product isolated from this procedure contains nearly-pure diphenyldiazomethane contaminated with trace quantities of DMSO, unreacted hydrazone, and benzophenone azine. The product is conveniently purified by simply dissolving the crude mixture in pentane and filtering the solution through a pad of basic alumina to provide analytically pure diphenyldiazomethane in 93% yield. This purification is simpler, and in our hands provides better product recovery, than recrystallization from petroleum ether.[6]

The dehydrogenation of hydrazones with chlorodimethylsulfonium chloride has also been used to prepare solutions of the resonance-stabilized diazo compounds shown in Table 1.[19] Due to the higher reactivity and the potential of these diazo species to explode, no attempts were made to isolate or purify them beyond the initial low-temperature filtration to remove triethylamine hydrochloride; the yields presented Table 1 were determined by nitrogen gas evolution measurements.

Table 1: Preparation of other diazo compounds via chlorodimethylsulfonium chloride mediated hydrazone dehydrogenation.[19]

hydrazone	diazo	N$_2$ yield
		97
		87
		88
		97 *
		51
		5 - 42

* Isolated yield

1. Department of Chemistry, The University of Vermont, Burlington, VT 05405. E-mail: matthias.brewer@uvm.edu

2. Pangborn, A. B.; Giardello, M. A.; Grubbs, R. H.; Rosen, R. K.; Timmers, F. J. *Organometallics* **1996**, *15*, 1518–1520.

3. Regitz, M.; Mass, G. *Diazo Compounds: Properties and Synthesis* Academic Press, INC.: Orlando, 1986.

4. Heydt, H. In *Science of Synthesis Houben-Weyl Methods of Molecular Transformations: Heteroatom Analogues of Aldehydes and Ketones*; 5th ed.; A., P., Ed.; Thieme: Stuttgart, 2004; Vol. 27, p 843—935.

5. Elliott, M. In *Comprehensive Organic Functional Group Transformations II*; 2 ed.; Katritzky, A. R., Taylor, R. J. K., Eds.; Elsevier Science: New York, NY, 2005; Vol. 3, p 469–523.

6. Smith, L. I.; Howard, K. L. *Org. Synth.. Coll. Vol. 3*, **1955**, 351.

7. Holton, T. L.; Shechter, H. *J. Org. Chem.* **1995**, *60*, 4725–4729.

8. Schroeder, W.; Katz, L. *J. Org. Chem.* **1954**, *19*, 718–720.

9. Ko, K. Y.; Kim, J. Y. *Bull. Korean Chem. Soc.* **1999**, *20*, 771–772.

10. Nakagawa, K.; Onoue, H.; Minami, K. *Chem. Commun.* **1966**, 730–731.

11. Ibata, T.; Singh, G. S. *Tetrahedron Lett.* **1994**, *35*, 2581–2584.

12. Curini, M.; Rosati, O.; Pisani, E.; Cabri, W.; Brusco, S.; Riscazzi, M. *Tetrahedron Lett.* **1997**, *38*, 1239–1240.

13. Morrison, H.; Danishefsky, S.; Yates, P. *J. Org. Chem.* **1961**, *26*, 2617–2618.

14. Guziec, F. S.; Murphy, C. J.; Cullen, E. R. *J. Chem. Soc. Perkin Transactions 1* **1985**, 107–113.

15. Adamson, J. R.; Bywood, R.; Eastlick, D. T.; Gallagher, G.; Walker, D.; Wilson, E. M. *J. Chem. Soc. Perkin Trans. 1* **1975**, 2030–2033.

16. Weiss, R.; Seubert, J. *Angew. Chem. Int. Ed.* **1994**, *33*, 891–893.

17. Lapatsanis, L.; Milias, G.; Paraskewas, S. *Synthesis* **1985**, 513–515.

18. Nishinaga, A.; Yamazaki, S.; Matsuura, T. *Chem. Lett.* **1986**, 505–506.

19. Javed, M. I.; Brewer, M. *Org. Lett.* **2007**, *9*, 1789–1792.

20. Mancuso, A. J.; Huang, S.-L.; Swern, D. *J. Org. Chem.* **1978**, *43*, 2480–2482.

Appendix
Chemical Abstracts Nomenclature (Collective Index Number); (Registry Number)

Diphenyldiazomethane (883-40-9)

Benzophenone hydrazone (5350-57-2)

Dimethyl sulfoxide: Methyl sulfoxide; Methane, sulfinybis-; (67-68-5)

Oxalyl chloride: HIGHLY TOXIC; Ethanedioyl dichloride; (79-37-8)

Triethylamine; Ethanamine, *N,N*-diethyl-; (121-44-8)

Matthias Brewer was born in 1974 and was raised in Holliston, Massachusetts. He received his undergraduate education at the University of Vermont where he carried out research with Professor A. Paul Krapcho. He received his Ph.D. in 2002 from the University of Wisconsin-Madison where he worked with Professor D. H. Rich on the synthesis of botulinum neurotoxin metalloproteinase inhibitors. Following an NIH Postdoctoral Fellowship with Professor L. E. Overman, he joined the faculty at the University of Vermont in 2005. His research interests include the development of new synthetic methods, the synthesis of natural products, and the synthesis of biologically active compounds.

Muhammad Irfan Javed received a B.S. in Applied Chemistry in 1998 from the University of Engineering & Technology, Lahore, Pakistan under the supervision of Dr. Inam-ul-Haque. He completed his M.S. in Chemistry in 2001 with Dr. T. K. Vinod at Western Illinois University, Illinois before joining Dr. Gregory Friestad's research group at the University of Vermont in 2003. Upon Dr. Friestad's move to the University of Iowa in 2005, Javed opted to remain at the University of Vermont and subsequently joined Dr. Matthias Brewer's research group. His research interests include the development of synthetic methods and green chemistry research.

Andy Tsai was born in Fujian, China in 1984 and immigrated to the United States in 1989. He obtained his undergraduate degree at the University of Michigan where he worked under Professor Richard Laine on functionalized silsesquioxanes. He is currently in the laboratories of Professors Jonathan A. Ellman and Robert G. Bergman pursuing a doctorate in chemistry.

SYNTHESIS OF 2-SUBSTITUTED BIARYLS VIA Cu/Pd-CATALYZED DECARBOXYLATIVE CROSS-COUPLING OF 2-SUBSTITUTED POTASSIUM BENZOATES: 4-METHYL-2'-NITROBIPHENYL AND 2-ACETYL-4'-METHYLBIPHENYL.

[Biphenyl, 4-Methyl-2'-nitro-]
[Biphenyl, 4-Methyl-2'-acetyl-]

Method A. 2-Nitrobenzoic acid derivatives.

Method B. Other ortho-substituted benzoic acids.

Submitted by Lukas J. Gooßen,[1] Nuria Rodríguez, Christophe Linder, Bettina Zimmermann, and Thomas Knauber.
Checked by Scott E. Denmark and Nathan S. Werner.

1. Procedure

Method A. 2-Nitrobenzoic acid derivatives.

A. *Potassium 2-nitrobenzoate.* A 1-L, two-necked, round-bottomed flask equipped with a magnetic stirring bar and a 250-mL dropping funnel is charged with 2-nitrobenzoic acid (33.4 g, 200 mmol) (Note 1) and 200 mL

196

of ethanol (Note 2). To this solution is added a solution of potassium hydroxide (85%) (13.2 g, 200 mmol, 1.00 equiv) (Note 3) in 200 mL of ethanol dropwise over 2 h. After complete addition, the reaction mixture is stirred for another 1 h at room temperature. A gradual formation of a white precipitate is observed. The resulting solid is collected by filtration through an 10-cm Büchner funnel (fitted with Whatman-1 filter paper, 90 mm), washed sequentially with ethanol, 2 x 50 mL, and 50 mL of cold (0 °C) diethyl ether, transferred to a 500 mL round-bottomed flask, and dried at 1.0 mmHg to provide potassium 2-nitrobenzoate (35.3 g, 86%) as a white powder (Note 4).

B. *4-Methyl-2'-nitrobiphenyl.* An oven-dried 500-mL, three-necked, round-bottomed flask equipped with a magnetic stirring bar, an internal thermometer and a reflux condenser fitted with a nitrogen gas inlet (Note 5), is charged with potassium 2-nitrobenzoate (22.6 g, 110 mmol, 1.10 equiv) (Note 6), (1,10-phenanthroline)bis(triphenylphosphine)copper(I) nitrate (1.25 g, 1.50 mmol, 0.015 equiv) (Note 7) and palladium acetylacetonate (45.7 mg, 0.15 mmol, 0.0015 equiv) (Note 8). After deoxygenating the flask with three alternating vacuum and nitrogen purge cycles, a solution of 4-bromotoluene (12.3 mL, 100 mmol) (Note 9) in 150 mL of anhydrous mesitylene (Note 10) is added. The flask is lowered into an oil bath preheated to 175 °C, so that the internal temperature remains constant at 150 °C (Note 11). After the reaction mixture is stirred and heated for 16 h, an aliquot checked by GC analysis indicates that the reaction is complete (Note 12). The heat source is removed and the mixture is allowed to stir for 2 h further while cooling to room temperature (Note 13). The nitrogen source is removed and any condensate is washed from the condenser into the reaction mixture with 10 mL of ethyl acetate. The mixture is then suction filtered through a 6-cm fritted glass funnel (coarse) containing a 1 cm bed of dry Celite (Note 14) into a 500-mL round-bottomed flask. The filter cake is rinsed with ethyl acetate, 3 x 20 mL (Note 15). The resulting dark brown filtrate and the washes are concentrated under reduced pressure (23 °C, 20 mmHg). The residue is purified by Kugelrohr distillation at 1.0 mmHg, gradually increasing the oven temperature from 25 °C to 155 °C. The fraction distilling at 155 °C oven temperature contains 4-methyl-2'-nitrobiphenyl (21.1 g, 99%) as a pale yellow oil (Note 16).

Method B. Other ortho-substituted benzoic acids.

A. *Potassium 2-acetylbenzoate.* A 500-mL, one-necked, round-bottomed flask equipped with a magnetic stirring bar and a 250-mL dropping funnel is charged with 2-acetylbenzoic acid (16.4 g, 100 mmol) (Note 17) and 100 mL of ethanol (Note 2). To this solution is added a solution of 85 % potassium hydroxide (6.6 g, 100 mmol, 1.00 equiv) (Note 3) in 100 mL of ethanol dropwise over 1 h. After complete addition, the reaction mixture is stirred for another 1 h at room temperature. The solvent is removed at 60 °C and 20 mmHg on a rotary evaporator to afford the crude product as a foam (Note 18). The crude product can be crystallized by dissolving it in 17 mL of boiling ethanol, slowly cooling the solution to room temperature, and then carefully layering it with 70 mL of diethyl ether. The biphasic mixture is then allowed to stand for 12 h at room temperature then 24 h at –20 °C. The resulting off-white crystals are collected by filtration through an 8-cm Büchner funnel (fitted with Whatman-1 filter paper, 70 mm), then are washed sequentially with 3 x 25 mL of diethyl ether, are transferred to a 250-mL round-bottomed flask, and are dried at 1.0 mmHg to provide potassium 2-acetylbenzoate (19.2 g, 95%) as colorless crystals (Note 19).

B. *2-Acetyl-4'-methylbiphenyl.* An oven-dried 250-mL, three-necked, round-bottomed flask equipped with a magnetic stirring bar, an internal thermometer and a reflux condenser fitted with a nitrogen gas inlet (Note 5), is charged with potassium 2-acetylbenzoate (8.9 g, 44 mmol, 1.10 equiv) (Note 20), copper(I) bromide (574 mg, 4.00 mmol, 0.10 equiv) (Note 21), 1,10-phenanthroline (721 mg, 4.00 mmol, 0.10 equiv) (Note 22), palladium acetylacetonate (122 mg, 0.40 mmol, 0.01 equiv) (Note 8), and bis(diphenylphosphino)methane (307 mg, 0.80 mmol, 0.02 equiv) (Note 23). After deoxygenating the flask with three alternating vacuum and nitrogen purge cycles, a solution of 4-bromotoluene (4.92 mL, 40 mmol) (Note 9) in a mixture of 60 mL of anhydrous 1-methyl-2-pyrrolidone (Note 24) and 20 mL of anhydrous quinoline (Note 25) is added. The flask is lowered into an oil bath preheated to 180 °C, so that the internal temperature remained constant at 160 °C (Note 26). After stirring the reaction mixture for 24 h (Note 27), the heat source is removed and the reaction mixture is allowed to stir while cooling to room temperature (Note 28). The nitrogen source is removed and any condensate is rinsed from the condenser into the reaction mixture with 10 mL of ethyl acetate. The reaction mixture is suction filtered through a 6-cm fritted glass funnel (coarse) containing a 1 cm bed of dry Celite (Note 14) into a 250-mL round-bottomed flask. The filter cake is rinsed with ethyl acetate, 3 x 15 mL (Note 15). The resulting dark brown filtrate and washings are transferred into a 250-mL separatory funnel and are washed with a 1 N aqueous hydrochloric acid solution (1 x 100 mL and 2 x 50 mL). Each aqueous layer is extracted with ethyl acetate, 2 x 35 mL. The combined organic phases are washed with brine (1 x 100 mL), dried over anhydrous magnesium sulfate (2-3 g) and filtered. The solvent is evaporated under reduced pressure (23 °C, 20 mmHg), and the residue is purified by Kugelrohr distillation at 1.0 mmHg, gradually increasing the oven temperature from 25 °C to 130 °C (Note 29). The fraction distilling at 130 °C oven temperature contains 2-acetyl-4'-methylbiphenyl (6.78 g, 81%) as a pale-green liquid (Note 30 and 31).

2. Notes

1.   2-Nitrobenzoic acid was purchased from Acros Organics, and used as received. *Avoid contact with skin and eyes. Use only in a chemical fume hood.*

2.   Ethanol was purchased from Pharmco-AAPER, and used as received.

3.   Potassium hydroxide was purchased from Aldrich Chemical Company, Inc., and was used as received.

4.   Potassium 2-nitrobenzoate exhibits the following physicochemical properties: ^{1}H NMR (500 MHz, methanol-d_4) δ: 7.91-7.92 (m, 1 H), 7.64-7.67 (m, 1 H), 7.59-7.61 (m, 1 H), 7.48-7.51 (m, 1 H); ^{13}C NMR (100 MHz, methanol-d_4) δ: 173.9, 148.0, 138.4, 134.3, 129.61, 129.59, 124.5; Anal. Calcd. for $C_7H_4KNO_4$: C, 40.97; H, 1.96; N, 6.83. Found: C, 40.69; H, 1.93; N, 6.77.

5.   The apparatus is maintained under an atmosphere of nitrogen during the course of the reaction.

6.   Potassium 2-nitrobenzoate was prepared according to Method A, Procedure A. Avoid contact with skin and eyes. Use only in a chemical fume hood.

7.   (1,10-Phenanthroline)bis(triphenylphosphine)copper(I) nitrate was purchased from Strem Chemicals and was used as received.

8.   Palladium acetylacetonate was purchased from Aldrich Chemical Company, Inc. and was used as received.

9.   4-Bromotoluene was purchased from Aldrich Chemical Company, Inc. and was used as received.

10.   Mesitylene, purchased from Aldrich Chemical Company, Inc., was dried over $CaCl_2$, and then was distilled from sodium benzophenone ketyl.

11.   The reaction mixture undergoes a gradual color change from dark orange to dark brown.

12.   The reaction can be monitored by quenching small aliquots (0.25 mL) with HCl (1 N, 2 mL) and extracting them with ethyl acetate (2 mL). The organic layer is dried over a mixture of $MgSO_4$ and $NaHCO_3$ and analyzed by GC. GC analyses were carried out using an HP-1 capillary column (30 m x 0.32 mm) and a time program beginning with 3 min at 100 °C followed by 40 °C/min ramp to 260 °C, then 7 min at this temp. The following retention times are observed for the compounds within the mixture: mesitylene (2.40 min), nitrobenzene (3.66 min), 4-bromotoluene (3.39 min), 4-methyl-2'-nitrobiphenyl (7.35 min).

Alternatively, the reaction can be monitored by quenching small aliquots with water and extracting them with a small amount of ethyl acetate. The organic layer is spotted onto an analytical silica gel TLC plate (0.20 mm thickness, obtained from Aldrich Chemical Company, Inc.) and eluted with hexane/EtOAc, 9/1 using 254 nm UV light to visualize the spots. The following R_f-values are observed for the compounds within the mixture: 4-bromotoluene (0.89), 4-methyl-2'-nitrobiphenyl (0.48).

13.   At room temperature, the excess of potassium 2-nitrobenzoate precipitates and is removed in the filtration step, together with the KBr formed during the reaction.

14.   Celite was purchased from Fischer Scientific and was used as received.

15.   The filter cake should be washed until the filtrate is colorless. Washing is necessary to avoid crystallization underneath the fritted filter during the filtration.

16.   The distilled 4-methyl-2'-nitrobiphenyl exhibits the following analytical data: ^{1}H NMR (500 MHz, CDCl$_3$) δ: 7.85 (dd, J = 8.1 and 1.0, 1 H), 7.62 (td, J = 7.6 and 1.3, 1 H), 7.50-7.45 (m, 2 H), 7.28-7.24 (m, 4 H), 2.43 (s, 3 H); ^{13}C NMR (126 MHz, CDCl$_3$) δ: 138.1, 136.2, 134.4, 132.2, 131.9, 129.4, 127.9, 127.7, 124.0, 21.2; IR (neat) cm^{-1}: 3026 (w), 2920 (w), 2866 (w), 1641 (w), 1565 (w), 1525 (s), 1476 (m), 1355 (s), 1306 (w), 1285 (w), 1187 (w), 1163 (w), 1112 (w), 1092 (w), 1043 (w), 1007 (w), 954 (w), 853 (m), 819 (m), 782 (m), 750 (s); MS (EI, 70 eV) m/z: 213 (M$^+$, 100), 196

(52), 185 (48), 168 (76), 152 (62), 139 (24), 129 (23), 115 (35), 63 (17); Anal. Calcd. for $C_{13}H_{11}NO_2$: C, 73.23; H, 5.20; N, 6.57. Found: C, 73.04; H, 5.12; N, 6.91.

17. 2-Acetylbenzoic acid was purchased from Acros Organics and was used as received. Avoid contact with skin and eyes. Use only in a fume hood.

18. The temperature for the evaporation of the ethanol is crucial. At lower temperatures, the water of neutralization is not removed and crystallization of the resulting salt is irreproducible.

19. Potassium 2-acetylbenzoate exhibits the following analytical data: ^{1}H NMR (500 MHz, methanol-d_4) δ: 7.68-7.66 (m, 1 H), 7.47-7.39 (m, 3 H), 2.54 (s, 3 H); ^{13}C NMR (126 MHz, methanol-d_4) δ: 206.8, 176.2, 141.2 140.5, 131.5, 129.8, 127.7, 30.3; Anal. Calcd. for $C_9H_7KO_3$: C, 53.45; H, 3.49; Found: C, 53.16; H, 3.18.

20. Potassium 2-acetylbenzoate was prepared according to Method B, Procedure A. *Avoid contact with skin and eyes. Use only in a fume hood.*

21. Copper(I) bromide was purchased from Acros Organics and was used as received.

22. 1,10-Phenanthroline was purchased from Aldrich Chemical Company, Inc., as 1,10-phenanthroline monohydrate. The water of hydration was removed by stirring a solution of the monohydrate in ethyl acetate over anhydrous $MgSO_4$. The solution was then filtered and the volatiles were removed under reduced pressure.

23. Bis(diphenylphosphino)methane was purchased from Aldrich Chemical Company, Inc. and was used as received.

24. 1-Methyl-2-pyrrolidone was purchased from Aldrich Chemical Company, Inc. and dried by removing water as a toluene azeotrope.

25. Quinoline was purchased from Aldrich Chemical Company, Inc. and was dried by removing the water by fractional distillation.

26. The reaction mixture undergoes a gradual color change from dark orange to dark brown.

27. The progress of the reaction can be monitored by quenching small aliquots (0.25 mL) with HCl (1 N, 2 mL) and extracting them with ethyl acetate (2 mL). The organic layer is dried over a mixture of $MgSO_4$ and $NaHCO_3$ and analyzed by GC. GC analyses were carried out using an HP-1 capillary column (30 m x 0.32 mm) and a time program beginning with 3 min at 100 °C followed by 40 °C/min ramp to 260 °C, then 7 min at this temp. The following retention times are observed for the compounds within

the mixture: 1-methyl-2-pyrrolidone (2.99 min), 4-bromotoluene (3.39 min), acetophenone (3.51 min), quinoline (4.97 min) and 2-acetyl-4'-methylbiphenyl (7.15 min).

Alternatively, the reaction can be monitored by quenching small aliquots with water and extracting them with a small amount of ethyl acetate. The organic layer is spotted onto an analytical silica gel TLC plate (0.20 mm thickness, obtained from Aldrich Chemical Company, Inc.) and eluted with 10 % ethyl acetate in n-hexane, using 254 nm UV light to visualize the spots. The following R_f-values are observed for the compounds within the mixture: 4-bromotoluene (0.89), 2-acetyl-4'-methylbiphenyl (0.44).

28. At room temperature, the excess potassium 2-acetylbenzoate precipitates and is removed in the filtration step, together with the KBr formed during the reaction.

29. Acetophenone is removed at 50 °C and 1.0 mmHg.

30. 2-Acetyl-4'-methylbiphenyl exhibits the following physicochemical properties: ^{1}H NMR (500 MHz, CDCl$_3$) δ: 7.54-7.48 (m, 2 H), 7.41-7.38 (m, 2 H), 7.24 (broad s, 4 H), 2.41 (s, 3 H), 2.02 (s, 3 H)); ^{13}C NMR (126 MHz, CDCl$_3$) δ: 205.2, 140.9, 140.5, 137.8, 130.7, 130.2, 129.4, 128.8, 128.7, 127.8, 127.2, 30.5, 21.2; IR (neat) cm^{-1}: 3022 (w), 1685 (s), 1595 (w), 1560 (w), 1517 (w), 1473 (w), 1441 (w), 1354 (m), 1266 (m), 1231 (m), 1110 (w), 1076 (w), 1041 (w), 1005 (w), 967 (w), 823 (m), 764 (m), 747 (m); MS (EI, 70 eV) m/z: 210 (M$^+$, 63), 195 (100), 182 (10), 165 (36). Anal. Calcd. for C$_{15}$H$_{14}$O: C, 85.68; H, 6.71; Found: C, 85.28; H, 6.72.

31. The products of homocoupling, 2,2'-diacetylbiphenyl and 4,4'-dimethylbiphenyl were detected by GC/MS.

Safety and Waste Disposal Information

All hazardous materials should be handled and disposed of in accordance with "Prudent Practices in the Laboratory"; National Academy Press; Washington, DC, 1995.

3. Discussion

The method described herein illustrates a safe and convenient cross-coupling strategy for the synthesis of unsymmetrical biaryls, often found as substructures in biologically active molecules or functional materials.[2]

Whereas in traditional cross-coupling reactions, the prior preparation of organometallic reagents is required,[3,4] in this procedure, the carbon nucleophiles are generated *in situ* from easily accessible arenecarboxylic acid salts using a copper-phenanthroline catalyst. In the absence of acidic protons, the arylcopper species thus generated transfer their aryl residue onto palladium complexes arising from the oxidative addition of aryl halides to Pd(0) precursors. During the subsequent release of the biaryls, the initial Pd(0) species are regenerated, closing the catalytic cycle for palladium. The phenanthroline-stabilized copper bromide released in the transmetalation step and the substrate, potassium arenecarboxylate, equilibrate under formation of more copper carboxylate along with the potassium bromide byproduct. This allows a process catalytic in both Pd and Cu.

To achieve good yields with this method, important precautions have to be taken: Firstly, water and other proton sources have to rigorously be excluded from the reaction medium. This precaution will prevent protonolysis of the arylcopper intermediate under formation of the corresponding arenes. Therefore, the preformed potassium salt has to be dried thoroughly along with all other reagents, or the water inherently formed during the deprotonation step has to be removed carefully by azeotropic distillation.

Secondly, the overall rate of formation of the arylcopper species has to be adjusted to the rate of cross-coupling for each given arenecarboxylate substrate, to avoid the formation of byproducts from the competing homocoupling. As the rate of decarboxylation varies vastly among arenecarboxylate substrates,[5] slight modifications in the reaction protocols are necessary to allow clean conversions. Variations in reactivity of the aryl halides are less marked, so that one general protocol for a given arenecarboxylate is applicable to a wide range of aryl halides.

Because of the high reactivity of the starting 2-nitrobenzoic acid derivatives, the protocol was improved in a number of ways: Only very low catalyst loadings are required for their coupling with a broad variety of aryl bromides and some chlorides (Method A). Moreover, an easy-to-handle, commercially available copper-phenanthroline-phosphine complex can be used as the decarboxylation catalyst. Finally, an inexpensive, non-polar solvent was employed, leading to a simplified work-up procedure: Removal of the solvent and Kugelrohr distillation gives the reaction products in pure form (Table 1). This optimized protocol could easily be scaled up from millimolar to molar quantities without a decrease in yield.

Table 1. Scope of the cross-coupling reaction with regard to the aryl halide substrate.[a]

Entry	Aryl Halide	Product no. Yield (%)
1	Br	6 99
2	Br, OMe	7 91
3	Br	8 87
4	Cl, NO₂	9 77

a) With the exception of entry 1 (100 mmol scale), all reactions were performed on a 40 mmol scale. For additional examples and analytical data of the products, see ref. 2b.

2-Acetylbenzoic acid represents a comparatively difficult substrate to cross-couple as the steric bulk of the acetyl group appears to hinder the cross-coupling step. In this case, the *in situ* formation of the copper-phenanthroline catalyst and the addition of a palladium-stabilizing ligand are recommended (Method B). This method is applicable to many other ortho-substituted or heterocyclic aromatic carboxylates (Table 2), although some fine-tuning is required to ensure optimal yields. For example, 2-fluorobenzoic acid, a sterically unhindered, moderately reactive derivative, is best coupled in the absence of palladium-stabilizing ligands.

When applying this method to the synthesis of 2-carboxy-substituted biaryl derivatives, the potassium carboxylates can be generated *in situ* by adding potassium isopropoxide to readily available phthalic anhydride.

Org. Synth. **2008**, *85*, 196-208

Again, it is crucial to carefully remove any traces of residual alcohols and water prior to the decarboxylative cross-coupling step, in order to suppress protodecarboxylation.

Table 2. Scope of the cross-coupling reaction with regard to the benzoate.[a]

Entry	Benzoate		Product no. Yield (%)
1	10		13 75
2[b]	11		14 85
3[b,c]	12		15 63

a) All reactions were performed on a 40 mmol scale. For additional examples and analytical data of the products, see ref. 2b. b) In the absence of phosphine. c) The carboxylate derivative was preformed in situ using 48.0 mmol of phthalic anhydride and a solution of 48.0 mmol of potassium tert-butoxide in 80 mL of isopropanol, at 70 °C; 0.80 mmol of palladium catalyst were required.

1. Fachbereich Chemie - Organische Chemie, Technische Universität Kaiserslautern, Erwin-Schrödinger-Straße Geb. 54, D-67663 Kaiserslautern, goossen@chemie.uni-kl.de
2. (a) Gooßen, L. J.; Deng, G.; Levy, L. M. *Science* **2006**, *313*, 662-664; (b) Gooßen, L. J.; Rodríguez, N.; Melzer, B.; Linder, C.; Deng, G.; Levy, L. M. *J. Am. Chem. Soc.* **2007**, *129*, 4824-4833.

3. For an overview of metal-catalyzed cross-couplings, see e.g.: de Meijere, A.; Diederich, F.; Eds., *Metal-Catalyzed Cross-Coupling Reactions*, 2nd Ed, Wiley-VCH: Weinheim, 2004.
4. (a) Miyaura, N.; Suzuki, A. *Chem. Rev.* **1995**, *95*, 2457-2483; (b) Kosugi, M.; Sasazawa, K.; Shimizu, Y.; Migita, T. *Chem. Lett.* **1977**, 301-302; (b) Stille, J. K. *Angew. Chem. Int. Ed. Engl.* **1986**, *25*, 508-524; (c) King, O.; Okukado N.; Negishi E. *J. Chem. Soc., Chem. Comm.* **1977**, 683-684; (d) Lipshutz, B. H.; Siegmann, K.; Garcia, E.; Kayser F. *J. Am. Chem. Soc.* **1993**, *115*, 9276-9282; (e) Corriu, R. J. P.; Masse, J. P. *J. Chem. Soc., Chem. Comm.* **1972**, 144a; (f) Kumada, M. *Pure Appl. Chem.* **1980**, *52*, 669-679.
5. (a) Cohen, T.; Schambach, R. A. *J. Am. Chem. Soc.* **1970**, *92*, 3189-3190; (b) Cohen, T.; Berninger, R. W.; Wood, J. T. *J. Org. Chem.* **1978**, *43*, 837-848; (c) Nilsson, M. *Acta Chem. Scand.* **1966**, *20*, 423-426; (d) Gooßen, L. J.; Thiel, W. R.; Rodríguez, N.; Linder, C.; Melzer, B. *Adv. Synth. Catal.* **2007**, *349*, 2241-2246.

Appendix
Chemical Abstracts Nomenclature; (Registry Number)

2-Nitrobenzoic acid; (552-16-9)

Potassium hydroxide; (1310-58-3)

(1,10-Phenanthroline)bis(triphenylphosphine)copper(I) nitrate; (33989-10-5)

Palladium acetylacetonate; (140024-61-4)

4-Bromotoluene; (106-38-7)

Mesitylene; (108-67-8)

4-Methyl-2'-nitrobiphenyl; (70680-21-6)

2-Acetylbenzoic acid; (577-56-0)

Copper(I) bromide; (7787-70-4)

1,10-Phenanthroline; (66-71-7)

Bis(diphenylphosphino)methane; (2071-20-7)

1-Methyl-2-pyrrolidone; (872-50-4)

Quinoline; (91-22-5)

2-Acetyl-4'-methylbiphenyl; (16927-79-0)

Prof. Dr. Lukas J. Gooßen, born 1969 in Bielefeld/Germany, studied chemistry at Universität Bielefeld/University of Michigan, conducted his diploma work with K. P. C. Vollhardt, UC Berkeley and his PhD research with W. A. Herrmann, TU München. After a postdoctoral stay with K. B. Sharpless, Scripps Research Institute and a position as a head of laboratory at Bayer AG, he started his academic career with a "Habilitation" at MPI für Kohlenforschung in the group of M. T. Reetz (2000-2003). In 2004, he moved to RWTH Aachen as a Heisenberg fellow, and in 2005, he was appointed professor at TU Kaiserslautern.

Nuria Rodríguez was born in 1978 in Valencia, Spain. After undergraduate studies at University of Valencia, she joined the group of G. Asensio and M. Medio-Simón in 2001. She obtained her PhD in 2006 on *Palladium-catalyzed cross-coupling reactions over a Csp³ at the alpha position of a sulfinyl group. Methodology and synthetic applications.* Since November 2005, she is conducting postdoctoral research in the group of L. J. Gooßen at TU Kaiserslautern.

Christophe Linder was born in 1977 in Mannheim, Germany. After undergraduate studies at TU Kaiserslautern, he joined the group of L. J. Gooßen in 2005 for diploma work on *Catalytic reductive etherification of ketones with alcohols at ambient hydrogen pressure.* In May 2006, he started his PhD research on *Extensions of the decarboxylative cross-coupling of arenecarboxylates.*

Bettina Zimmermann, née Melzer was born in 1980 in Rodalben, Germany. After undergraduate studies at TU Kaiserslautern, she received her diploma in May 2006 under the supervision of L. J. Gooßen with a thesis entitled *Catalytic cross-coupling reactions of aromatic carboxylic acids*. She started her PhD research in June 2006 in the same group on *Synthetic applications of decarboxylative cross-coupling reactions*.

Thomas Knauber was born in 1982 in Neustadt an der Weinstraße, Germany. In 2002, he started to study chemistry at TU Kaiserslautern. In spring of 2006, he joined the group of L. J. Gooßen for his diploma research on the *Synthesis of Telmisartan using decarboxylative cross-couplings of phthalic acid monoesters*.

Nathan Werner received a BS degree from Southern Utah University in 2005. After which, he began doctoral work at the University of Illinois with Scott Denmark. His thesis work focuses on the development of palladium-catalyzed cross-coupling reactions of allylic silanolate salts with aryl bromides.

208

SYNTHESIS OF AMINOARENETHIOLATO-COPPER(I) COMPLEXES

A.

1. t-BuLi, -80 °C, pentane

2. 1/8 S$_8$, THF, -80 °C to -20 °C, 2.5 h
3. Me$_3$SiCl, -20 °C to rt, 2.5 h

B.

CuCl, toluene, rt

Submitted by Elena Sperotto, Gerard P.M. van Klink and Gerard van Koten.[1]
Checked by Karla Bravo-Altamirano and Peter Wipf.[2]

1. Procedure

A. *Diethyl(2-[(trimethylsilanyl)sulfanyl]benzyl)amine*. A flame-dried, 250-mL Schlenk tube (Note 1), containing a magnetic stir bar, equipped with a rubber septum and a nitrogen inlet is charged with benzyldiethylamine (6.00 g, 36.7 mmol) (Note 2). The flask is cooled in a dry ice-acetone bath to –78 °C (temperature of the bath is checked with a thermometer, internal temperature is not monitored) (Note 3) and anhydrous, degassed pentane (80 mL) (Notes 4, 5) is added by syringe. A solution of *t*-butyllithium in pentane (1.7 M, 23.8 mL, 40.4 mmol, 1.1 equiv) (Note 6) is added dropwise by syringe over 5 min, and the reaction mixture is maintained at –78 °C for another 1.5 h. The mixture is slowly allowed to reach room temperature with stirring over 16 h. The solution is concentrated *in vacuo* at 0.01 mmHg. Cold anhydrous, degassed THF (70 mL at –78 °C, kept in a dry ice-acetone bath) (Notes 5, 7) is added by syringe to the Schlenk tube which is immersed in a cold bath at -78 °C. Sublimed sulfur (1.296 g, 40.43 mmol, 1.1 equiv) (Note 8) is added to the reaction flask by temporarily removing the septum (Note 9). The reation mixture is maintained at a bath temperature of –78 °C for 2 h, then the reaction mixture is allowed to slowly warm (by adding acetone to the cooling bath) to about –20 °C (temperature of the bath). Chlorotrimethylsilane (9.8 mL, 77.2 mmol, 2.1 equiv) (Note 10) is added over 1 min by syringe and the mixture is stirred for another 2 h while slowly warming up to room temperature. The solvent is then removed *in vacuo* at 0.01 mmHg and the residue is re-dissolved in anhydrous, degassed hexanes (50 mL) (Notes 4, 5). The solvent is then removed *in vacuo* at 0.01 mmHg, the residue is re-dissolved in hexanes (50 mL), and precipitation of LiCl is observed. The suspension is transferred to a flame-dried, 100-mL centrifuge vessel via cannula (Note 1), and is separated from the precipitated LiCl by centrifugation (7 min, 28,000 rpm). A suitable alternative to centrifugation is Schlenk filtration or any other technique that allows manipulation under inert atmosphere. Finally, the hexane supernatant is removed by cannula into a flame-dried, 100-mL Schlenk tube (Note 1). The solution is concentrated *in vacuo* at 0.01 mmHg to afford 9.32 g (95%) of the product as a yellow oil (Note 11).

B. 2-*[(Diethylamino)methyl]benzene thiolato-copper(I)*. A flame-dried, 100-mL Schlenk tube (Note 1), containing a magnetic stir bar, is equipped with a rubber septum and a nitrogen inlet, and is charged with diethyl(2-[(trimethylsilanyl)sulfanyl]benzyl)amine (2.32 g, 8.69 mmol). Anhydrous, degassed toluene (20 mL) (Notes 5, 12) is added by syringe resulting in a yellow solution. A second flame-dried 100-mL Schlenk tube (Note 1), containing a magnetic stir bar, equipped with a rubber septum and a nitrogen inlet, is charged with CuCl (0.774 g, 7.82 mmol, 0.9 equiv) (Note 13) and anhydrous, degassed toluene (20 mL), to provide a white suspension. The thiolate solution is transferred by cannula to the suspension of CuCl and the mixture is stirred under nitrogen until the reaction mixture is clear and free from visible solids (Note 14). The solvent is removed at 0.01 mmHg, anhydrous, degassed hexanes (50 mL) (Notes 4, 5, 15) is added and the reaction mixture is vigorously stirred. The solution is transferred to a flame-dried, 100-mL centrifuge vessel (Note 1) by cannula and separated from the precipitate by centrifugation (7 min, 28,000 rpm). Finally, the hexanes solution is removed by cannula. The precipitate is dried *in vacuo* at 0.01 mmHg to give the product as a fine off-white powder (1.63 g, 6.32 mmol, 81%) (Note 16).

2. Notes

1. The apparatus is first flame-dried under vacuum (using an inert gas manifold with a mercury bubbler) and once it cools to room temperature, it is placed under an atmosphere of nitrogen and carefully maintained under

this atmosphere during the entire course of the reaction because the product is highly air and moisture sensitive.

2. Benzyldiethylamine was purchased from TCI America at ≥98% purity and vacuum-distilled (60 °C, 2 mmHg) before use. The submitters purchased benzyldiethylamine from Acros Organics and used it as received.

3. The submitters performed the reaction using a liquid nitrogen-ethanol bath at –80 °C.

4. n-Pentane was purchased from Acros Organics at 99+% purity, and hexanes (technical grade) was purchased from Fischer Scientific. Both solvents were distilled under nitrogen over calcium hydride before use. The submitters purchased pentane and hexane from Interchema and distilled them under nitrogen over sodium sand.

5. All anhydrous solvents used in the procedure were freeze-thaw degassed (4 cycles) prior to use.

6. t-Butyllithium (1.7 M in pentane) was used as received from Sigma Aldrich. The submitters obtained t-butyllithium (1.5 M in pentane) from Acros Organics.

7. THF was obtained from Alfa Aesar at 99+% purity and distilled under nitrogen over sodium benzophenone ketyl. The submitters purchased THF from Sigma Aldrich and distilled it under nitrogen over sodium sand.

8. Sulfur (99.50%, sublimed) was purchased from Acros Organics and used as received. The submitters obtained sulfur from Sigma Aldrich (powder 100 mesh, sublimed) and used it as received.

9. The sulfur is added in a single portion via the use of weighing paper. The submitters found it more convenient to add the sulfur in a small glass container, which is dropped into the reaction mixture.

10. Chlorotrimethylsilane (TMSCl) was purchased from Sigma Aldrich at ≥99% purity, distilled under nitrogen from quinoline (TMSCl/quinoline, 9:1) (T = 57 °C) and stored over 4 Å molecular sieves. The submitters purchased TMSCl from Acros Organics, distilled it under nitrogen and stored it over activated 4 Å molecular sieves.

11. The product displayed the following physicochemical properties: ^{1}H NMR (400 MHz, C_6D_6) δ: 0.13 (s, 9H, SSi(CH_3)$_3$), 0.94 (t, 6 H, J = 7.0 Hz, N(CH$_2$CH$_3$)$_2$), 2.44 (q, 4 H, J = 7.0 Hz, N(CH_2CH$_3$)$_2$), 3.85 (s, 2 H, CH_2N), 6.92 (m, 1 H, ArH), 7.10 (m, 1 H, ArH), 7.44 (d, 1 H, J = 7.40 Hz, ArH), 7.73 (d, 1 H, J = 7.8 Hz, ArH); ^{13}C NMR (100.5 MHz, C_6D_6) δ: 1.5 (SSi(CH_3)$_3$), 12.9 (NCH$_2$CH$_3$)$_2$), 47.9 (NCH_2CH$_3$)$_2$), 57.2 (CH_2N), 127.1 (Aryl C), 127.8, 130.3, 131.5, 136.9, 144.9. Because of the high air and

moisture sensitivity of the compound, no elemental analysis or GC-MS analysis could be obtained. Preparation of NMR samples under a nitrogen atmosphere (NMR tube kept in a Schlenk tube during preparation) allows NMR analysis and determination of purity of the sample. The solvent C_6D_6 was distilled under nitrogen over calcium hydride at 80 °C, stored over activated 4 Å molecular sieves and freeze-thaw degassed prior to use.

12. Toluene was obtained from Fischer Scientific (99.9%) and was distilled under nitrogen over sodium benzophenone ketyl. The submitters purchased toluene from Interchema and distilled it under nitrogen over sodium sand.

13. CuCl was freshly prepared and kept under nitrogen according to the following procedure: To a preheated solution of $CuSO_4 \cdot 5H_2O$ (99.9 g, 0.40 mol) and NaCl (25.96 g, 0.40 mol, 1 equiv) in 400 mL of water at 65 °C (oil bath temperature) is slowly added over 1.5 h an aqueous solution of NaOH (14.40 g, 0.36 mol, 0.9 equiv) and $NaHSO_3$ (22.89 g, 0.22 mol, 0.55 equiv). The mixture is stirred for 1 h and the temperature is kept between 65 and 70 °C. After being cooled, the reaction mixture is transferred under nitrogen onto a 250-mL Schlenk-type filter funnel (25-50 μm porosity) where the solid is washed with: an aqueous solution of acetic acid (1 mL in 1 L of water), an aqueous solution of $NaHSO_3$ (1 g in 1 L of water), acetone (800 mL), technical diethyl ether (1 L), pre-dried (on KOH) diethyl ether (1 L) and dry diethyl ether (1 L). *Very important*: always keep the solid under a layer of solvent, and never in contact with air! The resulting solid is then transferred into a pre-dried Schlenk tube and dried under vacuum, to afford copper(I) chloride as a white powder (31.7 g, 0.32 mol, 80%). This reagent must be stored under an atmosphere of nitrogen.

14. The reaction time may vary from 2 to 4 h. Under a strictly air-free manifold, a suitable alternative is to allow the reaction to proceed overnight.

15. Hexanes were used to remove all soluble impurities.

16. The product is slightly air sensitive and is stored under a nitrogen atmosphere. The product displayed the following physicochemical properties: ^1H NMR (400 MHz, C_6D_6) δ: 0.99 (t, 6 H, J = 7.0 Hz, N(CH_2CH_3)_2), 2.44 (q, 4 H, J = 6.3 Hz, N(CH_2CH_3)_2), 3.68 (br s, 2 H, CH_2N), 6.78 (d, 1 H, J = 7.0 Hz, ArH), 6.88 (t, 1 H, J = 7.0 Hz, ArH), 7.02 (t, 1 H, J = 7.0 Hz, ArH), 7.91 (d, 1 H, J = 7.4 Hz, ArH); ^{13}C NMR (100.5 MHz, C_6D_6) δ: 11.4 (N(CH_2CH_3)_2), 49.9 (N(CH_2CH_3)_2), 61.6 (CH_2N), 124.0 (Aryl C), 127.9, 132.7, 135.5, 136.4, 143.3; Anal. Calc. for $C_{11}H_{16}CuNS$: C, 51.24; H, 6.25; N, 5.43; Found: C, 50.57; H, 6.24; N, 5.33. Although the

combustion data for carbon is outside the acceptable range, it is the best result obtained by the checkers.

Safety and Waste Disposal Information

All hazardous materials should be handled and disposed of in accordance with "Prudent Practices in the Laboratory"; National Academy Press; Washington, DC, 1995.

3. Discussion

Copper(I)-mediated reactions have recently become the choice for large industrial scale applications, since copper is environmentally friendly and cheaper than other transition metals already explored. However, most organocopper compounds still present several limitations including the sensitivity towards moisture and air, the low solubility in organic solvents and the thermal instability. Many studies were performed trying to improve the activity and selectivity of copper-catalyzed reactions.[3]

One of the problems met with copper-catalyzed reactions is the solubility of the catalyst, which influences the catalytic activity and often requires the use of high temperatures and/or harsh reaction conditions. Improvements can be found, using additives or external ligands or solvents, which render the copper salts more soluble, allowing the use of milder reaction conditions.[4,5] An obvious solution to these problems is the synthesis of well-defined copper complexes, which are soluble in a broad range of solvents.

The procedure reported herein describes the synthesis of such a well-defined copper(I) complex in which a copper(I) cation is bonded to a monoanionic, potentially *S,N*–bidentate coordinating amino-thiolate ligand. The synthesis proceeds through a three-step procedure; the first step involves the heteroatom assisted *ortho*-lithiation of the arene to the corresponding organolithium compound. In the second step, the insertion of sulfur into the lithium–carbon bond of the aryllithium reagent affords the *in situ* preparation of the corresponding lithium arenethiolate. Subsequent reaction with chlorotrimethylsilane gives the trimethylsilyl-protected arenethioether, which then is reacted with CuCl in toluene to afford the desired copper(I) arenethiolate complex.

A set of modified complexes (Fig.1) can be synthesized by changing the diethylamino functionality for, *i.e.*, dimethylamino, pyrrolidinyl, piperidinyl or morpholinyl substituents. Diverse substituents can be introduced directly onto the aromatic ring (*i.e.* TMS, *t*-Bu, CF_3, OMe) and a naphthalene backbone can be used instead of a benzene unit.

These well-defined copper(I) complexes[6] exhibit good thermal stability and solubility in common organic solvents. In addition, electronic and physical properties can easily be fine tuned by introducing substituents at the arene ring or the amino-functionality.[7]

Figure 1. Selected examples of copper(I) aminoarenethiolates.

86%[a]	75%[a]	95%[a]

| [a] overall yields of steps A,B | 81%[a] | 91%[a] | 70%[a] |

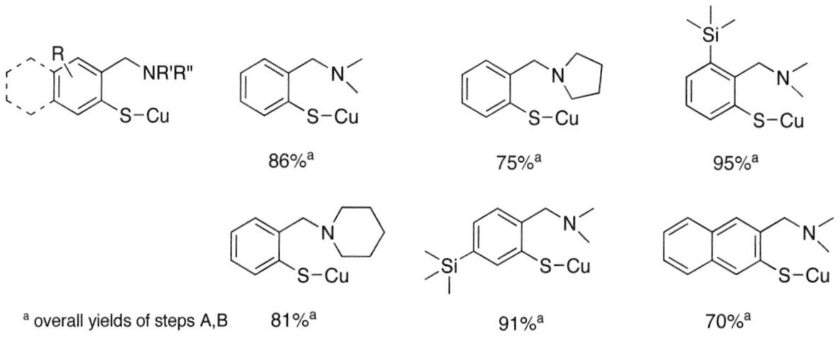

Characterization of copper(I) aminoarenethiolates by NMR and X-ray crystal structure determination, both in solution and in the solid state, showed highly aggregated species (trimers, tetramers or even nonamers).[8] The molecular structure is generally based on a core comprising a six-membered cyclohexane-like Cu_3S_3 ring, with alternating copper and sulfur atoms. The sulfur atom of each of the arenethiolate ligands bridges two copper atoms while the trigonal coordination geometry at each copper atom is attained by intramolecular N–Cu coordination.[7]

These copper-aminoarenethiolates have already been tested as catalyst (precursors) in allylic substitution,[9] 1,4-[10] and 1,6-[11] addition reactions, aromatic N-[12] and O-[13] arylation reactions, and have shown excellent catalytic properties (Figure 2).

Taking into account the actual performances and activities of these copper complexes, and their wide range of applicability, they can be considered as a useful and versatile tool in organic synthesis, and a valid soluble replacement of the common, insoluble copper(I) halide salts.

Figure 2. Application examples of aminoarenethiolato-copper(I) catalyzed reactions.

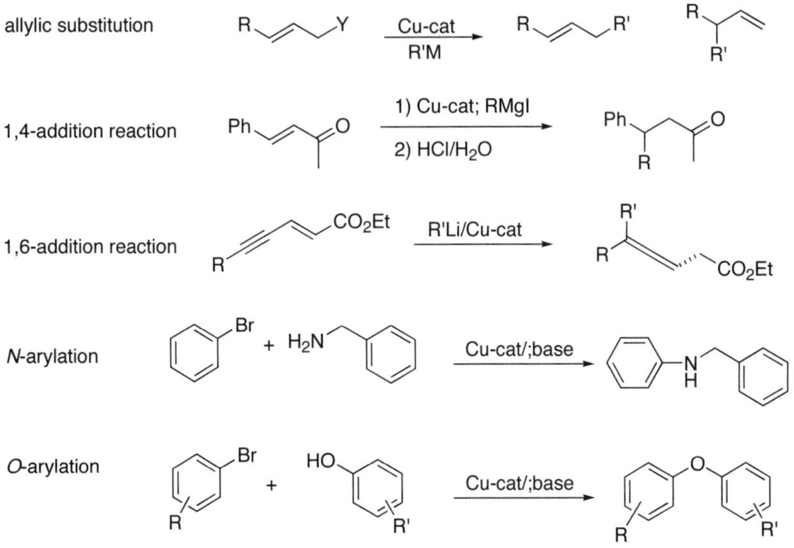

1. Organic Chemistry and Catalysis, Utrecht University, Padualaan 8, 3584 CH, Utrecht, The Netherlands. E-mail: g.vankoten@uu.nl
2. Department of Chemistry, University of Pittsburgh, Pittsburgh, PA 15260, USA. E-mail: pwipf@pitt.edu.
3. Ley, S. V.; Thomas, A. W. *Angew. Chem. Int. Ed.* **2003**, *42*, 5400-5449.
4. Beletskaya, I. P.; Cheprakov, A. V. *Coord. Chem. Rev.* **2004**, *248*, 2337-2364.
5. Kunz, K.; Scholz, U.; Ganzer, D. *Synlett.* **2003**, *15*, 2428-2439.
6. Posner, G. H.; Whitten, C. E.; Sterling, J. J. *J. Am. Chem. Soc.* **1973**, *95*, 7788-7800.
7. Knotter, D. M.; van Maanen, H. L.; Grove, D. M.; Spek, A. L.; van Koten, G. *Inorg. Chem* **1991**, *30*, 3309-3317.
8. Janssen, M. D.; Grove, D. M.; van Koten, G. *Prog. Inorg. Chem.* **1997**, *46*, 97-149.
9. (a) Persson, E. S. M.; van Klaveren, M.; Grove, D. M.; Bäckvall, J.-E.; van Koten, G. *Chem. Eur. J.* **1995**, *1*, 351-359; (b) van Klaveren, M.; Persson, E. S. M.; del Villar, A.; Grove, D. M.; Bäckvall, J.-E.; van Koten, G. *Tetrahedron Lett.* **1995**, *36*, 3059-3062; (c) Meuzelaar, G. J.;

Karlström, A. S. E.; van Klaveren, M.; Persson, E. S. M.; del Villar, A.; van Koten, G.; Bäckvall, J.-E. *Tetrahedron* **2000**, *56*, 2895-2903; (d) van Klaveren, M.; Persson, E. S. M.; Grove, D. M.; Bäckvall, J.-E.; van Koten, G. *Tetrahedron Lett.* **1994**, *35*, 5931-5934.

10. (a) Arink, A. M.; Braam, T. W.; Keeris, R.; Jastrzebski, J. T. B. H.; Benhaim, C.; Rosset, S.; Alexakis, A.; van Koten, G. *Org. Lett.* **2004**, *6*, 1959-1962; (b) van Klaveren, M.; Lambert, F.; Eijkelkamp, D. J. F. M.; Grove, D. M.; van Koten, G. *Tetrahedron Lett.* **1994**, *35*, 6135-6138; (c) Lambert, F.; Knotter, D. M.; Janssen, M. D.; van Klaveren, M.; Boersma, J.; van Koten, G. *Tetrahedron Asymmetry* **1991**, *2*, 1097-1100; (d) Knotter, D. M.; Grove, D. M.; Smeets, W. J. J.; Spek, A. L.; van Koten, G. *J. Am. Chem. Soc.* **1992**, *114*, 3400-2410.

11. Haubrich, A.; van Klaveren, M.; van Koten, G.; Handke, G.; Krause, N. *J. Org. Chem.* **1993**, *58*, 5849-5852.

12. Jerphagnon, T.; van Klink, G. P. M.; de Vries, J. G.; van Koten, G. *Org. Lett.* **2005**, *7*, 5241-5244.

13. Sperotto, E.; Jerphagnon, T.; van Klink, G. P. M.; de Vries, J. G.; van Koten, G. *In preparation.*

Appendix
Chemical Abstracts Nomenclature (Registry Number)

t-Butyllithium: (594-19-4)
Sulfur; (7704-34-9)
Trimethylsilyl chloride: Silane, chlorotrimethyl-; (75-77-4)
Copper chloride: Cuprous chloride; (7758-89-6)

Gerard van Koten retired in September 2007 and was appointed Emeritus Professor. He is currently active as Distinguished University Professor of Utrecht University and Distinguished Research Professor University of Cardiff (UK). He is known for his research on XCX-pincer metal complexes. The preparation and use of the first examples of homogeneous metallodendrimer catalysts demonstrate his interest for supramolecular systems with (organometallic) catalytically active functionalities. Recent developments involve the introduction of the XCX-pincer metal units in polypeptide chains, carbohydrates and in the active site of serine hydrolases. Currently systems are under development for cascade catalysis, which is connected to his interest in the development of sustainable (green) chemistry.

Elena Sperotto was born in 1978 in Sandrigo, Italy, and studied chemistry at the University of Padova where she received her Master degree in 2003. Afterwards, she spent one year at DSM Research in Geleen, The Netherlands, for a post-graduate research project on enzymatic resolution of amino acids. Since October 2004, she is working on her Ph.D. thesis at Utrecht University under the supervision of Prof. G. van Koten and she currently investigates copper-catalyzed carbon-heteroatom coupling reactions.

Gerard van Klink (27.8.1966) completed his Ph.D. under the supervision of Prof. F. Bickelhaupt at the Vrije Universiteit Amsterdam. He worked as a post-doctoral fellow on the design and synthesis of co-catalysts for Ziegler-Natta polymerization processes, and on low-valent catalysts for olefin polymerization in the group of Prof. J. Eisch at the State University of New York at Binghamton. From 1998 until 2007 he had a position as assistant professor in the Department of Organic Chemistry and Catalysis headed by Prof. G. van Koten. Presently, he is working in the field of metal-organic frameworks at the Delft University of Technology.

Karla Bravo-Altamirano obtained a B.S. degree in Chemistry in 2002 from Universidad de las Americas in Puebla, Mexico, where she conducted research under the supervision of Prof. Cecilia Anaya. She obtained her Ph.D. in 2007 from Texas Christian University with Prof. Jean-Luc Montchamp. Her research focused on the development of catalytic methodologies for the synthesis and functionalization of *H*-phosphinic acid derivatives and their *P*-chiral counterparts. Currently, she is a postdoctoral research associate in the group of Prof. Peter Wipf at the University of Pittsburgh where she is working on the dynamic combinatorial library generation as well as on the synthesis of suppressors of the peroxidase activity of cyt *c*/TOCL complexes and PKD inhibitors.

218

HIGH-YIELDING, LARGE-SCALE SYNTHESIS OF *N*-PROTECTED-β-AMINONITRILES: *TERT*-BUTYL (1*R*)-2-CYANO-1-PHENYLETHYLCARBAMATE

A. (*S*)-phenylglycine

1. NaBH₄ / I₂
2. (Boc)₂O, Et₃N

B.

CH₃SO₂Cl, Et₃N

C.

NaCN, DMSO

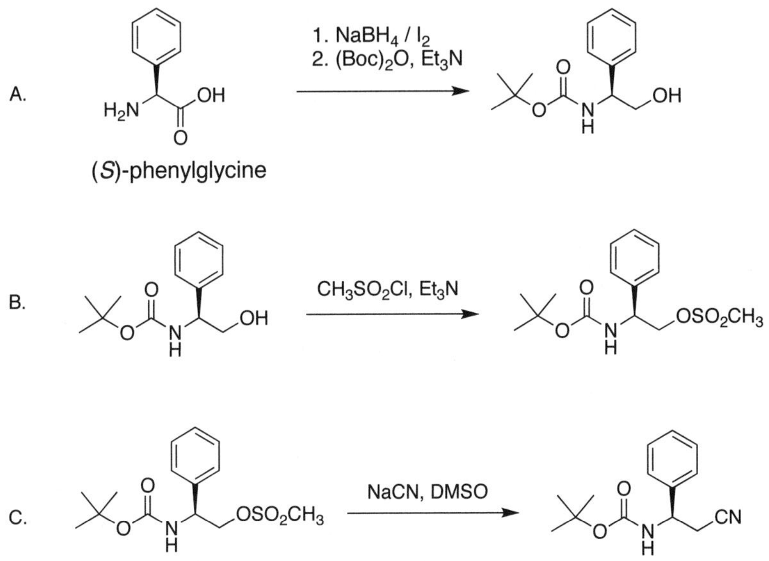

Submitted by Fathia Mosa,[1] Carl Thirsk,[1] Michel Vaultier,[2] Graham Maw, and Andrew Whiting.[1]
Checked by Robert B. Lettan II[3] and Peter Wipf.[3]

1. Procedure

A. tert-Butyl (1S)-2-hydroxy-1-phenylethylcarbamate. An oven-dried, 3-L, three-necked, round-bottomed flask is equipped with a nitrogen inlet adapter, a rubber septum, a 200-mL pressure equalizing dropping funnel (placed in the central neck of the flask and fitted with a rubber septum) and a 3.2 cm x 1.6 cm, egg shaped, Teflon-coated magnetic stirring bar. The flask is flushed with nitrogen and charged with sodium borohydride (30.0 g, 0.793 mol, 2.4 equiv) (Note 1) and 500 mL of anhydrous THF (Note 2). The resulting suspension is cooled in an ice bath and stirred rapidly. The pressure-equalizing dropping funnel is charged with a previously prepared solution of iodine (83.9 g, 0.331 mol, 1.0 equiv) (Note 3) in 150 mL of anhydrous THF (Note 2). The iodine-THF solution is added dropwise to the borohydride-THF suspension over 1.5 h (Note 4). Once addition is complete, the dropping funnel is removed and replaced by a water-cooled condenser fitted to the central neck of the flask. (S)-Phenylglycine (50.0 g, 0.331 mol) (Note 5) is then added to the mixture in small portions (Note 6). After the (S)-phenylglycine has been added, the ice-bath is replaced by an oil bath and the reaction mixture is heated at reflux under nitrogen for 18 h (Note 7). After this period, the solution removed from the oil bath and is placed in an ice-bath for 20 min, then 60 mL of methanol is very cautiously added (Note 8). After evolution of gas has ceased, the reaction mixture is diluted with 250 mL of anhydrous THF and triethylamine (48.6 mL, 35.3 g, 0.347 mol, 1.05 equiv) (Note 9) is added in one portion. With vigorous stirring, di-*tert*-butyl dicarbonate (72.9 g, 0.334 mol, 1.01 equiv) is added portionwise (Note 10), followed by 200 mL of anhydrous THF (Note 11). The mixture is allowed to warm to room temperature and was stirred for 3 h. After this time, the solvents are removed *in vacuo* (Note 12) and the resulting white solid is suspended in 400 mL of ethyl acetate and 300 mL of water and stirred vigorously. The white residues are dispersed by the gradual addition of 400 mL of a 1:1 solution of 1.2 N aqueous HCl and brine. After being stirred for 20 min, the mixture is poured into a 3-L separatory funnel and is shaken with 200 mL of ethyl acetate and the phases are separated. The aqueous phase is further extracted with ethyl acetate (2 x 200 mL) and the combined organic phases are then shaken with a 1:1 solution of 0.6 N HCl and brine (200 mL), followed by a 1:1 solution of sat. aq. sodium hydrogen carbonate and brine (200 mL), and finally, water (200 mL). The organic phase is dried over anhydrous sodium sulfate (100 g), filtered, and the solvent removed *in vacuo* (Note 12) to leave 100 mL of mother liquor. After addition of 300 mL of hexane, the solution is chilled at -20 °C for 12 h to induce precipitation. The mixture is then filtered through a 90 –mm Büchner funnel (fitted with 90-mm diameter Whatman 1 filter paper) to obtain crude *tert*-butyl (1S)-2-hydroxy-1-phenylethylcarbamate and the filtrate is kept. The crude product is purified by repeated recrystallization from dichloromethane-cyclohexane (3 crops) (Note 13) to afford 71.1 g (91%) of *tert*-butyl (1S)-2-hydroxy-1-phenylethylcarbamate as white needles after drying *in vacuo* (Note 12). The clear filtrate kept from each recrystallization is combined, concentrated *in vacuo* (Note 12), and the resulting mixture (usually an off-white solid or a sticky cream foam) purified by column chromatography (Note 14) yielding a further 2.68 g (3.4%) of *tert*-butyl

(1*S*)-2-hydroxy-1-phenylethylcarbamate after removal of solvents and drying *in vacuo* (Note 12). Total yield of *tert*-butyl (1*S*)-2-hydroxy-1-phenylethylcarbamate is 73.8 g (94%) obtained as white needles (Note 15).

   B.   *(2S)-2-[(tert-Butoxycarbonyl)amino]-2-phenylethyl   methane-sulfonate.* An oven-dried, 1-L, three-necked, round-bottomed flask is equipped with a nitrogen inlet adapter, a rubber septum, a 200-mL pressure equalizing dropping funnel fitted with a rubber septum, and a 3.2 cm x 1.6 cm, egg shaped, Teflon-coated, magnetic stirring bar. The flask is flushed with nitrogen and charged with *tert*-butyl (1*S*)-2-hydroxy-1-phenylethylcarbamate (20.0 g, 0.0843 mol), 200 mL of dry dichloromethane (Note 16), and freshly distilled triethylamine (17.4 mL, 12.6 g, 0.126 mol, 1.5 equiv) (Note 9), then is cooled in an ice bath. The dropping funnel is charged with a solution of methanesulfonyl chloride (6.86 mL, 10.1 g, 0.0885 mol, 1.05 equiv) (Note 17) and 100 mL of dry dichloromethane, which is then added to the flask dropwise over 1 h. The reaction mixture is stirred at <20 °C for 12 h (Notes 18, 19), after which time 150 mL of sat. aq. sodium hydrogen carbonate solution are added. After being stirred for a 20 min, the mixture is poured into a 1-L separatory funnel and the phases are separated. The aqueous phase is extracted with dichloromethane (2 x 100 mL) and the combined organic phases are then washed with saturated sodium chloride solution (2 x 100 mL). After drying the organic phase over anhydrous magnesium sulfate (20 g), the solvents are filtered, then are removed *in vacuo* (Notes 12 and 20) to yield 27.8 g (105%) of an off-white solid that is recrystallized from dichloromethane-hexane (Note 13) to give (2*S*)-2-[(*tert*-butoxycarbonyl)amino]-2-phenylethyl methanesulfonate (25.0 g, 94%) as a cream powder (Note 21).

   C.   *tert-Butyl (1R)-2-cyano-1-phenylethylcarbamate.* An oven-dried, 1-L, round-bottomed flask is equipped with a nitrogen inlet adapter and a 3.2 cm x 1.6 cm, egg shaped, Teflon-coated, magnetic stirring bar. The flask is charged with (2*S*)-2-[(*tert*-butoxycarbonyl)amino]-2-phenylethyl methanesulfonate (20.0 g, 0.0634 mol) and purged with nitrogen. After addition of 300 mL of anhydrous DMSO (Note 22), sodium cyanide (9.32 g, 0.190 mol, 3.0 equiv) (Note 23, **CAUTION!**) is added in one portion and the mixture is then stirred for 18 h in a thermostat-controlled oil bath set at 45 °C (Note 24). After this time, the reaction mixture is cooled to 0 °C in an ice bath, and 400 mL of water is added. The mixture is poured into a 2-L separatory funnel and the DMSO-H_2O phase is extracted with diethyl ether (3 x 200 mL) (Note 25). The combined ether extracts are washed with saturated sodium chloride solution (3 x 200 mL), water (2 x 200 mL) and then are dried over anhydrous magnesium sulfate (50 g) and then are filtered. Removal of the solvent *in vacuo* (Note 12) gives a white solid that is recrystallized from dichloromethane-hexane (Note 13) to yield *tert*-butyl (1*R*)-2-cyano-1-phenylethylcarbamate (12.0 g, 77%) as white crystals (Note 26).

2. Notes

1.   Sodium borohydride was obtained from Aldrich Chemical Company and was used as received.

2.   Tetrahydrofuran was obtained from Fisher Chemicals and was distilled under argon (at atmospheric pressure) from sodium benzophenone ketyl.

3.   Iodine (99.8%) was obtained from Fisher Chemicals and was used as received. The THF-iodine solution is best prepared by weighing the iodine into a conical flask fitted with a septum and containing a magnetic stirrer bar, flushing the flask with argon, charging with THF and stirring rapidly until the iodine has dissolved. The THF-iodine solution can be transferred to the dropping funnel by cannula, or by quickly pouring under a stream of argon.

4.   The THF-iodine solution should be added to the sodium borohydride suspension at such a rate that it reacts instantly with the borohydride and the reaction remains white. After all the iodine has been added, the mixture should have a milky white appearance.

5.   (*S*)-Phenylglycine (>98%) was obtained from Alfa Aesar and used as received.

6.   (*S*)-Phenylglycine is typically added portionwise over 1 h. Mild effervescence and a slight exotherm is observed. As a precaution, the rubber septum is removed from the condenser to allow direct venting.

7.   The nitrogen inlet is moved to the top of the condenser and a glass stopper is put in the neck of the flask. Upon heating the reaction mixture, effervescence and considerable volumes of hydrogen are evolved; adequate care should be taken to ensure this is safely vented; as a precaution, the rubber septum is removed from the condenser to allow direct venting.

8.   General reagent grade methanol can be used. Quenching with methanol is to be carried out very cautiously, adding 1 mL at a time with rapid stirring until the effervescence begins to abate. Again, a considerable

volume of hydrogen may be evolved so adequate precautions to vent should be taken.

9. Triethylamine was obtained from Fisher Chemicals and was distilled at atmospheric pressure from calcium hydride immediately prior to use.

10. Di-*tert*-butyl dicarbonate (>98%) was obtained from Advanced Asymmetrics, Inc. and was used as received. It was convenient to melt the reagent by placing the bottle in hot water, before weighing and adding it portionwise as a liquid to the reaction mixture.

11. Upon addition of the di-*tert*-butyl dicarbonate, the reaction mixture has a tendency to thicken considerably. Adding more THF should resolve any difficulties with stirring.

12. *In vacuo* denotes solvent evaporation using a Büchi rotary evaporator at a water temperature of ca. 30 °C and a ca. 15 mmHg vacuum, followed by drying under higher vacuum (ca. 0.5 mmHg).

13. As with all subsequent recrystallizations, the organic material is dissolved in the minimum quantity of the hot polar solvent (in this case refluxing dichloromethane), then is cooled, and the cold non-polar solvent (hexanes) added until the solution becomes cloudy. The solution is then refrigerated for 1-2 h, and the crystals collected by Büchner filtration and washed with cold non-polar solvent, keeping the filtrate for further re-concentration and subsequent recrystallizations, or column chromatography. Specifically, the crude *tert*-butyl (1*S*)-2-hydroxy-1-phenylethylcarbamate (Step A) was isolated from three subsequent recrystallizations: The crude material was taken up in 300 mL of hot dichloromethane, cooled to ambient temperature, and diluted with 60 mL of hexanes, to afford 61.6 g (78%) of the pure product after filtration. The remaining crude material was dissolved in 50 mL of hot dichloromethane, followed by 10 mL of hexanes, to afford 8.43 g (11%) of the pure product after filtration. Finally, the crude material was taken up again in 10 mL of hot dichloromethane, followed by 2 mL of hexanes, to afford 1.11 g (1%) of the pure product after filtration. The crude (2*S*)-2-[(*tert*-butoxycarbonyl)amino]-2-phenylethyl methanesulfonate (Step B) was dissolved in 100 mL of hot dichloromethane and then was diluted with 20 mL of hexanes, to afford 25.0 g (94%) of the pure product after filtration. The crude *tert*-butyl (1*R*)-2-cyano-1-phenylethylcarbamate (Step C) was dissolved in 30 mL of hot dichloromethane and then was diluted with 6 mL of hexanes, to afford 12.0 g (77%) of the pure product after filtration.

14. Column chromatography was performed under medium pressure. A 4.5-cm diameter column was slurry packed with 300 mL of SiO_2 (EMD silica gel 60, 230-400 mesh). The compound was loaded in the eluent and 20-mL fractions were collected. TLC was carried out on EMD silica gel 60 F_{254} glass backed plates, visualized using a potassium permanganate stain, which gave yellow stains in all cases. The elution solvent was 1:5 ethyl acetate/hexanes.

15. The product displayed the following physicochemical properties: mp 136-138 °C (cyclohexane/dichloromethane), lit. value 137-138 °C;[4a] $[\alpha]^{25}_D$ +36.1 (c 1.67, CHCl₃), lit. $[\alpha]^{22}_D$ +39.4 (c 1.67, CHCl₃);[4a] TLC R_f 0.19 (diethyl ether/hexane, 1:1); ¹H NMR (300 MHz, CDCl₃) δ: 1.44 (s, 9 H), 2.35 (bs, 1 H), 3.85 (bd, J = 3.9 Hz, 2 H), 4.78 (bs, 1 H), 5.25 (bd, 1 H, J = 6.3 Hz), 7.28-7.40 (m, 5 H); ¹³C NMR (125 MHz) δ: 28.4, 56.9, 66.9, 80.0, 126.6, 127.7, 128.8, 139.5, 156.2; IR (film) cm⁻¹: 3246, 3060-2900 (several bands), 1670, 1366, 1054; HRMS (ES+) m/z calcd for $C_{13}H_{19}NO_3Na$ (M⁺+Na): 260.1263, found 260.1273. Enantiomeric purity was determined by chiral stationary phase HPLC: er >99:1 with a Chiralpak AD-H column (hexane/2-propanol, 95:5), flow rate 1.0 mL/min; λ = 254 nm; retention times, (S)-enantiomer = 16.98 min, (R)-enantiomer = 18.77 min. Anal. Calcd. for $C_{13}H_{19}O_3N$: C, 65.80; H, 8.07; N, 5.90. Found: C, 65.30; H, 8.07; N, 5.87.

16. Anhydrous dichloromethane was obtained by atmospheric pressure distillation from calcium hydride under a nitrogen atmosphere.

17. Methanesulfonyl chloride was obtained from Sigma-Aldrich Chemical Company and was purified by atmospheric pressure distillation from calcium hydride (from Sigma-Aldrich). The purified reagent was used immediately following distillation.

18. The reaction temperature was conveniently maintained by simply allowing the ice-bath to melt and keeping the flask immersed in water. In this way, the bath temperature was found to remain at around 8-12 °C. Keeping the temperature low reduces the incidence of side-reactions (see discussion).

19. The progress of the reaction can be monitored by disappearance of the starting material TLC (R_f = 0.19, silica gel, Et₂O/hexane, 1:1)

20. Note that both organic solvent and the residual triethylamine need to be removed. Traces of triethylamine can be removed by placing the flask on a high-vacuum line for several hours after rotary evaporation.

224

21. The product displays the following physicochemical properties: mp 113-115 °C, lit. value 112-114 °C;[8] $[\alpha]^{25}_D$ +23.2 (c 1.0, CHCl$_3$), lit. value for (R)-enantiomer $[\alpha]^{25}_D$ -21 (c 1, CHCl$_3$);[9] TLC R$_f$ 0.43 (diethyl ether/hexane, 1:1); ^{1}H NMR (300 MHz, CDCl$_3$) δ: 1.44 (bs, 9 H), 2.88 (s, 3 H), 4.45, 4.48 (AB of ABX, 2 H, J_{AB} = 10.4 Hz, J_{AX} = 5.9 Hz, J_{BX} = 4.7 Hz), 5.02 (bs, 1 H), 5.19 (d, 1 H, J = 7.2 Hz), 7.31-7.41 (m, 5 H); ^{13}C NMR (75.5 MHz) δ: 28.3, 37.5, 46.0, 71.3, 80.3, 126.7, 128.3, 128.9, 137.7, 155.0; IR (film) cm^{-1}: 3381, 2979, 2937, 1699, 1520, 1391, 1357, 1173, 1052; HRMS (ES+) m/z calcd for C$_{14}$H$_{21}$NO$_5$SNa (M$^+$+Na): 338.1038, found 338.1015. Anal. Calcd. for C$_{14}$H$_{21}$NO$_5$S: C, 53.32; H, 6.71; N, 4.44. Found: C, 53.00; H, 6.75; N, 4.57.

22. Anhydrous DMSO was obtained from EMD Biosciences and was used as received.

23. Sodium cyanide was obtained from Fisher Chemicals and was used as received. Utmost care should be exercised when handling this reagent; maintain a supply of 10% aqueous bleach solution for washing any apparatus that comes into contact with it.

24. The progress of the reaction can be monitored by disappearance of the starting material TLC (R$_f$ = 0.43, silica gel, Et$_2$O/hexane, 1:1)

25. Great care should also be exercised when handling solutions (DMSO and aqueous) containing sodium cyanide, and any apparatus that has come into contact with sodium cyanide or solutions therefore. All glassware should be washed thoroughly with a 10% bleach solution prior to normal cleaning. Aqueous waste containing residual NaCN should also be treated with 10% bleach solution before disposal.

26. The product displays the following physicochemical properties: mp 110-112 °C, lit. value 112-113 °C;[4a] $[\alpha]^{25}_D$ +39.0 (c 0.5, EtOH), lit. value $[\alpha]^{25}_D$ +42.2 (c 0.45, EtOH);[4a] TLC R$_f$ 0.55 (diethyl ether/hexane, 1:1); ^{1}H NMR (300 MHz, CDCl$_3$) δ: 1.47 (s, 9 H), 2.98, 3.03 (AB of ABX, 2 H, J_{AB} = 16.4 Hz, J_{AX} = 4.5 Hz, J_{BX} = 5.8 Hz), 4.98 (bd, 1 H, J = 5.1 Hz), 5.05 (bs, 1 H), 7.34-7.42 (m, 5 H); ^{13}C NMR (75.5 MHz) δ: 25.2, 28.3, 51.3, 80.5, 117.1, 126.3, 128.6, 129.2, 138.6, 154.9; IR (film) cm^{-1}: 3377, 2980, 2940, 2249, 1685, 1519, 1367, 1274, 1252, 1170; HRMS (ES+) m/z calcd for C$_{14}$H$_{18}$N$_2$O$_2$: 246.1368, found 246.1363. Enantiomeric purity was determined by chiral stationary phase HPLC: er >99:1 with a Chiralpak AD-H column (hexane/2-propanol, 90:10), flow rate 1.0 mL/min; λ = 254 nm; retention times, (S)-enantiomer = 11.06 min, (R)-enantiomer = 13.00 min.

Anal. Calcd for $C_{14}H_{18}N_2O_2$: C, 68.27; H, 7.37; N, 11.37. Found C, 68.57, H, 7.28, N, 11.22.

Safety and Waste Disposal Information

All hazardous materials should be handled and disposed of in accordance with "Prudent Practices in the Laboratory"; National Academy Press; Washington, DC, 1995.

3. Discussion

To the best of our knowledge there are only two reported examples of one-pot amino acid reduction/N-protection sequences. Both employ lithium aluminum hydride (LAH) to achieve the reduction prior to N-protection with Boc$_2$O, affording the N-Boc-amino alcohols in 69-83% yield.[4] In all other instances, the amino alcohol, usually obtained *via* LAH, activated borohydride or mixed anhydride reductions, is isolated prior to N-protection and the often extensive aqueous work-up required following the reduction methods cited leads to diminished yields.[5] The protocol described herein employs the safer borohydride-iodine system[6] and avoids the isolation of the intermediate amino alcohol.

Documented conditions for the formation of O-sulfonylates of N-Boc-amino alcohols vary considerably, and in our hands proved unworkable or unreliable in most instances. Regarding the N-Boc-phenylglycinol system, it has to be noted that cyclization to the 4-phenyloxazolidin-2-one is a facile process occurring even at ambient temperature (particularly if less than totally pure reagents are used); indeed this is a documented method for the preparation of these useful chiral auxiliaries.[7] In this regard, we have found that use of up to two equivalents of sulfonyl chloride (methanesulfonyl of *p*-toluenesulfonyl chloride) in either pyridine or triethylamine with or without DMAP at ambient temperatures often leads to significant formation of this by-product; additionally yields are only moderate to good (55-75%).[4b] The added disadvantage is the need to remove the excess sulfonyl chloride, which is a more serious issue with *p*-toluenesulfonyl chloride. We found shorter reactions times (0.33-1.5 h) at lower temperatures (-15 to 0 °C) to be unsuitable since the reactions often fail to go to completion. Crucially, we noted that rigorous purification of reagents (particularly the sulfonyl

226

chlorides) is a prerequisite for obtaining the *O*-sulfonylates in high yields, allowing the reaction to be carried out at ambient temperature and with only one equivalent of sulfonyl chloride. Products are isolated by recrystallization.

Reported syntheses of *N*-protected β-amino nitriles are scarce; the best alternative to the route outlined above uses a β-amino iodide (obtained via reaction of iodine, imidazole and polystyryl-diphenylphospine with *N*-Boc-phenylglycinol), which undergoes displacement with tetramethylammonium cyanide.[5a] Given the cost of polystyryl-diphenylphosphine, this approach is economically unfeasible on large scale. The only reported example of displacement of an *O*-sulfonylate by cyanide in similar systems proved to be somewhat capricious in our hands.[5b]

N-Boc-β-amino-nitriles are useful intermediates affording, for example, chiral β-amino aldehydes upon reduction with DIBAL[8] (which may themselves be used to provide access to enantiopure 1,3-disubstituted *N*-Boc-1,3-aminoalcohols[9]) and chiral β-amino acids upon hydrolysis.[5a] Moreover, the *O*-sulfonylated precursors are themselves useful intermediates, with the leaving group abilities of –OMs and –OTs providing easy access to a variety of homologues.[10] The procedure described thus represents a safe and facile route toward these important chiral intermediates that should be widely applicable to other amino acids besides phenylglycine.

1. Department of Chemistry, Science Laboratories, Durham University, South Road, Durham DH1 3LE, U.K.
2. Institut de Chimie de Rennes, Campus de Beaulieu, Avenue du Général Leclerc 35042 Rennes Cedex, France.
3. Department of Chemistry, University of Pittsburgh, Pittsburgh, PA 15260, USA. E-mail: pwipf@pitt.edu.
4. (a) Correa, A.; Denis, J.-N.; Greene, A. E. *Synth. Commun.* **1991**, *21*, 1-9. (b) Cran, G. A.; Gibson, C. L.; Handa, S. *Tetrahedron: Asymmetry* **1995**, *6*, 1553-1556.
5. (a) Caputo, R.; Cassano, E.; Longobardo, L.; Palumbo, G. *Tetrahedron* **1995**, *51*, 12337-12350. (b) Sutherland, A.; Willis, C. L. *J. Org. Chem.* **1998**, *63*, 7764-7769.
6. McKennon, M. J.; Meyers, A. I.; Drauz, K.; Schwarm, M. *J. Org. Chem.* **1993**, *58*, 3568-3571.

7. (a) Lee, J.W.; Lee, J. H.; Son, H. J.; Choi, Y. K.; Yoon, G. J.; Park, M.
   H. *Synth. Commun.* **1996**, *26*, 83-88. (b) Agami, C.; Couty, F.; Hamon,
   L.; Venier, O. *Tetrahedron Lett.* **1993**, *34*, 4509-4512. (c) Agami, C.;
   Couty, F.; Hamon, L.; Venier, O. *Bull Chim. Soc. Fr.* **1995**, *132*, 808-
   814.
8. Toujas, J.-L.; Jost, E.; Vaultier, M. *Bull. Chim. Soc. Fr.* **1997**, *134*, 713-
   717.
9. Toujas, J.-L.; Toupet, L.; Vaultier, M. *Tetrahedron* **2000**, *56*, 2665-
   2672.
10. (a) Lazer, E. S.; Miao, C. K.; Wong, H.-C.; Sorcek, R.; Spero, D. M.;
    Gilman, A.; Pal, K.; Behnke, M.; Graham, A. G.; Watrous, J. M.;
    Homon, C. A.; Shah, A.; Guindon, Y.; Farina, P. R.; Adams, J. *J. Med.
    Chem.* **1994**, *37*, 913-923. (b) O'Brien, P. M.; Sliskovic, D. R.;
    Blankley, C. J.; Roth, B. D.; Wilson, M. W.; Hamelehle, K. L.; Krause,
    B. R.; Stanfield, R. L. *J. Med. Chem.* **1994**, *37*, 1810-1822.

Appendix
Chemical Abstracts Nomenclature; (Registry Number)

Sodium borohydride: Borate(1-), tetrahydro-, sodium (1:1); (16940-66-2)

Iodine; (7553-56-2)

(*S*)-Phenylglycine: Benzeneacetic acid, α-amino-, (αS)-; (2935-35-5)

Triethylamine: Ethanamine, *N*,*N*-diethyl-; (121-44-8)

Di-*tert*-butyl dicarbonate: Dicarbonic acid, *C*,*C*'-bis(1,1-dimethylethyl)
    ester; (24424-99-5

tert-Butyl (1*S*)-2-hydroxy-1-phenylethylcarbamate: Carbamic acid, *N*-[(1S)-
    2-hydroxy-1-phenylethyl]-, 1,1-dimethylethyl ester; (117049-14-6)

Methanesulfonyl chloride; (124-63-0)

(2*S*)-2-[(*tert*-Butoxycarbonyl)amino]-2-phenylethyl methanesulfonate:
    Carbamic acid, *N*-[(1S)-2-[(methylsulfonyl)oxy]-1-phenylethyl]-, 1,1-
    dimethylethyl ester; (110143-62-9)

Sodium cyanide; (143-33-9)

tert-Butyl (1*R*)-2-cyano-1-phenylethylcarbamate: Carbamic acid, [(1S)-2-
    cyano-1-phenylethyl]-, 1,1-dimethylethyl ester; (126568-44-3)

Andy Whiting carried out his undergraduate studies at the University of Newcastle upon Tyne (1978-81), followed by Ph.D. working on the synthesis β-lactam derivatives under the direction of Professor R. J. Stoodley (1981-84). After postdoctoral studies at Boston College, MA, USA working on natural product synthesis and asymmetric synthesis and developing chiral Diels-Alder Lewis-acid catalysts with Prof. T. R. Kelly (1984-86), he then worked for Ciba-Geigy plc. in Central Research UK. After a short while, he then moved into academia in the Chemistry Department at UMIST in 1989, before moving to a Readership at Durham University in 2001.

Mrs. Fathia A. Mosa obtained her B.Sc. in Chemistry in 1993 from Garyounis University (Libya) and carried out her M.Sc in 2000 at the same university. She did her M.Sc. thesis under the supervision of Professor Natiq A. R. Hatam on the isolation and spectroscopic elucidation of the constituents of Origanum majorana. She worked at Benghazi High Institute for Education (2001 to 2003) and Altahdi University (2003 to 2006) as an Assistance Lecturer and she is currently studying for her Ph.D. at Durham University under the supervision of Dr. Whiting in the area of natural product synthesis.

Carl Thirsk graduated from the University of Manchester Institute of Science and Technology (UMIST) in 2000 with an MChem in medicinal chemistry, before studying for his Ph.D. under the supervision of Dr. Whiting. His thesis title was 'Stereoselective routes to the polyene macrolide viridenomycin' and he was awarded a Ph.D. from Durham University in 2003. His research interests are focused on the concurrent application of synthetic and physical chemistry with chemical engineering science to aid the determination of reaction mechanism. Carl is a co-founder and CEO of LyraChem Limited.

Michel Vaultier was borned in Normandy (France). He received his Ph.D. degree from Rennes University in 1977 under the direction of Professor Robert Carrié. He was appointed at CNRS as an "Attaché de Recherche" in 1975. He joined Professor Barry Trost's group at the University of Wisconsin (Madison) for a one year postdoctoral stay in 1978-1979. He is currently Director of Research at CNRS. His main research interests at the moment are devoted to sustainable chemistry, the use of task specific ionic liquids in supported organic synthesis and the development of useful synthetic methods based on boron chemistry.

Graham Maw received a bachelor's degree in Chemistry from University of Leeds, UK in 1982. He then worked for three years in drug discovery at Hoechst Pharmaceutical Research Laboratories in Milton Keynes, UK before leaving to study for his Ph.D. in natural product synthesis with Professor Steven V. Ley CBE FRS at Imperial College, London (1989). He then went to University of Rochester as a postdoctoral fellow to study with Professor Robert K. Boeckman Jr., and subsequently joined the Sandwich laboratories of Pfizer Global Research & Development in 1991 where he is currently Senior Director, Head of Pain Chemistry.

Robert B. Lettan II did his undergraduate work at Otterbein College in Ohio and received his B. S. Chemistry degree in 2002. Subsequently, he obtained a Ph.D. degree at Northwestern University under the supervision of Prof. Karl Scheidt. His Ph.D. research involved the use of Lewis base activation of triethoxysilylalkynes and the development of a large-scale preparation of complex tertiary alcohols. Currently, he is a postdoctoral research associate in the group of Prof. Peter Wipf at the University of Pittsburgh where he is working on the use of bicyclobutanes in an approach toward the heterocyclic core structure of mubironine.

STEREOSELECTIVE SYNTHESIS OF (*E*)-2,3-DIBROMOBUT-2-ENOIC ACID
(2-Butenoic acid, 2,3-dibromo-, (2*E*)-)

1

Submitted by Samuel Inack Ngi, Elsa Anselmi, Mohamed Abarbri, Sandrine Langle, Alain Duchêne, Jérôme Thibonnet.[1]
Checked by Vikram Bhat and Viresh H. Rawal.

1. Procedure

Caution! All operations should be performed in a well-ventilated hood, and care should be taken to avoid skin contact with bromine.

An oven-dried, 100-mL, two-necked, round-bottomed flask protected from light (Note 1), equipped with a magnetic stir bar, a thermometer and a pressure-equalizing dropping funnel (fitted with a rubber septum and an argon inlet needle) is charged with tetrolic acid (3.0 g, 0.036 mol) (Note 2) and methanol (10 mL) (Note 3). The mixture is cooled to –10 °C (internal temperature) in an ice-salt bath.

Bromine (3.70 mL, 11.51 g, 0.072 mol, 2.0 equiv) (Note 4) is added dropwise via the addition funnel over 25 min (Note 5) while the reaction mixture is vigorously stirred. Complete bromine transfer is achieved by rinsing the addition funnel with 2 mL of methanol. The resulting dark-red solution is stirred for an additional 15 min while being cooled in the ice-salt bath. The reaction mixture is quenched by addition of 1.32 M aqueous sodium metabisulfite solution (30 mL) over 5-7 min (Note 6). The mixture is poured into a 250-mL separatory funnel and is rapidly extracted with Et_2O (4 x 50 mL) (Note 7). The combined organic layers are washed with brine (20 mL), dried over anhydrous $MgSO_4$ (7 g), filtered through a medium porosity fritted glass funnel, and concentrated on a rotary evaporator (room temperature, 25 mmHg) to provide a yellow solid (7.61 g, 0.031 mol, 87%). In a 200-mL beaker (Note 8) 7.61 g of crude product is dissolved in dichloromethane (10 mL) (Note 9). Hexanes (10 mL) is added and the solution is kept in a fume hood at room temperature for 15 h. Much of the solvent gradually evaporates over this period, leaving a white crystalline product in an orange liquid. The liquid (ca. 3 mL) is removed using a Pasteur pipette and the product is transferred to a Büchner funnel lined with Whatman filter paper (2 layers of Grade 1, 50 mm diameter) and is washed with 3 x 50 mL of petroleum ether (Note 9). The product is then dried for 2 h under vacuum (< 1 mmHg, room temperature) to yield 6.69 g (76%) of (E)-2,3-dibromobut-2-enoic acid (1) as colorless, monoclinic crystals (Note 10).

2. Notes

1.   The flask was protected from light with aluminum foil. All procedures were carried out in a room with overhead lights turned off.
2.   Tetrolic acid was purchased from Aldrich Chemical Co. and was used it as received.
3.   Methanol (99.8%) was purchased from Acros and was used as received.
4.   Bromine (>99.8%) was purchased from Acros and was used as received. *CAUTION: Bromine is extremely corrosive and must be handled with great care and always in a fume hood.*
5.   Because the bromination is exotherimic, the rate of addition of bromine was such that the internal temperature of the reaction mixture never exceeded –5 °C. Submitters found up to 6% of the Z-isomer was formed when the internal temperature was allowed to rise above 0 °C.
6.   Quenching the reaction was exothermic as well. Sodium metabisulfite solution was added dropwise using the addition funnel such that the internal temperature 7was allowed to rise gradually to 20 °C. The reaction mixture turned pale yellow when the excess bromine was quenched.
7.   The flasks, including the separatory funnel, were protected from light with aluminum foil. The product was found to be unstable under light when in solution.
8.   The 200 mL beaker (10.2 cm high with 5.8 cm diameter) was wrapped with aluminum foil and was covered with perforated aluminum foil (ca. 40 holes were made using a PrecisionGlide 16G1 ½ needle).
9.   Dichloromethane was purified by percolation through activated alumina. Hexanes (anhydrous, 99.9%) was obtained from Acros Petroleum

ether (bp range 36–60 °C) was obtained from Fisher Chemicals and was used as received.

10. (*E*)-2,3-Dibromobut-2-enoic acid (**1**), displayed the following physicochemical properties: mp 91–93 °C; IR (KBr) cm^{-1}: 3400-2300, 1703, 1600, 1274; ^{1}H NMR (500 MHz, CDCl$_3$ with 0.03% TMS) δ: 2.58 (3 H, s); 11.76 (1 H, bs); ^{13}C NMR (125 MHz, CDCl$_3$) δ: 30.2, 107.4, 126.2, 168.9; MS *m/z* (EI) 244 (M$^+$, 49), 165 (99), 163 (100), 135 (21), 79 (23), 67 (22), 55 (24), 45 (26); HRMS calcd for C$_4$H$_3$^{79}Br$_2$O$_2$ (M$^+$–H): 240.8494, found 240.8505; Anal. Calcd. For C$_4$H$_4$Br$_2$O$_2$: C, 19.70; H, 1.65; Br 65.53; O, 13.12. Found: C, 19.54; H, 1.73; Br, 65.24; O, 13.02.

Safety and Waste Disposal Information

All hazardous materials should be handled and disposed of in accordance with "Prudent Practices in the Laboratory"; National Academy Press; Washington, DC, 1995.

3. Discussion

The synthesis of α,β-dihalogeno-unsaturated systems has already been fully described and selectively achieved by addition of dihalogens on acetylenic systems.[2-14] On the other hand, little information is available regarding the synthesis of unsaturated α, β-dihalogeno carboxylic acids.[2-7, 15] Studies published up to now focus mainly on α, β-acetylenic ester dihalogenation by direct addition of dihalogen,[8-12] or in the presence of halonium ion sources.[6, 7] Direct site- and stereoselective dibromination of tetrolic acid is an easy procedure that can be carried out with good results and without secondary products.[16] The reaction is not seem effected by solvent, and similar yields were obtained when using MeOH, Et$_2$O, or CHCl$_3$. Maintaining the reaction temperature below 0 °C during the addition was imperative; above that temperature an *E/Z* mixture of the product was obtained. The crude dibrominated acid obtained before crystallization was pure enough for subsequent coupling reactions.

The same procedure can be applied to certain other α,β-acetylenic acids, and the results are summarized in Table 1.[16]

Table 1 - Stereoselective synthesis of (*E*)-2,3-dibromoalk-2-enoic acid

Entry	Alkyne	Solvent	Product	Yield, %
1	≡—COOH	MeOH or Et$_2$O or CHCl$_3$	**1**	87
2	C$_5$H$_{11}$—≡—COOH	MeOH or CHCl$_3$	**2**	97
3	MeO—≡—COOH	MeOH or neat	**3**	87

The *E* configuration of the double bond in the products obtained was proven by X-ray analysis.[16]

Unfortunately, this procedure does not give good results with propiolic acid, for which a mixture of brominated alkane and alkene products are formed.[9]

1. Laboratoire de Physicochimie des Matériaux et des Biomolécules, EA 4244, Faculté des Sciences de Tours, Parc de Grandmont, 37200 Tours (France).
2. Andersson, K. *Chem. Scr.* **1972**, *2*, 113-116.
3. Larson, S.; Luidhardt, T.; Kabalka, G. W.; Pagni, R. M. *Tetrahedron Lett.* **1988**, *29*, 35-36.
4. Rappe, C.; Andersson, K. *Ark. Kemi* **1965**, *24*, 303-313.
5. Berthelot, J.; Fournier, M. *Can. J. Chem.* **1986**, *64*, 603-607.
6. Castro, C. E.; Gaughan, E. J.; Owsley, D. C. *J. Org. Chem.* **1965**, *30*, 587-592.
7. Pagni, R. M.; Kabalka, G. W.; Boothe, R.; Gaetano, K.; Stewart, L. J.; Conaway, R.; Dial, C.; Gray, D.; Larson, S.; Luidhardt, T. *J. Org. Chem.* **1988**, *53*, 44774482.
8. Kishida, Y.; Nakamura, N. *Chem. Pharm. Bull.* **1969**, *17*, 2424-2435.

Org. Synth. **2008**, *85*, 231-237

9. Myers, A. G.; Alauddin, M. M.; Fuhry, M. A. M.; Dragovich, P. S.; Finney, N. S.; Harrington, P. M. *Tetrahedron Lett.* **1989**, *30*, 6997-7000.

10. Rossi, R.; Bellina, F.; Carpita, A.; Gori, R. *Gazz. Chim. Ital.* **1995**, *125*, 381-392.

11. Rossi, R.; Bellina, F.; Carpita, A.; Mazzarella, F. *Tetrahedron* **1996**, *52*, 4095-4110.

12. Rossi, R.; Bellina, F.; Mannina, L. *Tetrahedron Lett.* **1998**, *39*, 3017-3020.

13. Andersson, K. *Chem. Scr.* **1972**, *2*, 117-120.

14. Lu, X.; Zhu, G.; Ma, S. *Chin. J. Chem.* **1993**, *11*, 267-271.

15. Langle, S. Synthèse et réactivité des systèmes insaturés dihalogénes. Application en synthèse organique. Chemistry, Université François Rabelais, Tours, 2004.

16. Langle, S.; Ngi, S. I.; Anselmi, E.; Abarbri, M.; Thibonnet, J.; Duchêne, A. *Synthesis* **2007**, 1724-1728.

Appendix
Chemical Abstracts Nomenclature (Collective Index Number);
(Registry Number)

(*E*)-2,3-Dibromobut-2-enoic acid: (2-Butenoic acid, 2,3-dibromo-, (2*E*)- (9); (24557-17-3)

(*E*)-2,3-Dibromooct-2-enoic acid: (2-Octenoic acid, 2,3-dibromo- (4); (861041-43-2)

4-Methoxy-2-butynoic acid: 2-Butynoic acid, 4-Methoxy (9); (24303-64-8)

Tetrolic acid: 2-Butynoic acid (9); (590-93-2)

Pentylpropiolic acid: 2-Octynoic acid (6, 7, 8, 9); (5663-96-7)

Bromine (8, 9); (7726-95-6)

Jérôme Thibonnet, born in 1971 in Joinville (France), received his Ph.D. in 1999 from the University of Tours working on the total synthesis of retinoids under the supervision of Professor Alain Duchêne. After post-doctoral studies in the laboratory of Professor Paul Knochel at University of Munich, he joined the University of Aix-Marseille III as ingénieur de recherche at the laboratory of Professor Maurice Santelli (2000). He became Lecturer at the University of Tours in 2003. His research interests include new synthetic methods development of new organotin reagents and total synthesis of natural compounds.

Samuel Inack Ngi was born in Douala, Cameroon, 1977. He obtained his Master 2 of Organic Chemistry at the University of Orléans in 2005. He made his practical training under the guidance of Dr. Jérôme Thibonnet in the Organic Chemistry Laboratory of the Chemistry Department of Sciences at the University of Tours. During the same year he was allowed to begin his Ph.D. at the University of Tours under the direction of Pr. Alain Duchêne and Dr. Jérôme Thibonnet. His research focuses on the synthesis of heterocycles using reactions catalyzed by copper, in order to produce natural biologically active compounds or analogous.

Elsa Anselmi was born in Suresnes (France, Hauts-de-Seine) in 1974. She obtained her Ph.D. degree in 2000 from the "University of Versailles St-Quentin en Yvelines" under the guidance of Dr. Claude Wakselman and Jean-Claude Blazejewski on the development of new trifluoromethylation methods. After three years of post-doctoral studies on the synthesis of new fluorinated or deuterated monomers and polymers (CEA-Monts), she became Lecturer at the University of Tours in 2004. Her research interests include new synthetic methods development of new organotin reagents, new monomers and total synthesis of natural compounds.

Mohamed Abarbri was born in 1966 in Alhoceima (Marocco). He obtained, in 1995, his Ph.D. in Chemistry at the University of Tours. After post-doctoral studies in the laboratory of Professor Paul Knochel at the University of Marbourg, he became Lecturer (1996) and full professor in 2002 at the University of Tours. His research interests include new synthetic methods, development of new organotin reagents, ultra-sound chemistry and synthesis of perfluoro compounds.

Sandrine Langle was born in 1977 in Tours (France), obtained her Ph.D. in 2004 from the University of Tours working on the synthesis and reactivity of dihalogeno-unsaturated systems. After post-doctoral studies in the Service Hospitalier Frédéric-Joliot (CEA) at Orsay in radiochemistry, she became Lecturer at the University of Nancy in 2006. Her research interests include new prosthetic groups and radiotracers development for PET imaging and radiolabelling of molecules with 18-fluorine.

Alain Duchêne was born in Tours (France) in 1947. He studied organic chemistry at the Faculty of Sciences, Bordeaux I University. After post-doctoral studies in the laboratory of Professor Ryoji Noyori at the University of Nagoya (1992) he was promoted full professor in 1997 at the University of Tours. His research interests include new synthetic methods development of new organogermanium or tin reagents, ultra-sound chemistry and synthetic organic materials.

Vikram Bhat was born in 1981 in Ajmer, India. He received his undergraduate education at the Indian Institute of Technology, Bombay, Mumbai. In 2005 he entered the graduate program at the University of Chicago where he joined the research group of Professor Viresh H. Rawal. Currently, his graduate research focuses on the total synthesis of welwitindolinone alkaloids.

(R)-2,2'-BINAPHTHOYL-(S,S)-DI(1-PHENYLETHYL) AMINOPHOSPHINE. SCALABLE PROTOCOLS FOR THE SYNTHESES OF PHOSPHORAMIDITE (FERINGA) LIGANDS

Submitted by Craig R. Smith, Daniel J. Mans and T. V. RajanBabu.[1]
Checked by Scott E. Denmark and Son T. Nguyen.

1. Procedures

Caution! *Decomposition of excess PCl₃ (step A) should be carried out carefully in a fume-hood because hydrogen chloride is evolved in the highly exothermic hydrolysis reaction.*

    A. *(R)-(–)-(1,1'-Binaphthalene-2,2'-dioxy)chlorophosphine [1].*[2] A 250-mL, single-necked round-bottomed flask, equipped with a magnetic stirring bar, a reflux condenser, a three-way flow-controlled stopcock is flame-dried and purged with nitrogen. All joints are greased. The nitrogen outlet is connected to a gas scrubber trap (Note 1). The flask is charged with (R)-(+)-1,1'-bi(2-naphthol) (7.0 g, 24.5 mmol) (Note 2) and phosphorus trichloride (20.5 mL, 32.3 g, 235 mmol, 9.6 equiv) (Note 3) followed by 1-methyl-2-pyrrolidinone (20 µL, 0.2 mmol, 0.008 equiv) (Note 4). The flask is placed in an oil bath preheated to 92 °C. The solid dissolves within 5 min.

Org. Synth. **2008**, *85*, 238-247
Published on the Web 6/11/2008

The flask is swirled to dissolve the solid on the side. During this period, HCl evolves rapidly and the solution becomes yellow. The solution is kept at reflux for 10 min, then the oil bath is removed and the reaction mixture is allowed to cool to ambient temperature. The remaining phosphorus trichloride and HCl are removed under reduced pressure (7 mmHg) (Note 5) for 2 h, leaving a foamy wax. After venting the flask with nitrogen, diethyl ether (30 mL) (Note 6) is added to dissolve the solid. The solvent and traces of phosphorus trichloride are removed under reduced pressure (7 mm Hg) for 2 h, leaving a pale-yellow solid. The flask is connected to high vacuum (0.05 mm Hg) for 12 h to afford 9.08 g (106 %) of the desired product as off-white powder containing traces of diethyl ether (Note 7 and 8).

B.   *(R)-2,2'-Binaphthoyl-(S,S)-di-(1-phenylethyl)aminoylphosphine [3].*[3]   A 250-mL, three-necked flask equipped with a rubber septum, nitrogen inlet, and a temperature probe (Note 9) is flame-dried and purged with nitrogen. The flask is charged with anhydrous THF (35 mL) (Note 10) and (–)-bis-[(S)-1-phenylethyl]amine (**2**, 1.69 mL, 7.39 mmol) (Note 11). The solution is cooled to –78 °C in a dry ice-acetone bath. *n*-Butyllithium (4.62 mL, 7.39 mmol, 1.6 M solution in hexanes, 1.0 equiv) (Note 12) is added dropwise over a 15 min period maintaining the internal temperature below –68 °C until the addition is complete, resulting in a pale-pink solution (Note 13). The contents of the flask are warmed to –30 °C in about 30 min and immediately cooled to –78 °C for an additional 1 h, which results in the solution becoming dark pink. A solution of (*R*)-(-)-1,1'-binaphthyl-2,2'-dioxychlorophosphine (**1**, 2.85 g, 8.13 mmol, 1.1 equiv) in dry THF (10 mL) is then introduced dropwise *via* syringe maintaining the solution temperature below –68 °C. The solution is maintained at –78 °C for an additional 2 h before it is warmed to ambient temperature and allowed to stir for 12 h. The solvent is removed by rotary evaporation (23 °C, 20 mmHg) to afford a pale-yellow oil mixed with a white solid (Note 14). The residue is dissolved in 5 mL of dichloromethane, then is loaded onto a silica gel column (150 g, 50 cm x 5 cm) and is eluted with 2 L of pentane/dichloromethane, 4:1 (Note 15). Removal of the solvent from the eluate by rotary evaporation (23 °C, 20 mmHg) followed high vacuum (~ 0.05 mm Hg) affords 3.43 g (86%) of (*R*)-2,2'-binaphthoyl-(*S,S*)-di(1-phenylethyl)aminoyl-phosphine (**3**) as a white foam, mp 106-110 °C (Note 16). This product can be crushed into a powder and weighed as needed.

2. Notes

1. The gas scrubber trap contains 200 mL of aq. 6 N KOH solution.

2. (*R*)-(+)-1,1'-Bi-(2-naphthol), 99%, was purchased from Alfa Aesar and was used as received.

3. Phosphorus trichloride, Reagentplus®, 99%, was obtained from Sigma-Aldrich Company and was fractionally distilled under nitrogen before use. Excess PCl$_3$ ensures complete dissolution of the binaphthol.

4. 1-Methyl-2-pyrrolidinone 99.5% was purchased from Sigma-Aldrich Company and was used as received. The submitters employed a procedure that did not use 1-methyl-2-pyrrolidone and the reaction time was 16 h.

5. Three liquid nitrogen traps were used.

6. Diethyl ether (Fisher, BHT stabilized, HPLC grade) was dried by percolation through two columns packed with neutral alumina under a positive pressure of argon.

7. (*R*)-(-)-1,1'-Binaphthyl-2,2'-dioxychlorophosphine when stored in a dry nitrogen atmosphere is stable in excess of six months at ambient temperature.

8. The product displayed the following physicochemical properties: ^{1}H NMR (400 MHz, CDCl$_3$) δ: 8.02 (d, *J* = 8.80 Hz, 2 H), 7.97 (dd, J_1 = 8.20 Hz, J_2 = 3.40 Hz, 2 H), 7.55 (dd, J_1 = 8.80 Hz, J_2 = 0.80 Hz, 1 H), 7.52-7.45 (m, 3 H), 7.45-7.41 (m, 2 H), 7.35-7.29 (m, 2 H); ^{13}C NMR (100.6 MHz, CDCl$_3$) δ: 147.8, 147.2, 132.7, 132.4, 131.9, 131.5, 130.9, 130.1, 128.5, 127.0, 126.9, 126.7, 126.5, 125.7, 125.5, 124.44, 124.38, 123.1, 121.6, 121.1; ^{31}P NMR (101.33 MHz, CDCl$_3$) δ: 178.5.

9. A digital thermometer (Omega Instruments Digicator Model 400A with J-type thermocouple leads) was used.

10. Reagent-grade tetrahydrofuran (Fisher, BHT stabilized, HPLC grade) was dried by percolation through two columns packed with neutral alumina under a positive pressure of argon.

11. The submitters employed (-)-*bis*-[(*S*)-1-phenylethyl]amine from Sigma-Aldrich Company. Because the free base was not available, the checkers prepared (-)-*bis*-[(*S*)-1-phenylethyl]amine from (-)-*bis*-[(*S*)-1-phenylethyl]amine hydrochloride (97%; Sigma-Aldrich Company) using the following procedure. A 500-mL, round-bottomed flask equipped with a magnetic stir bar is charged with (-)-bis-[(*S*)-1-phenylethyl]amine hydrochloride (3.3 g, 12.6 mmol), aq. 6 N KOH solution (60 mL, 360 mmol,

Org. Synth. **2008**, *85*, 238-247

28 equiv) and methyl *tert*-butyl ether (MTBE, 60 mL, ACS grade, Sigma-Aldrich Company). The mixture is stirred vigorously for 6 h. The biphasic mixture is poured into a 500-mL separatory funnel and the aqueous layer is separated. The aqueous layer is extracted with MTBE (2 x 40 mL). The organic extracts are combined, dried over anhydrous $MgSO_4$ (10 g), filtered through a sintered glass funnel (6 cm x 6 cm), which is subsequently rinsed with MTBE (20 mL). The solvent is removed via rotary evaporation (23 °C, 20 mmHg). The product is purified by Kugelrohr distillation (air bath temperature 150 °C, 0.15 mmHg) to provide 2.8 g (98%) of the free base as a clear colorless liquid with the following physicochemical properties: ^{1}H NMR (400 MHz, $CDCl_3$) δ: 7.37-7.22 (m, 10 H), 3.52 (q, J = 6.6 Hz, 2 H), 1.59 (s (broad), 1 H), 1.29 (d, J = 6.6 Hz, 6 H); ^{13}C NMR (100.6 MHz, $CDCl_3$) δ: 145.8, 128.4, 126.6, 55.0, 25.0; IR ($CHCl_3$) cm^{-1}: 3082, 3061, 3025, 2960, 2924, 2863, 1602, 1492, 1469, 1451, 1368, 1202; MS (EI, 70 eV) *m/z* (%): 225 (2), 210 (58), 148 (3), 120 (10), 105 (100) 77 (17); HRMS (EI) calcd for $C_{16}H_{19}N$: 225.1517. Found: 225.1520. Anal. Calcd. for $C_{16}H_{19}N$: C, 85.28; H, 8.50; N, 6.22. Found: C, 84.90; H, 8.60; N, 6.36. $[\alpha]_D$ −170 (*c* 3.5, MeOH).

12. *n*-Butyllithium was obtained as a solution in hexanes from Sigma-Aldrich Company and was doubly titrated by the method of Gilman.

13. The initial addition of the *n*-BuLi may cause a rapid rise in temperature. Addition may need to be temporarily suspended upon onset of the exotherm.

14. The submitters removed the small amount of solid by filtering the reaction mixture through Celite (30 g) and rinsing with diethyl ether (3 x 80 mL) before the rotary evaporation.

15. Silica gel (Merck, grade 9385, mesh 230-400) was purchased from Sigma-Aldrich Company. Product was isolated from sixteen 50-mL fractions. The product, R_f = 0.10 (silica gel, pentane/dichloromethane, 4:1), was visualized by UV and $KMnO_4$.

16. The product displayed the following physicochemical properties: ^{1}H NMR (400 MHz, $CDCl_3$) δ: 7.93 (d, J = 8.80 Hz, 2 H), 7.88 (dd, J_1 = 7.80 Hz, J_2 = 4.20 Hz, 2 H), 7.58 (d, J = 8.80 Hz, 1 H), 7.43 (d, J = 8.80 Hz, 1 H), 7.40-7.34 (m, 3 H), 7.28 (d, J = 8.00 Hz, 1 H), 7.25-7.16 (m, 2 H), 7.15-7.06 (m, 10 H), 4.49 (m, 2 H), 1.72 (d, J = 6.8 Hz, 6 H); ^{13}C NMR (100.6 MHz, $CDCl_3$) δ: 150.1, 150.0, 149.5, 142.8, 132.8, 132.7, 131.4, 130.4, 130.2, 129.4, 128.3, 128.1, 127.9, 127.7, 127.1, 127.07, 126.6, 126.0, 125.9, 124.7, 124.4, 124.1, 124.0, 122.4, 122.3, 121.73, 121.71, 52.3, 52.2,

21.9; ^{31}P NMR (101.3 MHz, CDCl$_3$) δ: 145.6; IR (thin film) cm^{-1}: 3069, 2971, 1617, 1590, 1506, 1495, 1375, 1326, 1231, 1203, 1134, 1070. MS (ES, 70 eV) *m/z* (%): 539 (1), 538 (2), 450 (6), 434 (100), 315 (8), 268 (15), 239 (7), 105 (28). Anal. Calcd. for C$_{36}$H$_{30}$NO$_2$P: C, 80.13; H, 5.60; N, 2.60. Found: C, 80.37; H, 5.63; N, 2.89. [α]$_D$ –500.7 (*c* 1.8, EtOH).

Safety and Waste Disposal Information

All hazardous materials should be handled and disposed of in accordance with "Prudent Practices in the Laboratory"; National Academy Press; Washington, DC, 1995.

3. Discussion

Phosphoramidite ligands derived from biaryldiphenols were originally introduced by Feringa for the asymmetric Cu-catalyzed conjugate addition of dialkylzinc reagents to enones.[3,4a] These ligands have been found to have extensive applications in asymmetric catalysis. Representative examples since the publication of Feringa's review[4b] include, intramolecular Heck reactions,[5] Cu-catalyzed S$_N$2' substitutions of allylic halides with Grignard reagents[6] and addition of organo-aluminum reagents to enones (Cu-catalyzed)[7] and aldehydes (Ni-catalyzed)[8], Rh[9]- and Ir[10] -catalyzed asymmetric hydrogenation of enamides, Ir-catalyzed allylic alkylations,[11] aminations,[12] and etherifications,[13] Rh-catalyzed arylation of enones[14] and imines,[15] asymmetric Pd-catalyzed diamination of conjugated dienes,[16] Rh-catalyzed [2+2+2] cycloaddition of alkenyl isocyanates and terminal alkynes,[17] asymmetric hydrovinylation of vinylarenes,[18] strained alkenes,[18b,19] and 1,3-dienes.[20]

In general, two procedures starting with phosphorus(III) chlorides have been employed for the synthesis of phosphoramidites.[21] Feringa used a phosphoryl chloride [(RO)$_2$P-Cl], prepared by the base-mediated reaction of a diol with PCl$_3$ followed by nucleophilic substitution of the chloride with a secondary amine, or the corresponding lithium amide.[3] Alternatively a series of biphenol-based phosphoramidites have been prepared by the reaction of R$_2$NPCl$_2$ with the requisite biphenol in the presence of a tertiary amine.[4c] The procedure described here follows the former protocol with minor revision starting with a high-yielding synthesis of the phosphoryl chloride from the diol, neat phosphorus trichloride and catalytic amount of

1-methyl-2-pyrrolidinone,[2a,c] followed by the use of a lithium amide to effect the P-N bond formation. A number of other phosphoramidites, including several new ones, were prepared by this general procedure. These, along with the yields (starting from the diphenol) are shown in Figure 1. Use of one of these ligands for asymmetric hydrovinylation is illustrated in the following *Organic Syntheses* procedure.[18e]

Figure 1. Structures and yields (from the corresponding diphenol) of phosphoramidites synthesized by the two-step procedure (shown in brackets are the best reported yields from literature)

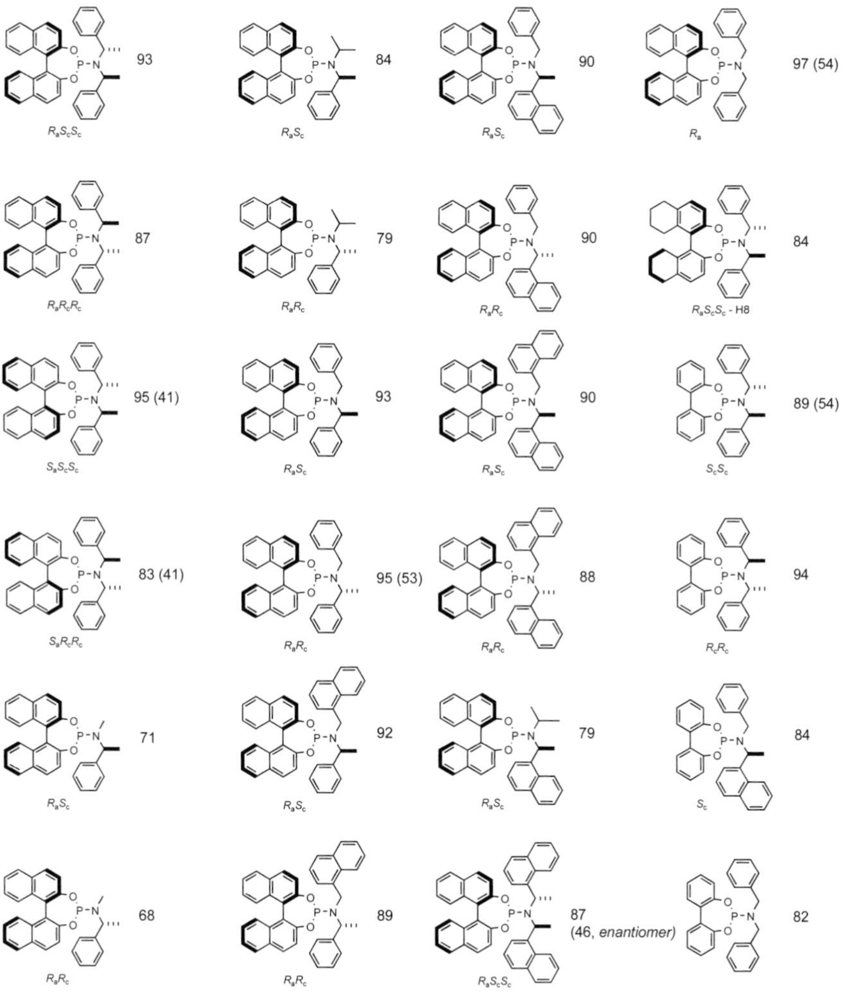

1. Department of Chemistry, The Ohio State University, 100 W. 18th Avenue, Columbus, OH 43210.

2. (a) Lucas, H. J.; Mitchell, F. W., Jr., Scully, C. N. *J. Am. Chem. Soc.* **1950**, *72*, 5491-5497. (b) Green, N.; Kee, T. P. *Synth. Commun.* **1993**, *23*, 1651-1657. (c) Cramer, N.; Laschat, S.; Baro, A. *Organometallics*, **2006**, *25*, 2284-2291.

3. Arnold, L. A.; Imbos, R.; Mandoli, A.; de Vries, A. H. M.; Naasz, R.; Feringa, B. L. *Tetrahedron*, **2000**, *56*, 2865-2878.

4. (a) Feringa, B. L.; Pineschi, M.; Arnold, L. A.; Imbos, R.; de Vries, A. *Angew. Chem., Int. Ed. Engl.* **1997**, 36, 2620-2623. (b) Feringa, B. *Acc. Chem. Res.* **2000**, 33, 346-353. For the use of structurally related biphenol-derived phosphoramidites, see: (c) Alexakis, A.; Polet, D.; Rosset, S.; March, S. *J. Org. Chem.* **2004**, *69*, 5660-5667. See also: (d) G. Franciò, F. Faraone and W. Leitner, *Angew. Chem., Int. Ed.*, **2000**, *39,* 1428-1430. (e) O. Huttenloch, J. Spieler and H. Waldmann, *Chem. Eur. J.*, **2001**, *7,* 671-675.

5. Imbos, R.; Minnaard, . J.; Feringa, B. L. *J. Am. Chem. Soc.* **2002**, *124*, 184-185.

6. Tissot-Croset, K.; Polet, D.; Alexakis, A. *Angew. Chem. Int. Ed.* **2004**, *43*, 2426-2428.

7. (a) d'Augustin, M.; Palais, L.; Alexakis, A. *Angew. Chem. Int. Ed.* **2005**, *44*, 1376-1378. (b) Alexakis, A.; Albrow, V.; Biswas, K.; Augustin, M.; Prieto, O.; Woodward, S. *Chem. Commun.* **2005**, 2843-2845.

8. Biswas, K.; Prieto, O; Goldsmith, P.J.; Woodward, S. *Angew. Chem. Int. Ed.* **2005**, *44*, 2232-2234.

9. (a) van den Berg, M.; Minnaard, A. J.; Schudde, E. P.; van Esch, J.; de Vries, A. H. M.; de Vries, J. G.; Feringa, B. L. *J. Am. Chem. Soc.* **2000**, *122*, 11539-11540. (b) Pena, D.; Minnaard, A. J.; de Vries, J. G.; Feringa, B. L. *J. Am. Chem. Soc.* **2002**, *124*, 14552-14553. (c) Bernsmann, H.; van den Berg, M.; Hoen, R.; Minnaard, A. J.; Mehler, G.; Reetz, M. T.; de Vries, J. G.; Feringa, B. L. *J. Org.Chem.* **2005**, *70*, 943-951. (d) Liu, Y.; Ding, K. *J. Am. Chem. Soc.* **2005**, *127*, 10488-10489.

10. Giacomina, F.; Meetsma, A.; Panella, L.; Lefort, L.; de Vries, A. H. M.; de Vries, J. G. *Angew. Chem. Int. Ed.* **2007**, *46*, 1497-1500.

11. (a) Bartels, B.; Helmchen, G. *Chem. Commun.* **1999**, 741-742. (b) Lipowsky, G.; Miller, N.; Helmchen, G. *Angew. Chem. Int. Ed.* **2004**, *43*, 4595-4597. (c) Streiff, S.; Welter, C.; Schelwies, M.; Lipowsky, G.; Miller, N.; Helmchen, G. *Chem. Commun.* **2005**, 2957-2959.

12. (a) Ohmura, T.; Hartwig, J. F. *J. Am. Chem. Soc.* **2002**, *124*, 15164-15165. (b) Weihofen, R.; Dahnz, A.; Tverskoy, O.; Helmchen, G. *Chem. Commun.* **2005**, 2957, 3541-3543. (c) Singh, O. V. ; Han, H. *J. Am. Chem. Soc.* **2007**, *129*, 774-775.

13. (a) Lopez, F.; Ohmura, T.; Hartwig, J. F. *J. Am. Chem. Soc.* **2003**, *125*, 3426-3427. (b) Fischer, C.; Defieber, C.; Suzuki, T. Carreira, E. *J. Am. Chem. Soc.* **2004**, *126*, 1628-1629.

14. Boiteau, J.-G.; Minnaard, A. J.; Feringa, B. L. *J. Org. Chem.* **2003**, *68*, 9481-9484.

15. (a) Jagt, R. B. C.; Toullec, P.Y.; Geerdink, D.; de Vries, J. G.; Feringa, B. L.; Minnaard, A. J. *Angew. Chem. Int. Ed.* **2006**, *45*, 2789-2791. (b) Marelli, C.; Monti, C.; Gennari, C.; Piarulli, U. *Synlett* **2007**, 2213-2216.

16. Du, H.; Yuan, W.; Zhao, B.; Shi, Y. *J. Am. Chem. Soc.* **2007**, *129*, 11688-11689.

17. Yu, R.; Rovis, T. *J. Am. Chem. Soc.* **2006**, *128*, 12370-12371.

18. (a) Franció, G.; Faraone, F.; Leitner, W. *J. Am. Chem. Soc.* **2002**, *124*, 736-737. (b) Park, H.; Kumareswaran, R.; RajanBabu, T. V. *Tetrahedron* **2005**, *61*, 6352-6367. (c) Zhang, A.; RajanBabu, T. V. *J. Am. Chem. Soc.* **2006**, *128*, 5620-5621. (d) Smith, C. R.; RajanBabu, T. V. *Org. Lett.* **2008**, *10*, 1657-1659. (e) Smith, C.; Zhang, A.; Mans, D.; RajanBabu, T. V. *Org. Synth.* **2008**, *85*, 248-266. (f) For the use of a different type of phosphoramidite, see: Shi, W.-J.; Zhang, Q.; Xie, J.-H.; Zhu, S.-F.; Hou, G.-H.; Zhou, Q.-L. *J. Am. Chem. Soc.* **2006**, *128*, 2780-2781.

19. Kumareswaran, R.; Nandi, N.; RajanBabu, T. V. *Org. Lett.* **2003**, *5*, 4345-4348.

20. (a) Zhang, A.; RajanBabu, T. V. *J. Am. Chem. Soc.* **2006**, *128*, 54-55. (b) Saha, B.; Smith C. R.; RajanBabu, T. V. *J. Am. Chem. Soc.* **2008**, *130*, xxxx-xxxx (in press).

21. Phosphoramidites can also be synthesized by treatment of the diol with hexamethylphosphorus triamide. Hulst, R.; de Vries, N. K.; Feringa, B. L. *Tetrahedron: Asymmetry* **1994**, *5*, 699-708.

Appendix
Chemical Abstracts Nomenclature; (Registry Number)

(*R*)-(+)-1,1'-bi(2-naphthol): (*R*)-Binol; (18531-94-7)

Phosphorus trichloride (7719-12-2)

1-Methyl-2-pyrrolidinone (872-50-4)

(-)-Bis[(*S*)-1-phenylethyl]amine (56210-72-1)

(-)-Bis[(*S*)-1-phenylethyl]amine hydrochloride (40648-92-8)

n-Butyllithium (109-72-8)

(*R*)-(1,1'-Binaphthalene-2,2'-dioxy)chlorophosphine: (*R*)-Binol-P-Cl;
    (155613-52-8)

(*R*)-2,2-Binaphthoyl-(*S*,*S*)-di(1-phenylethyl)aminoylphosphine; (11b*R*)-4-
    [*N*,*N*-bis[(1*S*)-1-Phenylethyl]amino]-dinaphtho[2,1-d:1',2'-
    f][1,3,2]dioxaphosphepin (415918-91-1)

T. V. (Babu) RajanBabu received his undergraduate education in India (Kerala University and IIT, Madras). He obtained a Ph.D. degree from The Ohio State University in 1978 under the direction of Professor Harold Shechter, and was a postdoctoral fellow at Harvard University with the late Professor R. B. Woodward. He then joined the Research Staff of Dupont Central Research, becoming a Research Fellow in 1993. He returned to OSU as a Professor of Chemistry in 1995. His research interests are in new practical methods for stereoselective synthesis focusing on enantioselective catalysis, free radical chemistry, and applications in natural product synthesis.

Craig R. Smith was born in 1983 in Columbus, OH, USA. He graduated with his B.S. in Chemistry in 2003 and M.S. in Chemistry in 2005 from Youngstown State University under the direction of Professor Peter Norris. The same year, he started his Ph.D. studies at The Ohio State University under the supervision of Professor T. V. RajanBabu. His current interests are the development of asymmetric carbon-carbon, carbon-nitrogen and carbon-sulfur bond forming reactions, and the synthesis of biologically active heterocyclic natural products and analogs thereof.

246

Dan Mans was born in St. Louis, MO and got his undergraduate education at St. Louis University in 2002. He joined Department of Chemistry at the Ohio State University for his Ph.D., which was completed in 2008 under the supervision of Professor RajanBabu. In his graduate work he was involved with the development of the hydrovinylation reaction and its applications for the synthesis of pseudopterosins and helioporins. His research interests are in development and optimization of organic reactions of broad applicability. In his current position in the pharmaceutical industry, he plans to use his synthetic skills in the area of medicinal chemistry.

Son T. Nguyen was born in 1975 in Namdinh, Vietnam. He received his M.S. degree in chemistry from Hanoi University in 1999 with Professor Ngo Thi Thuan. He then joined the laboratory of Professor Ronald Caple at the University of Minnesota-Duluth where he obtained a M.S. degree in 2001. He moved to the University of Minnesota-Twin Cities campus to continue graduate study with Professor Craig J. Forsyth working on the total synthesis of azaspiracids. He graduated with a Ph.D. degree in 2006. He is currently doing postdoctoral research in the laboratory of Professor Scott E. Denmark developing a catalytic allylation process for carbonyl compounds.

(R)-3-METHYL-3-PHENYL-1-PENTENE *VIA* CATALYTIC ASYMMETRIC HYDROVINYLATION

Submitted by Craig R. Smith, Aibin Zhang, Daniel J. Mans and T. V. RajanBabu.[1,2]

Checked by Scott E. Denmark and Min Xie.

Org. Synth. **2008**, *85*, 248-266
Published on the Web 6/11/2008

1. Procedures

Caution!

Grignard reagents similar to 3,5-bis(trifluoromethyl)phenylmagnesium bromide have been known to decompose violently,[3] hence it is suggested that the reaction be conducted behind a blast shield in an efficient fume hood.

A. *Sodium Tetrakis[(3,5-trifluoromethyl)phenyl]borate (NaBArF₂₄) (2).*[3-6] NaBArF$_{24}$ can be conveniently prepared following the *Inorganic Syntheses* procedure.[4b] Although this procedure is reproducible, the additional details and the minor modifications described below are recommended.

Following the procedure described in *Inorganic Syntheses,*[4b] a Grignard reaction employing Mg turnings (1.17 g, 48.1 mmol, 6.5 equiv), sodium tetrafluoroborate (813 mg, 7.40 mmol), 1,2-dibromoethane (0.57 mL, 1.25 g, 6.7 mmol, 0.9 equiv) and 3,5-bis(trifluoromethyl)bromobenzene (7.15 mL, 12.16 g, 41.5 mmol, 5.6 equiv) (Note 1) is carried out in anhydrous ether (Note 2) under an argon atmosphere, The reaction mixture is stirred at room temperature for 12 h to ensure the completion of reaction. After the workup and the subsequent processing of the reaction mixture as described ref 4b, the solvent is removed from the ether extracts using a rotary evaporator (20 °C, ca. 20 mmHg). The resulting heavy, brown oil is dried by azeotropic distillation using a Dean-Stark trap and 200 mL of dry benzene (Note 3) under argon for two hours. Most (>190 mL) of the benzene is then removed by a continuously draining the solvent through the stopcock on the Dean-Stark trap (Note 4). After being cooled to room temperature, the remaining oily, brown mixture is triturated with dry benzene (50 mL) and the benzene layer is then decanted. The trituration is repeated twice more or until the resulting benzene layer is colorless (Note 5). The residual solvent is removed from the mixture with a rotary evaporator (ca. 40 °C, 20 mmHg). The remaining tan solid is further dried under high vacuum (0.1 mmHg) at room temperature for 8 h to afford a light tan powder (6.15 g, 94%) (Note 6). To further purify the product, the tan powder is transferred to a ceramic Büchner funnel (60 mm x 100 mm (d x h)) fitted with a Whatman 1 filter paper (55 mm) where it is washed with three, 30-mL portions of a cold (–40 °C) mixture of dichloromethane/hexane, 2:1 (Note 7) in a Büchner funnel to remove colored impurities. The solid is then dried under vacuum (90 °C, 0.1 mmHg) for 16 h and then is cooled to ambient temperature to afford 4.96 g

(76%) of the sodium salt as a free-flowing, fine, white powder (Notes 8 and 9). Typically, the salt is stored at ambient temperature in a glove-box.

B. *Di(μ-bromo)bis(η-allyl)nickel(II) (4).*[7,8] A 250–mL, three-necked, round-bottomed flask equipped with a rubber septum, a Teflon-taped flow-controlled argon inlet, a thermometer, and a magnetic stirring bar is flame-dried, purged with argon and transferred into a glove-box. The flask is then charged with bis[1,2:5,6-η-(1,5-cyclooctadiene)]nickel (1.38 g, 5.00 mmol) (Note 10) and is removed from the glove-box. Under argon from a Schlenk line, the flask is charged with anhydrous diethyl ether (80 mL) (Note 11) and is cooled to –70 °C in a dry ice/2-propanol bath, at which time allyl bromide (0.440 mL, 0.615 g, 5.08 mmol, 1.0 equiv) (Note 12) is added *via* syringe with stirring. The cold bath is removed and replaced by a ice-water bath. The reaction mixture is allowed to warm to 0 °C gradually (ca. 15 min), then to ambient temperature over another 15 min by removing the cold bath (Note 13). The reaction mixture is stirred at ambient temperature for another 30 min and then is transferred into the glove-box. The volatile contents of the flask are removed under high vacuum (ca. 0.1 mmHg) by connecting the flask to an evacuation port inside the glove-box (Note 14). The flask is rinsed with anhydrous ether (3 x 10 mL) and the resulting suspension is passed through a plug of dry Celite (2.5 cm x 7.5 cm (d x h)) in a fritted glass filter funnel (2.5 cm x 18 cm, (d x h) coarse) (Note 15). The Celite pad is washed with an additional 40 mL of ether. The ether washings are transferred into a 250-mL pear-shaped flask equipped with a magnetic stir bar. The ether is removed under vacuum (ca. 0.1 mmHg) with vigorous stirring to afford 0.803 g (89%) of the title compound as a dark-red powder, which was used without further purification (Note 16, 17).

C. *2-Phenyl-1-butene (6).* A 2–L, two-necked, round-bottomed flask equipped with a condenser (fitted with a nitrogen inlet), a rubber septum, and a magnetic stir bar is evacuated, flame-dried, and purged with nitrogen. The flask is then charged with methyltriphenylphosphonium bromide (43.4 g, 121 mmol, 1.2 equiv) (Note 18) through one neck of the flask under a strong flow of nitrogen and then quickly sealed with the rubber septum. Anhydrous THF (500 mL) (Note 19) is added into the flask through the rubber septum. The contents of the flask are then treated with *n*-BuLi (76.0 mL, 1.6 M solution in hexanes, 121 mmol, 1.2 equiv) (Note 20) at ambient temperature and the mixture is stirred for 2 h. Propiophenone (13.4 mL, 101 mmol) (Note 21) is then added dropwise via syringe over a period of 15 min. The reaction mixture is heated to reflux with stirring for 8 h. The mixture is allowed to cool to ambient temperature and 400 mL of hexane/diethyl ether, 1:1 (Note 22) is added. The resulting suspension is filtered through a pad of Celite (ca. 15 g, 5 cm x 5 cm) in a fritted glass filter funnel [5.0 cm x 12 cm, (d x h) coarse] which is washed with 100 mL of hexane/diethyl ether, 1:1 (Note 23). The filtrate is concentrated by rotary evaporation (15 °C, ca. 20 mmHg). The resulting pale-yellow slurry is first taken up in dichloromethane (30 mL) and then is triturated with hexane (200 mL) and the supernatant liquid is decanted. The remaining slurry is treated in the same manner twice more (Note 24). The combined extracts are filtered through a silica plug (ca. 20 g, 3 cm long, 5 cm in diameter), which is washed with 50 mL of hexane. The combined eluates are concentrated by rotary evaporation (15 °C, ca. 20 mmHg). The light-yellow liquid residue is purified by column chromatography (Note 25) to afford 11.9 g (88%) of the **6** as a clear colorless liquid (Note 26).

D. *(R)-3-Methyl-3-phenylpentene (7).*[2] The pre-catalyst is prepared as follows *in a glove-box*: A 100–mL, pear-shaped Schlenk flask with one side-arm fitted with a rubber septum and equipped with a magnetic stirring bar is evacuated, flame-dried, and purged with argon. The flask is charged with anhydrous dichloromethane (50 mL) (Note 27) and is transferred into a glove-box (Note 28).   To the flask is quickly added [di(μ-bromo)bis(η-allyl)nickel(II) (180 mg, 0.50 mmol, 0.01 equiv), (*R*)-2,2'-binaphthoyl-(*S*,*S*)-di(1-phenylethyl)aminoylphosphine (539 mg, 1.00 mmol, 0.02 equiv) (Note 29) and NaBArF$_{24}$ (886 mg, 1.00 mmol, 0.02 equiv) in the order mentioned. The resulting suspension is stirred at ambient temperature for 2 h to afford a dark-brown solution containing a small amount of fine particles (NaBr).

A 1-L, three-necked, round-bottomed flask equipped with a rubber septum, a Teflon-taped flow-controlled argon inlet, a thermometer, and a magnetic stir bar is flame-dried and purged with argon. The flask is then charged with 150 mL of anhydrous dichloromethane (Note 27). The catalyst solution prepared above, now removed from the drybox, is introduced to the vessel *via* cannula.   The flask containing the catalyst solution is further rinsed with 10 mL of dichloromethane, and this solution is also transferred to the reaction mixture. Upon completion of pre-catalyst transfer, the system closed at the flow-controlled stopcock and then is cooled to –70 °C in a dry ice/acetone bath, creating a small vacuum. A strong flow of dry ethylene (Note 30) is introduced through a needle through the serum stopper to relieve the vacuum and then is adjusted to maintain a pressure of 1 atm by releasing excess gas through an oil bubbler. The introduction of the ethylene causes the internal temperature to rise. Within ca. 5 min, the internal temperature increases by 5 °C and the ethylene line is removed (Note 31). The solution is cooled back to –70 °C with vigorous stirring. A solution of 2-phenyl-1-butene (**6**) (6.60 g, 50.0 mmol) in 30 mL of dry dichloromethane (Note 27) is introduced as a weak stream into the solution of pre-catalyst over a two-minute period *via* syringe followed by a 10 mL rinse with dichloromethane. Ethylene is introduced again through a needle, first as a strong flow and then regulated to maintain a pressure of 1 atm. Under an ethylene atmosphere, the internal temperature of the reaction mixture is then maintained between –65 °C and –70 °C for a period of 4 h. At the end of this period the ethylene line is removed and the reaction mixture is slowly poured into an Erlenmeyer flask containing 500 mL of pentane (Note 32) is and combined with a 50-mL pentane rinse of the reaction vessel. After being warmed to ambient temperature, the resulting, cloudy solution is filtered through a plug of silica gel (Merck, grade 9385, mesh 230-400, 60 Å, 4 cm x 5 cm, (d x h)), which is eluted with 100 mL of pentane (Note 33). The combined eluates are concentrated by rotary evaporation (20 °C, 20 mmHg) to afford 7.99 g (99.7%) of (*R*)-3-methyl-3-phenylpentene (er 98.8:1.2) as a clear, liquid (Notes 34 and 35).

2. Notes

1.   Reagent-grade magnesium turnings (99.98%) were purchased from Reade Manufacturing Company and activated prior to use with 1,2-dibromoethane following the method described by Reger.[4b] Sodium tetrafluoroborate (98%, Sigma-Aldrich Company) was dried at 100 °C under vacuum (0.1 mmHg) for 12 h and then was cooled to ambient temperature before being transferred into a glove-box. It was kept at ambient temperature in a glove-box for prolonged storage. 1,2-Dibromoethane (Sigma Chemical Company) was dried over 4 Å molecular sieves overnight before usage. 3,5-Bis(trifluoromethyl)bromobenzene (98%) was purchased from Matrix Scientifics and was used as received.

2.   Diethyl ether (Fisher, BHT stabilized, HPLC grade) was dried by percolation through two columns packed with neutral alumina under a positive pressure of argon.

3.   Benzene (Sigma-Aldrich Company) was distilled from sodium ribbon under a nitrogen atmosphere.

4.   The benzene was not completely removed at this stage while the flask was still warm.

5.   Benzene soluble impurities removed at this stage and a beige solid is formed.

6.   On a similar scale (703 mg, 6.37 mmol of sodium tetrafluoroborate as the limiting reagent), 5.15 g (91%) of the salt was obtained at this stage.

7.   Reagent grade dichloromethane (Sigma-Aldrich Company) and reagent grade hexane (Fisher Scientific Company) were mixed and cooled to –40 °C using a dry ice-acetonitrile cold bath.

8.   On a similar scale (703 mg, 6.37 mmol of sodium tetrafluoroborate was employed as the limiting reagent), 4.07 g (72%) of the salt was obtained.

9.   The product displayed the following physicochemical properties: ^{1}H NMR (500 MHz, acetone-d_6) δ: 7.79 (br s, 8 H), 7.67 (br s, 4 H). ^{13}C NMR (125 MHz, acetone-d_6) δ: 162.3 (q, J = 49.8 Hz), 135.2 (s), 129.6 (qq, J = 31.3, 2.9 Hz), 125.0 (q, J = 272.4 Hz), 118.1 (s); ^{19}F NMR (470 MHz, acetone-d_6) δ: –63.6; IR (KBr) cm^{-1}: 1781, 1712, 1628, 1356, 1282, 1130, 945, 932, 887, 838, 743, 710, 682, 670. Although the melting point of the title compound has been reported,[4b,6] the checkers found that the product thus obtained did not melt, but only gradually darkened when heated to above 300 °C and rapidly decomposed at 400 °C. Obtaining correct elemental analysis of highly fluorinated compounds, which do not burn completely, is difficult. Although the purity of the title product could not be secured by elemental analysis, it was sufficiently pure to be employed in the generation of the pre-catalyst (allyl)Ni(phosphine) BARF.

10.   The bis[1,2:5,6-η-(1,5-cyclooctadiene)]nickel was purchased from Strem Chemicals, Inc. and was used as received.

11.   Reagent-grade diethyl ether (Fisher Scientific Company) was freshly distilled from sodium benzophenone ketyl under an atmosphere of nitrogen prior to use.

12.   Allyl bromide (99%) was purchased from Sigma-Aldrich Company and was freshly distilled before use.

13.   The yellow Ni(COD)$_2$ crystals dissolve as the temperature approaches –10 °C, accompanied by the solution becoming pale orange. At 5 °C the solution turns deep red and becomes homogenous.

14.   To ensure the removal of all volatile components, the evacuation was continued for a period more than 4 h, during which time the vacuum

was released temporarily every hour and the reaction vessel was weighed. The evacuation was resumed until the loss of mass was less than 10 mg between two consecutive measurements.

15. The Celite (Fisher Scientific Company) was first stirred with concentrated HCl overnight, and then was washed with deionized water until neutral and finally with reagent grade MeOH (Sigma-Aldrich Company). The wet Celite was dried at 200 °C in an oven for 8 h and then was cooled to ambient temperature under vacuum (~0.1 mmHg). The dry Celite thus obtained was transferred into a glove-box before use.

16. To ensure the complete removal of ether, the evacuation was continued for ca. 4 h, during which the weight of flask containing the product was checked every hour until the loss of mass was less than 5 mg.

17. The allylnickel bromide dimer [di(μ-bromo)bis(η-allyl)nickel(II)] decomposes upon standing in air for several minutes. In an inert atmosphere, it is stable at 20 °C for several hours, however, it should be prepared in a drybox and stored at –20 °C in the freezer compartment if prolonged storage is required. No criterion of purity for this compound was established. This is a known compound[7,8] and several phosphine complexes of the allylnickel halides have been prepared by treatment with phosphines, and these complexes fully analyzed by ^{31}P NMR, ^{1}H NMR and X-ray crystallography.[14b] However, experience validates the use of the crude material for further reactions as prescribed in this procedure.

18. Methyltriphenylphosphonium bromide (99%) was purchased from Fisher Scientific Company and was used without further purification.

19. THF (Fisher, BHT stabilized, HPLC grade) was dried by percolation through two columns packed with neutral alumina under a positive pressure of argon.

20. n-BuLi was obtained as a solution in hexanes from Acros Organics and was titrated with N-benzylbenzamide, 99%, (Acros Organics) before use.

21. Propiophenone (99%) was purchased from Sigma Chemical Company and was used as received.

22. Reagent grade diethyl ether (Fisher Scientific Company) and ACS grade hexane (Fisher Scientific Company) were used as received.

23. The precipitated lithium bromide was thus removed.

24. The title product isolated at this stage is only slightly contaminated by triphenylphosphine oxide, as confirmed by ^{1}H NMR analysis of crude product.

254

25. Column chromatography was performed with silica gel (Merck, grade 9385, mesh 230-400, 60 Å), column size, 9 cm x 6 cm (h x d). The product (**6**) was isolated from 8 fractions (50 mL). UV and KMnO₄ detection were employed to visualize the product by TLC (silica gel); R_f value in pentane is 0.61. The product was sufficiently pure for use in the next stage. An analytically pure sample was obtained by a fractional distillation of a portion of the product at 45 mmHg and collecting the fraction boiling at 92-93 °C.

26. The distilled product displayed the following physicochemical properties: ^{1}H NMR (500 MHz, CDCl₃) δ: 7.43-7.41 (m, 2 H), 7.34-7.31 (m, 2 H), 7.29-7.24 (m, 1 H), 5.28 (s, 1 H), 5.12 (s, 1 H), 2.53 (q, J = 7.3 Hz, 2 H), 1.17 (t, J = 7.3 Hz, 3 H); ^{13}C NMR (125 MHz, CDCl₃) δ: 150.0, 141.5, 128.2, 127.2, 126.0, 110.9, 28.0, 12.9; IR (neat) cm^{-1}: 3081, 3056, 3030, 2967, 2934, 2876, 1945, 1877, 1799, 1628, 1600, 1573, 1495, 1463, 1443, 894, 776, 702. Anal. Calcd. for C₁₀H₁₂: C, 90.85; H, 9.15. Found: C, 90.77; H, 9.15.

27. Dichloromethane (Fisher, HPLC grade) was dried by percolation through two columns packed with neutral alumina under a positive pressure of argon.

28. Because halogenated solvents can poison the catalysts in the glove box, it is recommended that the purifier of the glove box is temporarily turned off during the preparation of the dichloromethane solution of the pre-catalyst. A minimum exposure of dichloromethane to the atmosphere in the glove box is also recommended. The residual dichloromethane vapor can be removed by purging the glove box with argon before turning the purifier back on.

29. See previous procedure in this volume.

30. Ethylene (99%) was purchased from Matheson Tri-gas. The gas flow was controlled by a regulator. The ethylene line was split into two lines by a three way stopcock, one connected to an oil bubbler, and the other to a 10 cm x 1.3 cm column of Drierite® in front of a needle outlet.

31. The checkers observed that if the exotherm was allowed to continue, the internal temperature of the reaction mixture could not be lowered to –70 °C. This undesired and uncontrolled exotherm is associated with the rapid generation of *cis*- and *trans*-2-butene as confirmed by ^{1}H NMR analysis of the reaction mixture. These side processes caused significant increase in the reaction volume and a significant attenuation of the rate of the desired reaction. It was thus necessary to remove the ethylene

line from the vessel and allow for the solution to cool. After the introduction of the substrate **6** and reintroduction of ethylene (maintained at a pressure of 1 atm), the exotherm and increase in volume were negligible. A low internal temperature was easily maintained and the desired reaction was facile and highly selective.

32. Reagent grade pentane was purchased from Fisher Scientific Co. and was used as received.

33. The more polar, colored impurities were effectively removed.

34. Based on ^{1}H NMR analysis, reaction conversion was higher than 99% and compound **7** was the only constitutional isomer observed. The product thus obtained was of very high purity. Analytically pure sample was prepared by a fractional distillation of a portion of the product at 15 mmHg and collecting the fraction boiling at 82-83 °C.

35. The product displayed the following physicochemical properties: $[\alpha]_D^{22} = -22.3$ (c 1.05, CHCl$_3$), lit.[10b] $[\alpha]_D^{20} = -12.5$ (c 0.8, CHCl$_3$, 96:4 er); ^{1}H NMR (500 MHz, CDCl$_3$) δ: 7.34-7.28 (m, 4 H), 7.21-7.16 (m, 1 H), 6.03 (dd, J_{trans} = 17.4 Hz, J_{cis} = 10.4 Hz, 1 H), 5.10 (dd, J_{cis} = 10.7 Hz, J_{gem} = 1.2 Hz, 1 H), 5.04 (dd, J_{trans} = 17.6 Hz, J_{gem} = 1.2 Hz, 1 H), 1.89-1.72 (ABX$_3$, v_A = 1.84, v_B = 1.77, J_{AB} = 13.7 Hz, J_{AX} = 7.4 Hz, J_{BX} = 7.4 Hz, 2 H), 1.35 (s, 3 H), 0.77 (t, J = 7.4 Hz, 3 H); ^{13}C NMR (125 MHz, CDCl$_3$) δ: 147.4, 146.9, 128.0, 126.7, 125.7, 111.7, 44.5, 33.4, 24.3, 8.9; IR (neat) cm^{-1}: 3083, 3058, 3023, 2968, 2935, 2878, 1944, 1873, 1830, 1801, 1635, 1600, 1493, 1456, 1445, 1379, 1370, 1030, 1003, 914, 760, 700; GC: t_R 10.61 min (poly(dimethylsiloxane), 25 m x 0.25 mm, 1.0 μm film thickness, 1.0 mL helium/min (1:100 split), 5 min at 100 °C, 5°C/min, 5 min at 200 °C); CSP–GC: t_R (R)-**7**, 24.28 min (98.8), (S)-**7**, 25.16 min (1.2). (Cyclodex B (J & W Scientific, 30 m x 0.25 mm, 0.25 μm film thickness) hydrogen (1.40 bar), 1:1 split, 30 min at 65 °C). Anal. Calcd. for C$_{12}$H$_{16}$: C, 89.94; H, 10.06. Found: C, 90.03; H, 10.16.

Safety and Waste Disposal Information

All hazardous materials should be handled and disposed of in accordance with "Prudent Practices in the Laboratory"; National Academy Press; Washington, DC, 1995.

3. Discussion

The procedure described here for the asymmetric hydrovinylation of 2-phenyl-1-butene (**6**) with the generation of a quaternary carbon center[2] has also been used with minor modifications for related asymmetric hydrovinylations of various other vinylarenes, 1,3-dienes and norbornene. The product **7** has been prepared previously by (i) stoichiometric γ-alkylation of an allylic Grignard reagent using a methylating agent with a chiral leaving group (38% yield, 32% ee)[9] and, (ii) Cu-catalyzed asymmetric alkylation of an allylic phosphate with an organozinc reagent (94% yield, 97% ee).[10] Advantages of the present method[2,11] include the use of simpler starting materials (2-phenyl-1-butene and ethylene) and nearly salt-free reaction conditions, still retaining the high yield and enantioselectivity.

Among the Ni(II)-catalyzed olefin dimerization reactions, the hydrovinylation reaction, viz., the addition of a vinyl group and a hydrogen across a double bond (Scheme 1), has attracted the most attention.[12] Since the branched product is chiral, a regio- and stereoselective version of this reaction could provide easy access to a variety of olefin-derived products including carboxylic acid derivatives. For example, the hydrovinylation of vinylarene derivatives that leads to 3-arylbutenes could be used for the synthesis of widely used antiinflammatory 2-arylpropionic acids. Even though the hydrovinylation reaction has had a long history dating back to 1965, until recently no catalyst system gave satisfactory yield and selectivity to be of practical value. Most often the reaction is complicated by isomerization of the primary products, and oligomerization of the vinylarene and ethylene (Scheme 1). Besides, use of high pressures of ethylene and metal components incompatible with sensitive organic groups often limited the utility of many of the procedures reported earlier.

Scheme 1. Hydrovinylation of Vinylarenes

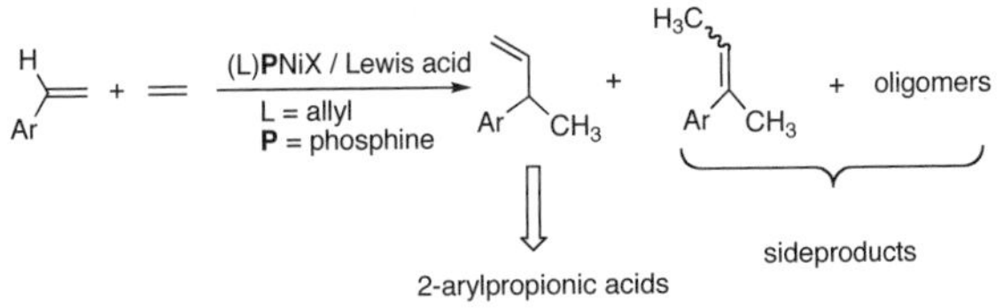

2-arylpropionic acids

Since our disclosure[13] of new protocols for this reaction, its scope has been considerably broadened. Application of old (Figure 1, **8**, **9**, **12**) and new (**10, 11, 13, 14**) ligands have enabled successful asymmetric hydrovinylation of vinyl arenes,[14] 1,3-dienes[14g,15] and strained bicyclic olefins such as norbornene.[14f,16] Several examples are illustrated in Eq. 1-6.

Figure 1. Assorted Ligands Used in Asymmetric Hydrovinylation Reactions

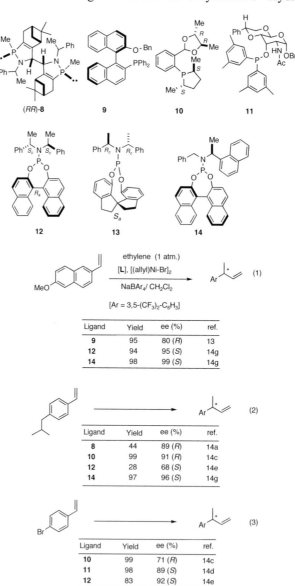

Ligand	Yield	ee (%)	ref.
9	95	80 (R)	13
12	94	95 (S)	14g
14	98	99 (S)	14g

Ligand	Yield	ee (%)	ref.
8	44	89 (R)	14a
10	99	91 (R)	14c
12	28	68 (S)	14e
14	97	96 (S)	14g

Ligand	Yield	ee (%)	ref.
10	99	71 (R)	14c
11	98	89 (S)	14d
12	83	92 (S)	14e

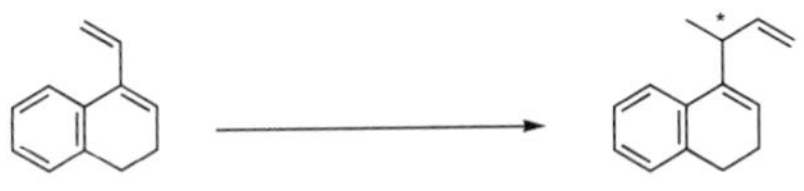

$(4)^{15}$

Ligand	Yield	ee (%)
10	99	93 (*R*)
12	99	99 (*S*)

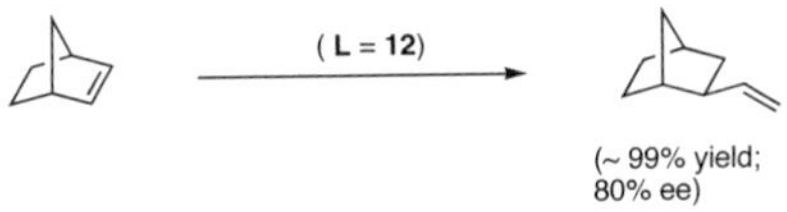

(L = 12)

$(5)^{16}$

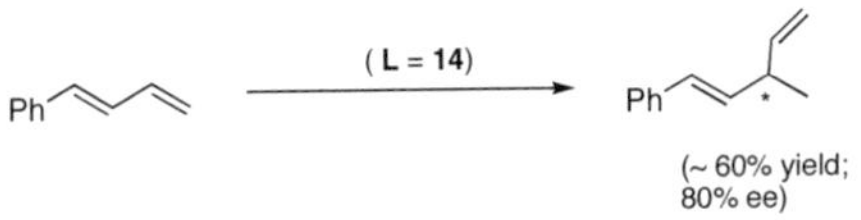

(~ 99% yield;
80% ee)

(L = 14)

$(6)^{14g}$

(~ 60% yield;
80% ee)

The search for new methods for stereoselective generation of quaternary carbon centers is a subject of considerable topical interest.[17] Several important pharmaceutically relevant compounds, among them, analgesic (-)-eptazocine,[18] protein kinase C activator lyngbyatoxin,[19] cognitive enhancing agent (−)-phenserine,[20] and serotonin antagonist LY426965,[21] contain an quaternary carbon center at the benzylic position. Hydrovinylation of 2-aryl-1-alkenes[13b] generate a quaternary center at the benzylic position (Eq. 7) and introduces a highly versatile latent functionality in the form of a vinyl group. The resulting intermediates could be quite valuable for further synthetic elaboration. An asymmetric variant of this reaction (Eq. 8) forms the basis of this submission.[2]

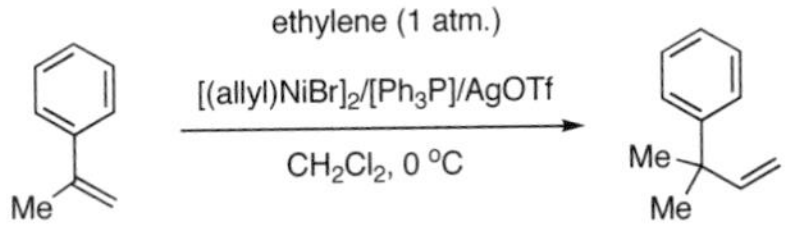

ethylene (1 atm.)

[(allyl)NiBr]$_2$/[Ph$_3$P]/AgOTf

CH$_2$Cl$_2$, 0 °C

(7)

ethylene (1 atm.)

[(allyl)NiBr]$_2$/**12**/NaBAr$_4$

(1 mol% cat.), CH$_2$Cl$_2$

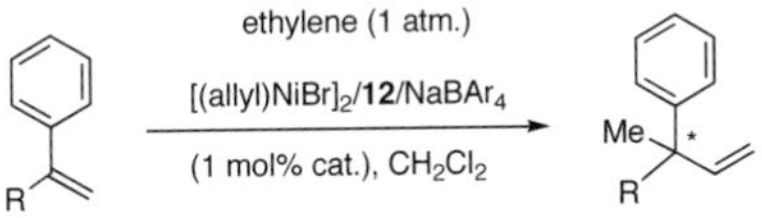

(8)

In scouting studies using 2-phenyl-1-butene (**6**) as substrate, catalysts derived from the MOP ligand (Fig. 1, **9**) show no reactivity while those derived from phospholane ligand **10**, which gave high ee's and turnover numbers in the hydrovinylation of a number of styrene derivatives[14c] and

1,3-dienes,[15] show only moderate reactivity under similar conditions.[2] Among the chiral ligands examined, the phosphoramidite **12** is found to provide the best results. Ligand **12**, when treated with [(allyl)NiBr]₂ followed by NaBARF gave a very active pre-catalyst that effects the hydrovinylation of 1-ethylstyrene at –78 °C (4 h), with as little as 1 mol% of catalyst to give a nearly quantitative reaction.[2] Under these conditions, no isomerization or oligomerization products is detected, as judged by careful GC analysis and ¹H NMR spectroscopy. The yields and selectivities are highly reproducible, and as expected, best selectivity is observed at low temperatures. They are independent of the catalyst loading or extent of reaction, clearly indicating the total absence of non-selective reactions.

Results of asymmetric hydrovinylation of several 2-aryl-1-alkenes under the optimal conditions are tabulated in Table 1. While the 4-methyl substrate **15** gave excellent selectivity for the formation of the expected product, the 4-chloro derivative **16** gave up to 5% isomerization of the starting olefin (entry 3). A similar minor side reaction was also observed for the substrates **18** and **20**. An isopropyl group at the 1-position of the styrene (**17**) retards the reaction (entry 4), and it is best accomplished at 24 °C with 10 mol% catalyst. Even though the yield of the reaction is only moderate, very high ee (~97%) was observed for the isolated product. The 2-naphthyl derivative **19** gave excellent yield (>98%) and selectivity (>99%) for the expected product. The tetralin derivative **20** represents a different class of substrates that underwent the hydrovinylation reaction giving >95% ee. Significant isomerization (~30%) of the starting material to an endocyclic olefin is a major distraction of this otherwise useful reaction. Compounds (e.g., **21b**) structurally related to the HV product **21a** from **20** have been synthesized previously via intramolecular asymmetric Heck reactions (~93% ee),[18] stoichiometric oxazoline directed alkylation (~99% ee),[22a] and enzyme-catalyzed desymmetrization of a chiral malonate (97% ee).[22b] By comparison, the asymmetric hydrovinylation route is significantly shorter, and operationally simpler.

Table 1. Asymmetric Hydrovinylation of 2-Aryl-1-alkenes[a]

entry	vinylarene	T (°C)/t (h)	conv./yield	sel.[b]	ee (%)[c]
1	**6**	–70/4	>99/>95	>99	>95
2	**15**	–60/12	>99/>90	>99	90
3	**16**	–70/11	>94/>90	>95[d]	90
4	**17**[e]	24/20	61[f]/60	>97	>95
5	**18**	–70/8	>99/ 93	>96[d]	>50[g]
6	**19**	–70/14	>99/>98	>99	93
7	**20**	–70/4	>98/ 70	71[d]	>95

[a] See Eq. 8 for details. [b] Selectivity for HV product. [c] Determined by GC, (R) isomer of **7**; all others assigned by analogy to **7**; Product from **20** (entry 7) assigned by comparison of $[\alpha]_D^{25}$ with that of a related compound.[18] [d] Rest isomerized product from starting material. [e] 10 mol% catalyst used. [f] Rest starting material. [g] ee determined via Mosher esters of hydroboration product.

Among the other olefins **22-24**, only the acyclic diene **24** undergoes hydrovinylation, and the product **25** is formed in nearly racemic form, contaminated with product of ethylene addition at the benzylic position.

Figure 2

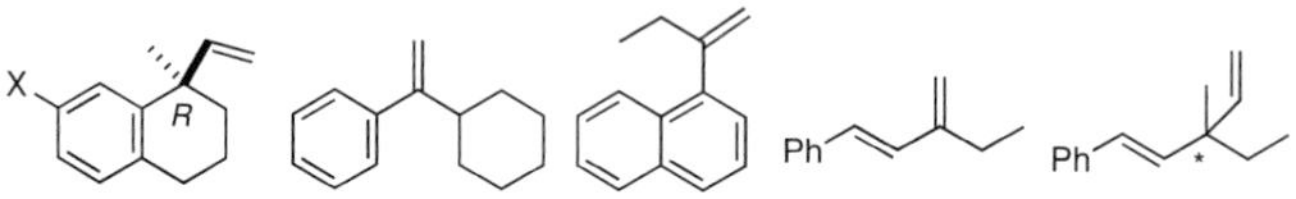

21a X = H
21b X = OMe   22   23   24   25 (<5% ee)

1. Department of Chemistry, The Ohio State University, 100 W. 18th Avenue, Columbus, OH 43210.

2. Zhang, A.; RajanBabu, T. V. *J. Am. Chem. Soc.* **2006**, *128*, 5620-5621.

3. (a) Appleby, I. C.; *Chem. Ind. (London),* **1971**, 120. (b) Leazer, J. L., Jr.; Cvetovich, R.; Tsay, F.-R.; Dolling, U.; Vickery, T.; Bachert, D. *J. Org. Chem.* **2003**, *68*, 3695-3698.

4. (a) Reger, D. L.; Wright, T. D.; Little, C. A.; Lamba, J. J. S.; Smith, M. D. *Inorg. Chem.* **2001**, *40*, 3810-3814. (b) Reger, D. L.; Little, C. A.; Brown, K. J.; Lamba, J. J. S.; Krumper, J. R.; Bergman, R. G.; Irwin, M.; Fackler, J. P. *Inorg. Synth.* **2004**, *34*, 5-8.

5. (a) Kobayashi, H.; Sonoda, A.; Iwamoto, H.; Yoshimura, M. *Chem. Lett.* **1981**, *10*, 579. (b) Brookhart, M.; Grant, B.; Volpe, A. F., Jr. *Organometallics*, **1992**, *11*, 3920-3922.

6. Yakelis, N. A.; Bergman, R. G. *Organometallics*, **2005**, *24*, 3579-3581.

7. *"Synthetic Methods of Organometallic and Inorganic Chemistry"*, Herrmann, W. A.; Salzer, A. Eds, Georg Thieme Verlag: Stuttgart, 1996, Vol. 1, p. 156.

8. Wilke, G.; Bogdanović, B.; Hardt, P.; Heimbach, P.; Keim, W.; Kröner, M.; Oberkirck, W.; Tanaka, K.; Steinrücker, E.; Walter, D.; Zimmerman, H. *Angew. Chem. Int. Ed. Engl.* **1966**, *5*, 151-164.

9. Yanagisawa, A.; Nomura, N.; Yamada, Y.; Hibino, H.; Yamamoto, H. *Synlett* **1995**, 841-842.

10. (a) Van Veldhuizen, J. J.; Campbell, J. E.; Giudici, R. E.; Hoveyda, A. H. *J. Am. Chem. Soc.* **2005**, *127*, 6877-6882. (b) Kacprzynski, M. A.; Hoveyda, A. H. *J. Am. Chem. Soc.* **2004**, *126*, 10676-10681.

11. Zhou et al. have reported the use of spirophosphoramidite **13** to carry out Ni-catalyzed asymmetric hydrovinylation of 2-aryl-1-alkenes,

262

including that of the substrate **6** (71% yield, 72% ee, *S*-**7**) described in this submission. Shi, W.-J.; Zhang, Q.; Xie, J.-H.; Zhu, S.-F.; Hou, G.-H.; Zhou, Q.-L. *J. Am. Chem. Soc.* **2006**, *128*, 2780-2781.

12. For a history of the reaction, scope, limitations and proposed mechanism, see: RajanBabu, T. V. *Chem. Rev.* **2003**, *103*, 2845-2860.

13. (a) Nomura, N.; Jin, J.; Park, H.; RajanBabu, T. V. *J. Am. Chem. Soc.* **1998**, *120*, 459-460. (b) RajanBabu, T. V.; Nomura, N.; Jin, J.; Nandi, M.; Park, H.; Sun, X. *J. Org. Chem.* **2003**, *68*, 8431-8446.

14. Use of Wilke's *Azaphospholene (8)*: (a) Wegner, A.: Leitner, W. *Chem. Commun.* **1999**, 1583-1584 and references cited therein. *Phospholanes* as ligands: (b) Nandi, M.; Jin, J.; RajanBabu, T. V. *J. Am. Chem. Soc.* **1999**, *121*, 9899-9900. (c) Zhang, A.; RajanBabu, T. V. *Org. Lett.* **2004**, *6*, 1515-1517. *Phosphinites*: (d) Park, H.; RajanBabu, T. V. *J. Am. Chem. Soc.* **2002**, *124*, 734-735. *Phosphoramidites:* (e) Franciò, G.; Faraone, F.; Leitner, W. *J. Am. Chem. Soc.* **2002**, *124*, 736-737. (f) Park, H.; Kumareswaran, R.; RajanBabu, T. V. *Tetrahedron* **2005**, *61*, 6352-6367. (g) Smith, C. R.; RajanBabu, T. V. *Org. Lett.* **2008**, *10*, 1657-1659.

15. (a) Zhang, A.; RajanBabu, T. V. *J. Am. Chem. Soc.* **2006**, *128*, 54-55. (b) Saha, B.; Smith, C. R.; RajanBabu, T. V. *J. Am. Chem. Soc.* **2008**, *130*, xxxx-xxxx. (in press)

16. Kumareswaran, R.; Nandi, N.; RajanBabu, T. V. *Org. Lett.* **2003**, *5*, 4345-4348.

17. For reviews, and history of the problem, see: (a) Douglas, C. J.; Overman, L. E. *Proc. Natl. Acad. Sci. U.S.A.* **2004**, *101*, 5363-5367 and references cited therein. (b) Denissova, I.; Barriault, L. *Tetrahedron* **2003**, *59*, 10105-10146. (c) Corey, E. J.; Guzman-Perez, A. *Angew. Chem., Int. Ed.* **1998**, *37*, 388-401. (d) Romo, D.; Meyers, A. I. *Tetrahedron*, **1991**, *47*, 9503-9569. (e) Martin, S. F. *Tetrahedron*, **1980**, 36, 419-460.

18. Takemoto, T. Sodeoka, M.; Sasai, H.; Shibasaki, M. *J. Am. Chem. Soc.* **1993**, *115*, 8477-8478 (corrections: ibid. *J. Am. Chem. Soc.* **1994**, *116*, 11207).

19. For a leading reference, see: Edwards, D. J.; Gerwick, W. H. *J. Am. Chem. Soc.* **2004**, *126*, 11432-11433.

20. Huang, A.; Kodanko, J. J.; Overman, L. E. *J. Am. Chem. Soc.* **2004**, *126*, 14043-14053.

21. Denmark, S. E.; Fu, J. *Org. Lett.* **2002**, *4*, 1951-1953.

22. (a) Hulme, A, N.; Henry, S. S.; Meyers, A. I. *J. Org. Chem.* **1995**, *60*, 1265-1270. (c) Fadel, A.; Arzel, P. *Tetrahedron: Asymmetry*, **1997**, *8*, 371-374.

Appendix
Chemical Abstracts Nomenclature; (Registry Number)

Benzophenone; (119-61-9)

Iodine; (7553-56-2)

3,5-Bis(trifluoromethyl)bromobenzene; (328-70-1)

Sodium tetrafluoroborate; (13755-29-8)

Bis[1,2:5,6-η-(1,5-cyclooctadiene)]nickel: [bis(1,5-cyclooctadiene)nickel
     (0)]; (1295-35-8)

Allyl bromide; (106-95-6)

Ethylene; (74-85-1)

n-Butyllithium; (109-72-8)

Methyltriphenylphosphonium bromide; (1779-49-3)

Triphenylphosphine oxide; (791-28-6)

Sodium tetrakis[(3,5-trifluoromethyl)phenyl]borate; (79060-88-1)

Di(*μ*-bromo)bis(*η*-allyl)nickel(II): [allylnickel bromide dimer]; (12012-90-7)

2-Phenyl-1-butene; (2039-93-2)

(*R*)-3-Methyl-3-phenylpentene: [(1*R*)-1-ethyl-1-methyl-2-propenyl]-
     benzene]; (768392-48-9)

T. V. (Babu) RajanBabu received his undergraduate education in India (Kerala University and IIT, Madras). He obtained a Ph.D. degree from The Ohio State University in 1978 under the direction of Professor Harold Shechter, and was a postdoctoral fellow at Harvard University with the late Professor R. B. Woodward. He then joined the Research Staff of Dupont Central Research, becoming a Research Fellow in 1993. He returned to OSU as a Professor of Chemistry in 1995. His research interests are in new practical methods for stereoselective synthesis focusing on enantioselective catalysis, free radical chemistry, and applications in natural product synthesis.

Craig R. Smith was born in 1983 in Columbus, OH, USA. He graduated with his B.S. in Chemistry in 2003 and M.S. in Chemistry in 2005 from Youngstown State University under the direction of Professor Peter Norris. The same year, he started his Ph.D. studies at The Ohio State University under the supervision of Professor T. V. RajanBabu. His current interests are the development of asymmetric carbon-carbon, carbon-nitrogen and carbon-sulfur bond forming reactions, and the synthesis of biologically active heterocyclic natural products and analogs thereof.

Aibin Zhang was born in China in 1974. After completing his B.S. in Chemistry at Anhui Normal University in 1996, he entered Shanghai Institute of Organic Chemistry, where he got his Ph.D. under the guidance of Dr. Shengming Ma in 2001. After a brief postdoctoral stint with Dr. Sean M. Kerwin at The University of Texas-Austin, he joined Professor RajanBabu's research group at The Ohio State University. His research at Ohio State focused on the development and application of asymmetric hydrovinylation reaction. Currently he is working for PharmaCore, NC.

Dan Mans was born in St. Louis, MO and got his undergraduate education at St. Louis University in 2002. He joined Department of Chemistry at the Ohio State University for his Ph.D., which was completed in 2008 under the supervision of Professor RajanBabu. In his graduate work he was involved with the development of the hydrovinylation reaction and its applications for the synthesis of pseudopterosins and helioporins. His research interests are in development and optimization of organic reactions of broad applicability. In his current position in the pharmaceutical industry, he plans to use his synthetic skills in the area of medicinal chemistry.

Min Xie received his bachelor's degree in Chemistry from Peking University in China. During that time he performed research with Professor Kai Wu on the fabrication of nanoporous anodic aluminium oxide (AAO) film and its application in the preparation of carbon nanotubes. In 2003 he began his graduate studies in Organic Chemistry at the University of Illinois, under the mentorship of Professor Scott E. Denmark. His current research focuses on the synthesis of rationally designed quaternary ammonium salts and investigation of phase transfer catalysis (PTC) by quantitative structure activity relationships (QSAR).

EFFICIENT OXIDATIVE SYNTHESIS OF (-)-2-*TERT*-BUTYL-(4*S*)-BENZYL-(1,3)-OXAZOLINE

Submitted by Björn T. Hahn, Kirsten Schwekendiek and Frank Glorius.[1]
Checked by Björn Gschwend and Andreas Pfaltz.

1. Procedure

(-)-2-tert-Butyl-(4S)-benzyl-(1,3)-oxazoline (2). In a two-necked, 500-mL round-bottomed flask are placed 12 g of molecular sieves (Note 1). (2*S*)-(-)-2-Amino-3-phenyl-1-propanol ((*S*)-phenylalaninol) (5.20 g, 0.0344 mol) (Note 2) and trimethylacetaldehyde (2.96 g, 0.0344 mol, 1.0 equiv) (Note 3) are dissolved in 150 mL of dichloromethane (Note 4). The resulting slightly turbid solution is added to the flask containing the molecular sieves. The flask is fitted with a glass stopper and a CaCl$_2$-filled drying tube. The reaction mixture is stirred at room temperature for 18 h (Note 5). Subsequently, the glass stopper is replaced by a thermometer, the drying tube is removed and *N*-bromosuccinimide (6.19 g, 0.0348 mol, 1.01 equiv) (Note 6) is added (Note 7). The resulting yellow-orange reaction mixture is stirred for 60 min at room temperature (Note 8). The orange mixture is filtered (Note 9) and the filtrate was transferred to a 500-mL separatory funnel where it is extracted with 50 mL of 0.5 M aq. Na$_2$S$_2$O$_3$ solution, 50 mL of sat. aq. Na$_2$CO$_3$ solution and 50 mL of water. The combined aqueous phases are then extracted with 50 mL of CH$_2$Cl$_2$ in a 500-mL separatory funnel and the combined organic phases are washed with 50 mL of brine in a 500-mL separatory funnel, then are dried over MgSO$_4$ (7 g) and are filtered through a glass funnel with a cotton wool plug. The filtrate is transferred to a 500-mL, round-bottomed flask and is treated with Celite (10 g) and the solvent is removed on a rotary evaporator under reduced pressure (Note 10). The resulting solid is transferred to a column (17 cm x 5 cm, h x d) containing 150 g of silica gel. Elution with pentane/MTBE

Org. Synth. **2008**, *85*, 267-277
Published on the Web 6/11/2008

(10:1 to 7:1 to 5:1) (Note 11) affords 5.57-5.72 g (74-76%) of a pale yellow liquid (Notes 12, 13, 14).

2. Notes

1. The 4 Å molecular sieves (beads, 1.7-2.4 mm, purchased from Fluka Chemie GmbH, Buchs) were activated by microwave irradiation (4 times 2 min at 700 Watts) and then cooled under an argon atmosphere in the sealed reaction flask.

2. (S)-Phenylalaninol was prepared by reduction of the corresponding amino acid according to the procedure of Abiko and Masamune.[2] Another convenient method has also been reported by McKennon and Meyers,[3a] as well as Gawley and Meyers in *Organic Syntheses*.[3b] Alternatively, (S)-phenylalaninol can be purchased from numerous vendors in enantiomerically pure form, e.g. Aldrich Chemical Co., and can be used without further purification.

3. Trimethylacetaldehyde was purchased from Aldrich Chemical Co. and was distilled prior to use.

4. Methylene chloride was of technical grade (Brenntag Schweizerhall AG) and was distilled under air prior to use. The submitters reported the use of CH_2Cl_2 distilled over CaH_2 which resulted in a slightly increased yield of 79%.

5. The stir rate was adjusted to about 200 rpm to minimize grinding of the molecular sieves.

6. N-Bromosuccinimide was purchased from Fluka and was recrystallized from water. The colorless plates were dried thoroughly under vacuum for 1 h.

7. The reaction flask was placed in a water bath (23 °C), and the NBS was added in portions, keeping the internal temperature below 35 °C.

8. The progress of the reaction can be followed by GC-MS analysis (see Note 13).

9. The reaction mixture was filtered through a 2 cm layer of Celite using a sintered glass funnel (porosity 3, 6.5 cm diameter).

10. The pressure was reduced to 15 mmHg and the heating bath kept at 50 °C. Upon addition of the eluent to the crude product, the formation of a sticky insoluble solid was observed. The crude product was therefore adsorbed on Celite.

11. Flash chromatography: Fluka silica gel 60 (0.040-0.063 mm). Pentane and MTBE were of technical grade (Brenntag Schweizerhall AG) and distilled prior to use. Three different pentane/MTBE mixtures were used successively: 750 mL of a 10:1, 1600 mL of a 7:1 and 1100 mL of a 5:1 mixture. The solvent was collected in 250 mL fractions (from the start) and the product was obtained in fractions 4 to 12.

12. The product is >95% pure, as judged by ^{1}H NMR spectroscopy. If necessary it can be further purified by Kugelrohr distillation (100 °C, 0.06 mmHg, > 95% yield) to give **2** as a colorless liquid (Note 13).

13. Oxazoline **2** displays the following physicochemical data: R_f = 0.3 (0.2 mm silica gel on plastic foil, pentane/MTBE 5:1). $[\alpha]_D^{20}$ = −35.8 (c 1.39, CHCl$_3$ (0.75% EtOH)), [lit.[4] $[\alpha]_D^{23}$ = −38.2 (c = 1.31, CHCl$_3$)]. ^{1}H NMR (400 MHz, CDCl$_3$) δ: 1.19 (s, 9 H), 2.63 (dd, 1 H, $^2J_{HH}$ = 13.7 Hz, $^3J_{HH}$ = 8.6 Hz), 3.09 (dd, 1 H, $^2J_{HH}$ = 13.7 Hz, $^3J_{HH}$ = 4.4 Hz), 3.96 (dd, 1 H, $^2J_{HH}$ = 8.5 Hz, $^3J_{HH}$ = 6.8 Hz), 4.11 (dd, 1 H, $^2J_{HH}$ = 8.5 Hz, $^3J_{HH}$ = 9.3 Hz), 4.35 (dddd, 1 H, $^3J_{HH}$ = 9.2 Hz, $^3J_{HH}$ = 8.7 Hz, $^3J_{HH}$ = 6.8 Hz, $^3J_{HH}$ = 4.5 Hz), 7.16 – 7.23 (m, 3 H), 7.24 – 7.31 (m, 2 H). ^{13}C NMR (100 MHz, CDCl$_3$) δ: 27.7, 33.0, 41.4, 66.8, 71.3, 126.3, 128.3, 129.4, 137.7, 174.2. IR (NaCl) cm^{-1}: 3062, 3028, 2970, 2928, 1949, 1881, 1807, 1658, 1604, 1482, 1456, 1393, 1361, 1284, 1219, 1140, 1097, 1026, 979, 926, 749, 703; GC-MS: t_R (**2**) = 11.6 min, EI m/z (%): 126 (100), 91 (18), 70 (46), 57 (56), 41 (31); t_R (**1**) = 13.5 min, EI m/z (%): 188 (23), 162 (76), 128 (100), 117 (53), 91 (79), 57 (15), 41 (43) (Marcherey-Nagel Optima 5 5% PhMeSi, 25 m x 0.2 mm, 0.35 μm film, 1.4 bar He, 100 °C/5 min/10 °C min^{-1}/ 270 °C/10 min); CSP-HPLC: t_R = 10.1 min (Daicel, Chiralpak AS, 25 cm, heptane/2-propanol, 400:1, 20 °C, 0.5 mL/min, 220 nm; UV: l$_{max1}$ = 208 nm, l$_{max2}$ = 259 nm. Anal. Calcd. for C$_{14}$H$_{19}$NO: C, 77.38; H, 8.81; N, 6.45. Found: C, 77.24; H, 8.80; N, 6.42.

14. The lability of oxazolines strongly depends on their substitution pattern. The product **2** is moisture sensitive. Standing at room temperature in a closed flask led to the formation and precipitation of a small amount of the corresponding hydrolysis product, amido alcohol **3** (a colorless, crystalline solid), within a few weeks. Storing at low temperature and exclusion of moisture is therefore recommended.

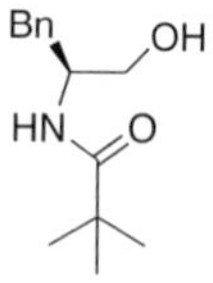

3

Safety and Waste Disposal Information

All hazardous materials should be handled and disposed of in accordance with "Prudent Practices in the Laboratory"; National Academy Press; Washington, DC, 1995.

3. Discussion

Chiral 2-oxazolines (IUPAC: 4,5-dihydrooxazoles)[5] have attracted increasing interest in the last decades, because of their multiple roles in biologically active compounds,[6] chiral ligands for asymmetric catalysis[7] like bisoxazolines and phosphinooxazolines and many other areas of chemistry.[8] Consequently, numerous methods have been developed for the efficient synthesis of oxazolines.[5] Most often, amino alcohols are reacted with either acid chlorides,[5] carboxylic acids,[9] nitriles[10] or imino ethers[11] (Scheme 1). However, despite their broad applicability, some disadvantages of these methods can be identified. Starting from acid chlorides, three steps are needed to form the oxazoline products. The direct use of carboxylic acids under Appel conditions most often leads to the formation of a number of byproducts requiring elaborate purification steps. Starting from nitriles, rather forcing conditions (Lewis acid catalysis in refluxing chlorobenzene) have to be used. Finally, imino ethers are generally not commercially available and have to be prepared in an additional step, either by using primary amides in conjunction with Meerwein's salt or from nitriles employing HCl gas in an alcohol.

270

Scheme 1

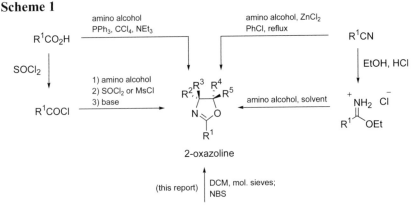

2-oxazoline

(this report) | DCM, mol. sieves;
NBS

R^1CHO + amino alcohol

Several oxidative methods for the formation of benzoxazoles,[12] benzimidazoles[12] and imidazolines[13] from aldehydes are described. Thus, recently, we and others have developed complementary, oxidative procedures for the formation of 2-oxazolines starting from the corresponding aldehydes.[14,15] In a typical procedure, (-)-2-*tert*-butyl-(4*S*)-benzyl-(1,3)-oxazoline can be formed from pivaldehyde and L-phenylalaninol using NBS as the oxidizing agent. Under optimized conditions, the in situ formation of the oxazolidine **1** is followed by the addition of one equivalent NBS, resulting in the formation of the oxazoline hydrobromide salts. Following the standard procedure, this salt was deprotonated subsequently using an aqueous basic work-up, also allowing for the separation of the oxazoline from the water-soluble succinimide byproduct. The product **2** was obtained with unchanged high enantiomeric excess of above 99%, as determined by HPLC using a chiral column (see above).

The substrate scope of this transformation is rather broad.[14a] Differently substituted aliphatic and aromatic aldehydes and various 1,2-aminoalcohols can be successfully employed (Table 1). A number of different benzaldehyde derivatives provide oxazolines in good yields. However, electron-rich aromatic aldehydes like salicylaldehyde, 4-hydroxybenzaldehyde, or pyrrole-2-carbaldehyde are unsuitable, since they were found to undergo an undesired electrophilic aromatic bromination. Nevertheless, moderately electron-rich aldehydes like methyl-substituted benzaldehydes are suitable substrates, although the reactions proceed with a significantly reduced rate (entries 2,3). Benzaldehyde derivatives bearing electron-withdrawing groups (entries 4-6,9) and other electron-poor substrates (entries 7,8) are generally good substrates. However, due to the increased acid-sensitivity of the oxazolines obtained from these latter substrates, slightly modified reaction conditions were required. Namely, these reactions were run in the presence of a solid inorganic base (K_2CO_3 or K_3PO_4) in toluene as the solvent. Aliphatic aldehydes were also used with good success (entries 16-20).

No limitation in the scope of the amino alcohol component was found. Unsubstituted (entry 9), mono- (entries 1-8,10,11,16-18) or disubstituted (entries 12-15,19,20) amino alcohols were converted into the oxazoline products with the method described. In addition, it seems that the enantiomeric purity of enantiomerically pure aminoalcohols (entries 10-13, 16-19) generally remains intact.[14a] This can be deduced from the following observations: First, in cases where values for optical rotation of optically pure 2-oxazolines were published in the literature, these values were found to be comparable to the ones obtained for the products of this study (entries 10-13,16,17). The deviations observed were less than 4% for these samples. Second, in cases where the oxazoline product bears two stereocenters an epimerization would lead to a mixture of diastereomers (entries 12,13,18,19). However, GC-MS and ^{1}H NMR analysis of these products revealed only a single diastereomer, indicating that no loss of enantiomeric purity had occured. Most importantly, however, for (-)-2-*tert*-butyl-(4*S*)-benzyl-(1,3)-oxazoline, the title compound of this publication, the enantiomeric purity was unequivocally established by an HPLC measurement using a chiral column.

In conclusion, this one-pot synthesis is characterized by mild reaction conditions, broad scope, high yield and its preparative simplicity.

272

Table 1. Oxidative formation of substituted oxazolines starting from various aldehydes and aminoalcohols.

entry	product	yield (%)	entry	product	yield (%)
1	Et, Ph (oxazoline)	91	11	Ph, Ph (oxazoline)	80
2	Et (oxazoline with o-tolyl)	93	12	Ph, Me (oxazoline with Ph)	81
3[b]	Et (oxazoline with mesityl)	42	13	indane-fused oxazoline, Ph	65
4[b]	Et (oxazoline with 2-Cl-phenyl)	76	14	gem-dimethyl oxazoline, Ph	82
5	Et (oxazoline with 4-CO$_2$Me-phenyl)	83	15	dioxolane-substituted oxazoline, Ph	88
6[b]	Et (oxazoline with 4-NO$_2$-phenyl)	68	16	Ph, t-Bu (oxazoline)	89
7[b]	Et (oxazoline with 2-pyridyl)	77	17	Ph, t-Bu (oxazoline)	85
8[b]	Et (oxazoline with CH=CH-Ph)	30	18[c]	Ph, t-Bu (oxazoline)	76
9	oxazoline with 4-CO$_2$Me-phenyl	88	19	Ph, Me, t-Bu (oxazoline)	91
10	isopropyl oxazoline, Ph	70	20[c]	gem-dimethyl oxazoline, (CH$_2$)$_8$	75

[a] General reaction conditions: 3 (1 mmol), CH$_2$Cl$_2$ (6 mL), 2 (1 mmol), mol. sieves (4 Å, 1.5 g), 14 h, rt; NBS (1 mmol), 0.5 h, rt. [b] Modified reaction conditions: 3 (1 mmol), toluene (6 mL), 2 (1 mmol), mol. sieves (4 Å, 1.5 g), 14 h, rt; K$_3$PO$_4$ (3 mmol), NBS (1 mmol), 1.5 h, rt.
[c] Large scale reaction according to the procedure of this publication.

1. Organisch-Chemisches Institut, Westfälische Wilhelms-Universität, Corrensstraße 40, 48149 Münster, Germany; glorius@uni-muenster.de
2. Abiko A.; Masamune S., *Tetrahedron Lett.* **1992**, *33*, 5517-5518.
3. (a) McKennon, M. J.; Meyers, A. I. *J. Org. Chem.* **1993**, *58*, 3568-3571. (b) Dickman, D. A.; Meyers, A. I.; Smith, G. A.; Gawley, R. E. *Org. Synth.* **1985**, *63*, 136-139.
4. Bates, G. S.; Varelas, M. A. *Can. J. Chem.* **1980**, *58*, 2562-2566.
5. (a) Gant, T. G.; Meyers, A. I. *Tetrahedron* **1994**, *50*, 2297-2360 and references cited therein. (b) Ager, D. J.; Prakash, I.; Schaad, D. R. *Chem. Rev.* **1996**, *96*, 835-876.
6. (a) Nicolaou, K. C.; Lizos, D. E.; Kim, D. W.; Schlawe, D.; de Noronha, R. G.; Longbottom, D. A.; Rodriguez, M.; Bucci, M.; Cirino, G. *J. Am. Chem. Soc.* **2006**, *128*, 4460-4470. (b) Pirrung, M. C.; Tumey, L. N.; McClerren, A. L.; Raetz, C. R. H., *J. Am. Chem. Soc.* **2003**, *125*, 1575-1586. (c) Bode, H. B.; Irschik, H.; Wenzel, S. C.; Reichenbach, H.; Müller, R.; Höfle, G. *J. Nat. Prod.* **2003**, *66*, 1203-1206. (d) Kline, T.; Andersen, N. H.; Harwood, E. A.; Bowman, J.; Malanda, A.; Endsley, S.; Erwin, A. L.; Doyle, M.; Fong, S.; Harris, A. L.; Mendelsohn, B.; Mdluli, K.; Raetz, C. R. H.; Stover, C. K.; Witte, P. R.; Yabannavar, A.; Zhu, S. *J. Med. Chem.* **2002**, *45*, 3112-3129.
7. (a) Ghosh, A. K.; Mathivanan, P.; Cappiello, J. *Tetrahedron: Asymmetry* **1998**, *9*, 1-45. (b) Pfaltz, A. *Acc. Chem. Res.* **1993**, *26*, 339-345. (c) Helmchen, G.; Pfaltz, A. *Acc. Chem. Res.* **2000**, *33*, 336-345.
8. (a) Greene, T. W.; Wuts, P. G. M. *Protective Groups in Organic Synthesis*, 3rd ed., John Wiley & Sons, New York, **1999**. (b) Snieckus, V. *Chem. Rev.* **1990**, *90*, 879-933. (c) Gnas, Y.; Glorius, F. *Synthesis* **2006**, 1899-1930. (d) Huang, H.; Hoogenboom, R.; Leenen, M. A. M.; Guillet, P.; Jonas, A. M.; Schubert, U. S.; Gohy, J.-F. *J. Am. Chem. Soc.* **2006**, *128*, 3784-3788.
9. (a) Vorbrüggen, H.; Krolikiewicz, K. *Tetrahedron* **1993**, *49*, 9353-9372. (b) Appel, R. *Angew. Chem.* **1975**, *87*, 863-874; *Angew. Chem., Int. Ed. Engl.* **1975**, *14*, 801-811.
10. (a) Witte, H.; Seeliger, W. *Liebigs Ann. Chem.* **1974**, 996-1009. (b) Bolm, C.; Weickhardt, K.; Zehnder, M.; Glasmacher, D. *Helv. Chim. Acta* **1991**, *74*, 717-726.
11. Neilson, D. G. in *The chemistry of amidines and imidates* (ed.: Patai, S.), Wiley, London, **1975**, 389.

12. (a) Kawashita, Y.; Nakamichi, N.; Kawabata, H.; Hayashi, M. *Org. Lett.* **2003**, *5*, 3713-3715. (b) Chang, J.; Zhao, K.; Pan, S. *Tetrahedron Lett.* **2002**, *43*, 951-954. (c) Varma, R. S.; Kumar, D. *J. Heterocycl. Chem.* **1998**, *35*, 1539-1540 (d) Varma, R. S.; Saini, R. K.; Prakash, O. *Tetrahedron Lett.* **1997**, *38*, 2621-2622. (e) Park, K. H.; Jun, K.; Shin, S. R.; Oh, S. W. *Tetrahedron Lett.* **1996**, *37*, 8869-8870.

13. Very recently, the first oxidative synthesis of imidazolines has been reported: (a) Fujioka, H.; Murai, K.; Ohba, Y.; Hiramatsu, A.; Kita, Y. *Tetrahedron Lett.* **2005**, *46*, 2197-2199. See also these related oxidation reactions: (b) H. Fujioka, K. Murai, Y. Ohba, H. Hirose, Y. Kita, *Chem. Commun.* **2006**, 832-834; (c) Ishihara, M.; Togo, H. *Synlett* **2006**, 227-230; (d) Gogoi, P.; Konwar, D. *Tetrahedron Lett.* **2006**, *47*, 79-82.

14. (a) Schwekendiek, K.; Glorius, F. *Synthesis* **2006**, 2996. (b) Sayama, S. *Synlett* **2006**, 1479.

15. In addition to the methods reported in references 13, a few less practical oxidative oxazoline formations appeared, requiring undesirable substrates like 2-azidoethanol derivatives, being unselective or having a very limited substrate scope: (a) Intramolecular redox reaction of 2-formylbenzaldehyde: Shipchandler, M. T. *J. Heterocyclic. Chem.* **1977**, *14*, 305-306. (b) Reaction of aldehydes with 1,2-hydroxyazides: Badiang, J. G.; Aubé, J. *J. Org. Chem.* **1996**, *61*, 2484-2487. (c) Synthesis of 3-oxazolines (with the formation of 2-oxazoline as minor byproducts): Favreau, S.; Lizzani-Cuvelier, L.; Loiseau, M.; Duñach, E.; Fellous, R. *Tetrahedron Lett.* **2000**, *41*, 9787-9790.

Appendix
Chemical Abstracts Nomenclature (Collective Index Number); (Registry Number)

(-)-2-*tert*-Butyl-(4*S*)-benzyl-(1,3)-oxazoline: 4,5-Dihydrooxazole, (4*S*)-
    benzyl, 2-*tert*-butyl; (75866-75-0)
(2*S*)-(-)-2-Amino-3-phenyl-1-propanol: Phenylalaninol: Propanol, 2-amino-,
    3-phenyl, (*S*); (3182-95-4)
Trimethylacetaldehyde: Pivaldehyde: Propanal, 2,2-dimethyl; (630-19-3)
N-Bromosuccinimide; (128-08-5)

Frank Glorius was educated in chemistry at the Universität Hannover, Stanford University (Prof. Paul A. Wender), Max-Planck-Institut für Kohlenforschung and Universität Basel (Prof. Andreas Pfaltz) and Harvard University (Prof. David A. Evans). In 2001 he began his independent research career at the Max-Planck-Institut für Kohlenforschung in Germany (Mentor: Prof. Alois Fürstner). In 2004 he became Assoc. Prof. at the Philipps-Universität Marburg and since 2007 he is a Full Prof. for Organic Chemistry at the Westfälische-Wilhelms-Universität Münster, Germany. He is a dedicated teacher and his research program focuses on the development of new concepts for catalysis and their implementation in organic synthesis.

Björn Hahn was born in Braunschweig (Germany) in 1980 and studied chemistry at the Technische Universität Braunschweig where he obtained his diploma in 2005. He joined the group of Prof. Frank Glorius at the Philipps-Universität Marburg in 2006 as a Ph.D. student. Currently he is working at the Westfälische-Wilhelms-Universität Münster on new catalytic reactions in organic synthesis.

Kirsten Schwekendiek was born in Emden (Germany) in 1981. She studied Chemistry at the Philipps-Universität of Marburg and the University of Melbourne. In 2006 she obtained her diploma degree from Marburg University, where she carried out her graduate work in the field of oxazoline synthesis under the direction of Prof. Frank Glorius. Currently she is a Ph.D. student in the group of Prof. Thisbe K. Lindhorst at the Christian-Albrechts-Universität of Kiel, dealing with the synthesis of photoswitchable glycoconjugates and their evaluation in the field of Biological Chemistry.

Björn Gschwend was born in Basel (Switzerland) in 1980. He studied Chemistry at the University of Basel where he obtained his diploma in 2005 under the supervision of Prof. Edwin C. Constable and Prof. Catherine E. Housecroft. He joined the group of Prof. Andreas Pfaltz in fall 2005 as a Ph.D. student and is currently working on the synthesis of new chiral ligands for metal-catalyzed reactions.

TERT-BUTYL *TERT*-BUTYLDIMETHYLSILYLGLYOXYLATE: A USEFUL CONJUNCTIVE REAGENT

A.

B.

C.

Submitted by David A. Nicewicz, Guillaume Brétéché, and Jeffrey S. Johnson.[1]

Checked by Christopher Bryan and Mark Lautens.

1. Procedure

A. tert-*Butyl diazoacetate* (Note 1). A modification of Dailey's procedure was employed.[2] A 1-L, two-necked, round-bottomed flask equipped with a Teflon-coated magnetic stir bar (4.1 x 1.9 cm) and a thermometer (one neck is left open) is charged with 4-acetamidobenzenesulfonyl azide (25.55 g, 106 mmol, 1.0 equiv) (Note 2), *tert*-butyl acetoacetate (17.6 mL, 17.1 g, 106 mmol), tetrabutylammonium bromide (0.68 g, 2.12 mmol, 0.02 equiv) and pentane (300 mL) (Note 3). The suspension is stirred and cooled to 0 °C in an ice-bath. An aqueous solution (100 mL) of sodium hydroxide (12.0 g, 300 mmol, 2.8 equiv) is added in several portions over 15 min, to create a suspension of a white solid in a yellow supernatant. The empty neck is fitted with a gas inlet adaptor

278

with stopcock and is vented to the atmosphere. The suspension is stirred for 20 h at ambient temperature. The white solid is removed by filtration through Celite (Note 4) and the pad is washed with 50 mL of pentane. The filtrate is transferred to a 1-L separatory funnel, and the aqueous phase is separated (Note 5) and extracted with three 70-mL portions of pentane. The combined organic extracts are washed with two 150 mL-portions of deionized water and 150 mL of brine, then are dried over $MgSO_4$ (ca. 7 g) for 10 min, filtered through a funnel packed with glass wool, and concentrated by rotary evaporation (Note 6) to yield 13.18 – 13.24 g (88 %) of *tert*-butyl diazoacetate as a yellow liquid that was used directly in the next step (Note 7).

B. tert-*Butyl* tert-*butyldimethylsilyldiazoacetate*. A modification of Bolm's method was employed.[3] A flame-dried, 500-mL, two-necked, round-bottomed flask equipped with a Teflon-coated magnetic stir bar (2.5 cm x 1.3 cm), a thermometer and a rubber septum is maintained under an inert atmosphere of argon and charged with *N,N*-diisopropylethylamine (19.3 mL, 14.7 g, 111 mmol, 1.2 equiv), *tert*-butyl diazoacetate (13.18 g, 92.8 mmol) and anhydrous diethyl ether (125 mL) (Note 8). The solution is stirred and cooled to –30 °C in a cryocooler. *tert*-Butyldimethylsilyl trifluoromethanesulfonate (25.6 mL, 29.4 g, 111 mmol, 1.2 equiv) is added dropwise via syringe over 15 min (Note 9). The yellow mixture is stirred for 24 h at –25 °C in a cryocooler (Note 10). The progress of the reaction is monitored by thin layer chromatography (Note 11). The mixture is filtered through Celite (Note 4) and the pad is washed with 50 mL of diethyl ether. The filtrate is concentrated by rotary evaporation (≤30 °C bath temperature, ≥70 mmHg) to yield 25.68 – 26.65 g of crude *tert*-butyl *tert*-butyldimethylsilyldiazoacetate as a yellow oil (Note 12).

C. tert-*Butyl* tert-*butyldimethylsilylglyoxylate*. A 1-L, three-necked, round-bottomed flask fitted with an overhead mechanical stirrer (fitted with a 6.0 x 2.0 cm Teflon paddle), a thermometer is charged with sodium bicarbonate (62.3 g, 742 mmol, 8.0 equiv), deionized water (210 mL) and acetone (150 mL) through the open neck. The mixture is mechanically stirred (600 rpm) and cooled to 0 °C in an ice-bath. Oxone® (114 g, 185 mmol, 1.85 equiv) is added in several portions (Note 13). A solution of crude *tert*-butyl *tert*-butyldimethylsilyldiazoacetate from Step B (25.68 – 26.65 g, 100 – 104 mmol) in dichloromethane (180 mL) is added to the reaction mixture all at once (Note 14). The resulting yellow mixture is stirred and cooled in the ice-bath for approximately 2 h. The progress of the reaction is monitored by TLC until no starting material is present (Note 15). The insoluble salts are removed by filtration through Celite (Note 4) and the pad is washed with two 25-mL portions of dichloromethane. Deionized water (150 mL) is added to the filtrate and the biphasic mixture is transferred to a 1-L separatory funnel. The organic layer is separated and the aqueous phase is extracted with two 50-mL portions of dichloromethane. The combined organic extracts are dried with Na_2SO_4 (ca. 50 g) for 10 min, then are filtered through a funnel packed with glass wool, and concentrated by rotary evaporation ($\leq$30 °C bath temperature, $\geq$70 torr) to give 21.1 – 23.4 g of a yellow oil (Note 16). The product is purified by flash column chromatography (Note 17), to yield 12.27 – 12.62 g (50.2 – 51.1 mmol, 47 – 49 % for three steps) of pure *tert*-butyl *tert*-butyldimethylsilylglyoxylate as a yellow oil (Note 18).

2. Notes

1. This compound is commercially available from Aldrich Chemical Company.

2. 4-Acetamidobenzenesulfonyl azide (97%) was purchased from Alfa Aesar and used as received. The submitter's material was synthesized using the procedure of Davies.[4]

3. *tert*-Butyl acetoacetate (98%) (d = 0.954 g/mL) was purchased from Aldrich Chemical Company. Tetrabutylammonium bromide (99+%) was purchased from Acros Organics. Sodium hydroxide and pentane were purchased from Fisher Scientific. All reagents were used as received.

4. All filtrations with Celite were performed using a slurry of Celite in the reaction solvent in a 10.5-cm diameter fritted funnel. The height of the Celite pad was 1.0 – 1.5 cm.

5. The organic phase is yellow whereas the aqueous is red.

6. The temperature of the bath should be kept below 20 °C to prevent distillation of the *tert*-butyl diazoacetate. The pressure should be $\geq$70 mmHg.

7. ^{1}H NMR data for *tert*-butyl diazoacetate: (400 MHz, $CDCl_3$) δ: 1.46 (s, 9 H), 4.61 (s, 1 H).

8. *N,N*-Diisopropylethylamine (98%) (d = 0.742 g/mL) was purchased from Fluka and distilled from calcium hydride before use (80 °C, 50 mm Hg). Diethyl ether was purified using an MBRAUN solvent purification system by percolation through a column of activated alumina.

280

9. *tert*-Butyldimethylsilyl trifluoromethanesulfonate (d = 1.151 g/mL) was purchased from Oakwood Products, Inc. and was used as received. During the addition of *tert*-butyldimethylsilyl trifluoromethanesulfonate the temperature increases from –30 °C to –25 °C. A white precipitate forms as well.

10. A Neslab CB-80 cryocooler was employed to maintain reaction temperature.

11. The yellow mixture is analyzed by thin layer chromatography (TLC; silica gel plates cut from 20 x 20 aluminum sheets (Silica gel 60 F_{254} from EMD Chemicals)), eluting with petroleum ether/EtOAc, 9:1 (R_f = 0.95 for *tert*-butyl *tert*-butyldimethylsilyldiazoacetate, R_f = 0.67 for *tert*-butyl diazoacetate). Visualization was accomplished with UV light and by staining with aqueous ceric ammonium molybdate solution followed by heating.

12. The submitters reported a yield of 22.48 g of 80 – 85 % purity. The checkers were unable to determine the purity of the crude material; 100% conversion was assumed. The silanol derived from hydrolysis of the silyl triflate is the main impurity (the submitters report that this accounts for around 9% of weight). Some of the ammonium salt produced during the reaction may be present as well. Purifying the compound at this stage is not trivial on this scale and there is no advantage to doing so since the impurities are easily rejected in the last step. A sample was purified by flash chromatography using petroleum ether/Et$_2$O, 99:1 in order to obtain spectroscopic data. Analytical data for *tert*-butyl *tert*-butyldimethylsilyldiazoacetate: IR (thin film), cm^{-1}: 2951, 2928, 2858, 2087, 1685, 1470, 1391, 1367, 1280, 1248, 1159, 1084, 838, 827, 803, 778, 675; ^{1}H NMR (400 MHz, CDCl$_3$) δ: 0.21 (s, 6 H), 0.96 (s, 9 H), 1.48 (s, 9 H). ^{13}C NMR (100 MHz, CDCl$_3$) δ: –6.2, 18.8, 26.4, 28.3, 81.0, 169.1. Anal. Calcd. for $C_{12}H_{24}N_2O_2Si$: C, 56.21; H, 9.43; N, 10.92. Found: C, 56.08; H, 9.59; N, 10.68.

13. Sodium bicarbonate (certified ACS) and acetone (99+%) were purchased from Fisher Scientific. Oxone® monopersulfate was purchased from Alfa Aesar, and it was used as received. As the Oxone® monopersulfate is added to the reaction mixture, gas is evolved.

14. Dichloromethane (99.5%) was purchased from Fisher Scientific and used as received. As the *tert*-butyl *tert*-butyldimethylsilyldiazoacetate is added, the mixture turns bright yellow.

15. Since it is a two phase reaction, the reaction time depends on the strength of the mechanical agitation. The yellow mixture is analyzed by thin

layer chromatography (TLC), eluting with petroleum ether/Et$_2$O, 9:1 (R$_f$ = 0.83 for *tert*-butyl *tert*-butyldimethylsilylglyoxylate, R$_f$ = 0.95 for *tert*-butyl *tert*-butyldimethylsilyldiazoacetate). Visualization is accomplished with UV light and on staining with aqueous ceric ammonium molybdate solution followed by heating.

16. The submitters reported a yield of 19.8 g. The compound is somewhat volatile and considerable amounts can be lost by extended exposure to high vacuum.

17. Flash column chromatography was performed on a silica gel column (23 cm length x 9 cm width, 720 g of silica gel) (Silicycle, grade 40 – 63μ). The compound was dissolved in 25 mL of petroleum ether. The silanol by-product was eluted with 1 L of petroleum ether and then the product was eluted with petroleum ether/Et$_2$O, 24:1 (ca. 2.7 L of eluent are required before collecting fractions) Collected fractions (eight 200-mL yellow fractions) were analyzed by thin layer chromatography (TLC), eluting with petroleum ether/Et$_2$O, 24:1 (R$_f$ = 0.49 for *tert*-butyl *tert*-butyldimethylsilylglyoxylate). Visualization was accomplished with UV light and on staining with aqueous ceric ammonium molybdate solution followed by heating.

18. *tert*-Butyl *tert*-butyldimethylsilylglyoxylate displayed the following physicochemical properties: IR (thin film) cm^{-1} 2931, 2860, 1716, 1657, 1464, 1369, 1252, 1159, 993, 841, 785; ^{1}H NMR (300 MHz, CDCl$_3$) δ: 0.27 (s, 6 H),), 0.96 (s, 9 H), 1.55 (s, 9 H; ^{13}C NMR (75 MHz, CDCl$_3$) δ: –6.6, 17.1, 26.6, 28.1, 83.6, 163.0, 232.9. Anal. Calcd. for C$_{12}$H$_{24}$O$_3$Si: C, 58.97; H, 9.90. Found: C, 58.81; H: 9.98.

Waste Disposal Information

All hazardous materials should be handled and disposed of in accordance with "Prudent Practices in the Laboratory"; National Academy Press; Washington, 1955.

3. Discussion

Silylglyoxylates were first synthesized by Ando's laboratory via dye-sensitized photooxidation of α-silyl α-diazo esters. Yields were ca. 40%.[5] Ozonolysis of ethyl trimethylsilyl diazoacetate has also been reported to afford the corresponding silyl glyoxylate in 30% yield with 15% over-oxidation to the silyl ester.[6] Recently silylglyoxylates were used by Bolm in the preparation of α-silyl α-hydroxy acetic acids.[3] The synthetic route involved *C*-silylation of benzyl diazoacetate and subsequent Rh(II)-catalyzed oxo transfer using propylene oxide. We have modeled the route presented herein on the Ando/Bolm route, incorporating an alternative oxo transfer that is more broadly applicable to sterically hindered substrates and does not require the use of $Rh_2(OAc)_4$. The diazo to oxo conversion is based on a recent report by Wang that uses Oxone® to transform α-diazo esters to α-keto esters.[7] The use of this oxidation does indeed work for the synthesis of silyl glyoxylates and moreover, allows for the synthesis of some highly hindered substrates (Table 1).[8] Since *tert*-butyl diazoacetate is fairly expensive, it is more cost effective to begin from *tert*-butyl acetoacetate. The diazo transfer and retro-Claisen condensation (deacylation) is conveniently achieved in one step using Davies's reagent[4] and applying a procedure adapted from Dailey.[2] Intermediates are not purified in the sequence. Impurities are efficiently rejected in the final purification by silica gel chromatography, the progress of which can be easily monitored due to the bright yellow color of the product.

Table 1.

1) $R_3SiOSO_2CF_3$, *i*-Pr_2NEt, Et_2O

2) **Method A:** $Rh_2(OAc)_4$ (2-5 mol %)
   propylene oxide, toluene, 110 °C
   or
   Method B: Oxone®, $NaHCO_3$
   acetone/CH_2Cl_2/H_2O

entry	R^1	SiR$_3$	method	yield (%)
1	*t*-Bu	SiMe$_3$	A	69
2	*t*-Bu	SiMe$_3$	B	55
3	*t*-Bu	Si*i*-Pr$_3$	A	no reaction
4	*t*-Bu	Si*i*-Pr$_3$	B	26
5	2,6-(*i*-Pr)$_2$Ph	SiMe$_3$	A	13
6	2,6-(*i*-Pr)$_2$Ph	SiMe$_3$	B	59

In addition to the hydrogenation work of Bolm, α-silyl-α-keto esters (silyl glyoxylates) react with alkynyl, vinyl, and hydridic nucleophiles to afford alkoxides that rearrange to glycolate enolates.[9,10] The [1,2]-Brook rearrangements[11] common to these reactions are presumably facilitated by the ester group. The nascent nucleophiles can be trapped with aldehydes to give aldol products wherein the two hydroxyl groups are differentiated. Silyl glyoxylates may prove useful in other reactions involving complementary nucleophilic and electrophilic reactants and the route described herein provides flexibility in both the ester group and silyl group of the reagent.

Figure 1.

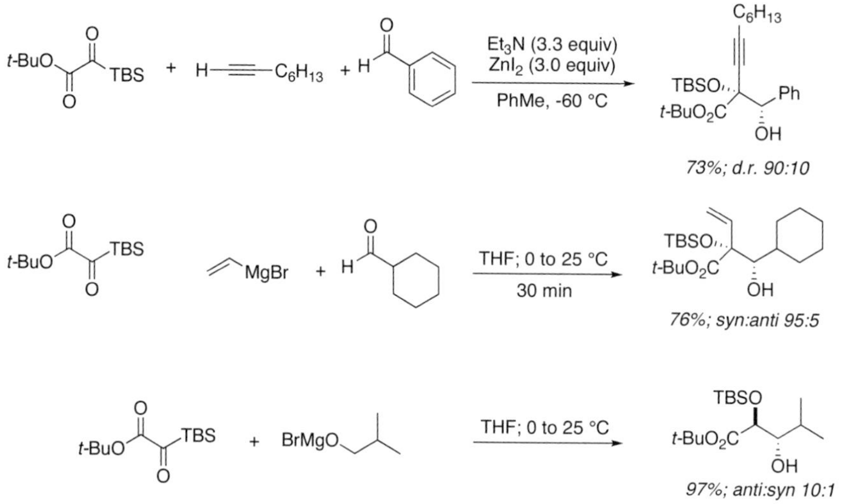

1. Department of Chemistry, University of North Carolina, Chapel Hill, NC   27599-3290, E-mail : jsj@unc.edu
2. O'Bannon, P. E.; Dailey, W. P. *Tetrahedron* **1990**, 46, 7341-7358.
3. Bolm, C.; Kasyan, A.; Heider, P.; Saladin, S.; Drauz, K.; Günther, K.; Wagner, C. *Org. Lett.* **2002**, 4, 2265-2267.
4. Davies, H. M. L.; Cantrell, W. R.; Romines, K. R.; Baum, J. S. *Org. Synth.* **1992**, 70, 93-100.

5. (a) Sekiguchi, A.; Kabe, Y.; Ando, W. *Tetrahedron Lett.* **1979**, *20*, 871-872. (b) Sekiguchi, A.; Kabe, Y.; Ando, W. *J. Org. Chem.* **1982**, *47*, 2900-2903.

6. Sekiguchi, A.; Ando, W. *J. Chem. Soc. Chem. Commun.* **1979**, 575.

7. Yao, W.; Wang, J. *Org. Lett.* **2003**, *5*, 1527-1530.

8. Nicewicz, D. A. Ph.D. Thesis, University of North Carolina at Chapel Hill, 2006.

9. Nicewicz, D. A.; Johnson, J. S. *J. Am. Chem. Soc.* **2005**, *127*, 6170-6171.

10. Linghu, X.; Satterfield, A. D.; Johnson, J. S. *J. Am. Chem Soc.* **2006**, *128*, 9302-9303.

11. Brook, A. G. *Acc. Chem. Res.* **1974**, *7*, 77-84.

Appendix
Chemical Abstracts Nomenclature; (Registry Number)

Sodium azide; (26628-22-8)

N-Acetylsulfanilyl chloride; (121-60-8)

4-Acetamidobenzenesulfonyl azide; (2158-14-7)

tert-Butyl acetoacetate; (1694-31-1)

Tetrabutylammonium bromide; (1643-19-2)

tert-Butyl diazoacetate; (35059-50-8)

tert-Butyldimethylsilyl trifluoromethanesulfonate; (69739-34-0)

N,N-Diisopropylethylamine; (7087-68-5)

Oxone® monopersulfate; (37222-66-5)

Jeffrey Johnson received his B.S. in Chemistry from the University of Kansas in 1994 (Highest Distinction and Honors in Chemistry). He obtained his Ph.D. from Harvard University in the laboratory of Professor David Evans in 1999 and was an NIH Postdoctoral Fellow with Professor Robert Bergman at the University of California at Berkeley from 1999-2001. He has been at the University of North Carolina since 2001.

David Nicewicz received his B.S. (2000) and M.S. (2002) in Chemistry from the University of North Carolina at Charlotte. He joined the laboratory of Professor Jeffrey Johnson at the University of North Carolina at Chapel Hill in 2001 where he studied enantioselective catalysis and natural product synthesis and was funded in part by an ACS Division of Organic Chemistry fellowship (Novartis Pharmaceuticals, 2004-2005). David completed his Ph.D. studies in 2006 and is currently a NIH Ruth L. Kirschstein NRSA postdoctoral fellow in the laboratory of Professor David MacMillan at Princeton University.

Guillaume Brétéché was born in 1984 in Nantes, France. He graduated from the Chemistry School of Toulouse with a M.S degree in 2007. His last year of undergraduate studies was completed at the University of North Carolina, Chapel Hill where he joined Professor Jeffrey Johnson's group. His research focused on the synthesis of silyl glyoxylates. He is currently employed by Arkema and works on synthesis of organic monomers for application in fuel cell membranes.

Christopher Bryan was born in Winnipeg, Canada in 1982. He received his B.Sc. degree with distinction in 2005 from the University of Victoria, where he worked in the laboratory of Scott McIndoe. While an undergraduate, he worked as a Co-op student in the medicinal chemistry department at Boehringer-Ingelheim Pharmaceuticals in Laval, PQ. He is currently pursuing his Ph.D. at the University of Toronto under the supervision of Professor Mark Lautens. His research is focused on the synthesis of heterocycles via metal-catalyzed tandem processes.

BENZYL ISOPROPOXYMETHYL CARBAMATE – AN AMINOMETHYLATING REAGENT FOR MANNICH REACTIONS OF TITANIUM ENOLATES

A.

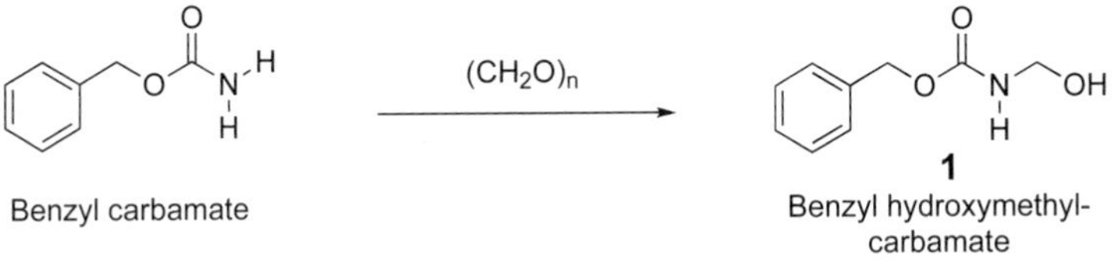

Benzyl carbamate

(CH$_2$O)$_n$

1

Benzyl hydroxymethyl-carbamate

B.

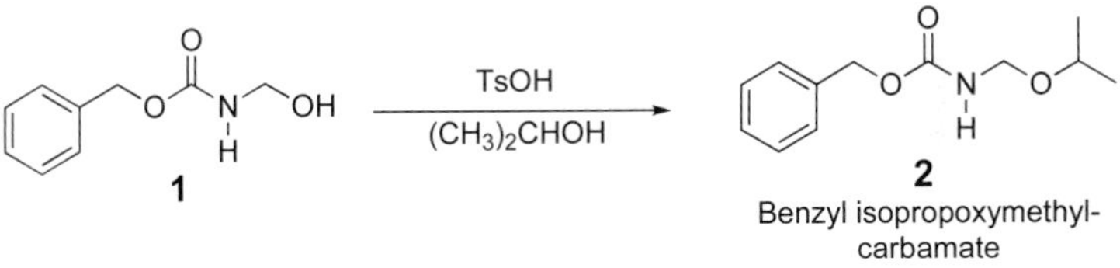

1

TsOH

(CH$_3$)$_2$CHOH

2

Benzyl isopropoxymethyl-carbamate

Submitted by Hartmut Meyer,[1] Albert K. Beck,[2] Radovan Sebesta,[2] and Dieter Seebach.[2]
Checked by Scott E. Denmark and Tyler W. Wilson.

1. Procedure

A. Benzyl hydroxymethyl carbamate (**1**). A 500-mL, four-necked, round-bottomed flask, fitted with a thermocouple, a reflux condenser, an adapter with a nitrogen inlet and an overhead mechanical stirrer (Note 1) is charged with benzyl carbamate (24.2 g, 0.16 mol) (Note 2) and dist. water (120 mL). Aqueous formaldehyde solution (37% in water, 30 mL, *ca.* 0.4 mol) (Note 3) and anhydrous potassium carbonate (0.44 g, 3.2 mmol) are added with stirring (400 rpm) at room temperature. The flask is then immersed into an oil bath, preheated to 65 °C, and the mixture is vigorously stirred until complete reaction has taken place (Note 4). The reaction flask is then transferred into an ice-water bath and vigorous stirring is continued until the product precipitates. After 2 h, water (120 mL) and anhydrous potassium carbonate (0.44 g, 3.2 mmol) are added, and the resulting mixture is stirred (600 rpm) at room temperature for 24 h until all of the product precipitates (Note 5). The solids are collected by suction filtration through a porcelain Büchner funnel (100 mm, Whatman #1, 90 mm), then are washed carefully with water (120 mL) and dried, first over night in air, then under vacuum (100 mmHg) at room temperature, to afford 20.1 g (69%) of **1**. This product is used without further purification (Note 6) in the next step.

B. *Benzyl isopropoxymethyl carbamate* (**2**). A 1-L, single-necked, round-bottomed flask fitted with a stir bar and nitrogen inlet is charged with benzyl hydroxymethyl carbamate (**1**) (19.6 g, 108 mmol), *tert*-butyl methyl ether (TBME, 200 mL) and isopropyl alcohol (50 mL). Next, *p*-toluenesulfonic acid hydrate (200 mg) (Note 7) is added to the solution and stirring is continued at room temperature for 24 h (Notes 8 and 9). Sodium hydrogen carbonate (2.0 g) and anhydrous magnesium sulfate (12 g) are added, and stirring is continued for another 2 h to remove the acid and the water formed. After removal of the solids by filtration through a glass-sintered Büchner funnel (45 mm, coarse) and washing the residue with TBME (50 mL), most of the solvent is removed from the filtrate under reduced pressure (Note 10). The crude product is purified by chromatographic filtration (Note 11) to give 23.5 g (97%) of **2** as a colorless solid (Notes 12 and 13).

2. Notes

1.   The submitters reported using a 45 g rugby-ball-shaped stir bar. However, more reproducible results were found by employing a mechanical stirrer (7.6 cm Teflon paddle).

2.   Benzyl carbamate is commercially available (Fluka Chemie AG) or can be easily prepared by modification of a literature procedure:[3] A 2-L, three-necked round-bottomed flask fitted with an overhead stirrer, dropping funnel and a reflux condenser with a gas adapter is charged with *conc.* aqueous ammonium hydroxide solution (1 L). Under ice-bath cooling benzyl chlorocarbonate (200 mL, 1.42 mol) is added over a period of 30 min. Stirring is continued for 2 h, after which the precipitated product is filtered off, then is washed carefully with water (2 L) and dried in air for two days. The crude product thus obtained is dissolved, under slight warming, in ethyl acetate (600 mL). Anhydrous magnesium sulfate is added for drying, then is filtered off and part of the solvent (*ca.* 400 mL) is evaporated until

precipitation ensues. Hexane (600 mL) is added to complete the precipitation, the solids are isolated to give 180.5 g (84%) benzyl carbamate. Checkers obtained benzyl carbamate from Sigma-Aldrich (99%) and it was used as received.

3.   Aqueous formaldehyde solution (37 wt. % solution in water, ACS-grade) was obtained from Sigma-Aldrich and was used as received. The submitters used aqueous formaldehyde solution (36.5% stabilized with 10% methanol) obtained from Fluka Chemie AG.

4.   At the beginning of the reaction the solid ester floats on the surface. After ca. 10 min, the mixture reaches a temperature of 45 °C, whereupon the solid material melts and the reaction begins. After 20 min, the internal temperature is 58 °C, and almost complete conversion occurs, as indicated by TLC.. Benzyl carbamate R_f = 0.28, **1**, R_f = 0.1 (silica gel, hexane/ethyl acetate, 3:2).

5. This protracted stirring is needed to cleave water soluble multi-addition products of formaldehyde. The yield of **1** increases from 45% (without this step) up to about 65%. The checkers noted that without mechanical stirring during this step a yield of 55% was typically obtained.

6.   Higher purity product can be obtained by recrystallization: 56.0 g of compound **1** are dissolved, with warming, in 60 mL of ethyl acetate; 220 mL of hexane are added in portions. After being cooled to 5 °C (refrigerator), the product **1** is collected and washed with an ice-cold solution of hexane/ethyl acetate, 2:1 to give 50.2 g (90%) of **1**. (Products containing up to 15% starting material from incomplete reaction may be effectively purified by adding only half the volume of hexane relative to the used ethyl acetate.) The checkers obtained an analytically pure sample by recrystallizing 1.0 g of **1** from 15 mL of hot *tert*-butyl methyl ether (TBME). Filtration and washing with cold TBME afforded 0.56 g of **1** as white needles (56% recovery). mp 87.5-88.5 °C; ^{1}H NMR (500 MHz, CDCl$_3$, 70 mM) δ: 7.38-7.31 (m, 5 H), 5.9 (br s, 0.1 H) and 5.75 (br s, 0.8 H), 5.13 (s, 2H), 4.72 (br d, J = 6.6 Hz), 3.16 (br s, 0.7 H) and 2.63 (br d, 0.05 H) equilibrium mixture of rotamers. ^{13}C NMR (125 MHz, CDCl$_3$, 470 mM) rotamers observed, δ for major rotamer: 156.8, 135.9, 128.5, 128.2, 128.0, 68.8, 67.0; minor signals observed δ: 66.9 and 65.9. IR (KBr) cm^{-1}: 3350, 3033, 2956, 2899, 1698, 1524, 1458, 1294, 1230, 1138, 1082, 1041, 973, 906, 747, 697; HRMS (ESI) *m/z*: calcd. for C$_9$H$_{11}$NO$_3$Na [M+Na]$^+$, 204.0637; found [M+Na]$^+$, 204.0636. Anal. Calcd for C$_9$H$_{11}$NO$_3$: C, 59.66; H, 6.12; N, 7.73. Found: C, 59.91; H, 6.32; N, 7.81.

7. Isopropyl alcohol (99.5%, ACS reagent), *tert*-butyl methyl ether (99%, ACS reagent), *p*-Toluenesulfonic acid monohydrate (98.5%, Reagent plus®) and sodium bicarbonate (99.7%, ACS reagent) were obtained from Sigma-Aldrich. Magnesium sulfate (certified anhydrous) was obtained from Fisher.

8. The reaction progress was monitored by TLC: **2**, $R_f = 0.40$ (silica gel, hexane/ethyl acetate, 3:2)

9. If the reaction is not complete at this point, more *p*-toluenesulfonic acid hydrate (100 mg) is added and stirring at room temperature is continued for 24 h.

10. Rotary evaporation at 45 °C and 20 mmHg. The checkers noted that after removing ca. 85% of the solvent the product would precipitate. Redissolution prior to chromatography was then achieved by adding 10 mL of dichloromethane and heating the mixture to 40 °C.

11. A chromatography column (6 cm diameter) is prepared with a 6 cm high bed of silica gel (ca. 100 g) packed with TBME. The crude product is placed on the column as a concentrated solution in TBME (ca. 25 mL) and is eluted with the same solvent (1000 mL in 25-mL fractions). The product-containing fractions are combined, evaporated and dried under vacuum (100 mm Hg) at room temperature. An analytical sample was obtained after drying 0.5 g the product in an Abderhalden at 55 °C at 100 mmHg for 12 h.

12. The product has the following physicochemical properties: mp 71.5-72.0 °C; ^{1}H NMR (500 MHz, CDCl$_3$) δ: 7.36-7.32 (m, 5 H), 5.51 (br s, 0.9 H) and 5.35 (br s, 0.1 H), 5.18 (br s, 0.2 H) and 5.13 (s, 1.8 H), 4.70 (d, J = 7 Hz, 1.8 H) and 4.63 (br s, 0.2 H), 3.8 (sept, J = 6.1 Hz, 0.9 Hz) and 3.7 (br s, 0.2 H), 1.17 (d, J = 6.1 Hz, 5.6 H) and 1.12 (br s, 0.8 H) equilibrium mixture of rotamers; ^{13}C NMR (125 MHz, CDCl$_3$) δ: 156.3, 136.2, 128.5, 128.2, 128.1, 70.1, 69.0, 66.9 22.3; IR (KBr) cm^{-1}: 3331 (br), 3066, 3028, 2967, 2926, 1718, 1522, 1458, 1374, 1312, 1240, 1176, 1126, 1096, 1042, 993 958, 754, 696; HRMS (ESI) *m/z*: calcd. for C$_{12}$H$_{17}$NO$_3$Na [M+Na]$^+$, 246.1106; found {M+Na]$^+$, 246.1100. Anal. Calcd. for C$_{12}$H$_{17}$NO$_3$: C, 64.55; H, 7.67; N, 6.27. Found: C, 64.56; H, 7.86; N, 6.37.

13. In contrast to the analogous, oily, methoxy-derivative **3** (see Discussion),[4] the product **2** can be stored at room temperature for months without decomposition.

3. Discussion

Benzyl methoxy-carbamate (**3**) was first prepared in 1951 by oxidative electrochemical decarboxylation (Hofer-Moest-electrolysis) of *N*-benzyloxycarbonyl-glycine (**4**) in methanol,[5,6] a method which is also applicable to other amino acid derivatives[7-11] and to peptides.[8,11-13]

The use of carbamate **3** as a carbamidomethylating reagent for titanium enolates (Mannich reaction) was reported 46 years later.[4] The authors of this report have prepared compound **3** by an alternative method, similar to that described herein: reaction of benzyl carbamate with formaldehyde to give the *N*-hydroxymethyl derivative **1** and acid-catalyzed methanolysis. Reagent **3** was employed extensively for the enantioselective preparation of β²-amino-acids[14] of type **6**, using the titanium enolates of *N*-acyl-5,5-diphenyl-4-isopropyl-1,3-oxazolidin-2-one (**5**, DIOZ derivatives)[15] as nucleophiles[16-19] (*cf.* accompanying procedure).

The isopropoxy derivative **2** had not been reported before. This derivative has the practical advantage of being solid and stable on storage under exclusion of moisture at room temperature, whereas the methoxy derivative **3** is an oil and has to be stored in a deep freezer (-15 °C). Other preparative advantages of **2** over **3** are described in the accompanying procedure employing the benzyl isopropoxymethyl carbamate.

1. Fachbereich Chemie, Institut für Organische Chemie, Leibniz Universität Hannover, Schneiderberg 1B, D-30167 Hannover, Germany.

2. Departement für Chemie und Angewandte Biowissenschaften, Laboratorium für Organische Chemie, ETH-Hönggerberg HCI, Wolfgang-Pauli-Str. 10, CH-8093 Zürich, Switzerland

3. Carter, H. E.; Frank, R. L.; Johnston, H. W. *Org. Synth., Coll. Vol. 3*, **1955**, 167-169.

4. Barnett, C. J.; Wilson, T. M.; Evans, D. A.; Somers, T. C. *Tetrahedron Lett.* **1997**, *38*, 735-738.

5. Linstead, R. P.; Shephard, B. R.; Weedon, B. C. L. *J. Chem. Soc.* **1951**, 2854-2858.

6. Shono, S.; Matsumura, Y.; Tsubata, K. *Nippon Kagku Kaishi* **1984**, *11*, 1782-1787 (*Chem. Abstr.* **1985**, *102*, 61636h).

7. Renaud, P.; Seebach D. *Synthesis* **1986**, 424-426.

8. Renaud, P.; Seebach, D. *Angew. Chem.* **1986**, *98*, 836-837; *Angew. Chem. Int. Ed.* **1986**, *25*, 843-844.

9. Renaud, P.; Seebach D. *Helv. Chim. Acta* **1986**, *69*, 1704-1710.

10. Stucky, G.; Seebach D. *Chem. Ber.* **1989**, *122*, 2365-2375.

11. Seebach, D.; Charczuk, R.; Gerber, C.; Renaud, P.; Berner, H.; Schneider, H. *Helv. Chim. Acta* **1989**,*72*, 401-425.

12. Gerber, C.; Seebach D. *Helv. Chim. Acta* **1991**, *74*, 1373-1385.

13. Sommerfeld, T. L.; Seebach D. *Angew. Chem.* **1995**, *107*, 622-623; *Angew. Chem. Int. Ed. Engl.* **1995**, *34*, 553-554.

14. Hintermann, T.; Ph.D.-thesis ETH No. 12964, Zürich 1998.

15. Brenner, M.; La Vecchia, L.; Leutert, T.; Seebach D. *Org. Synth.* **2003**, *80*, 57-61.

16. Hintermann, T.; Seebach D. *Helv. Chim. Acta* **1998**, *81*, 2093-2126.

17. Seebach, D.; Schaeffer, L.; Gessier, F.; Bindschädler, P.; Jäger, C.; Josien, D.; Kopp, S.; Lelais, G.; Mahajan, Y.; Micuch, P.: Sebesta, R.; Schweizer, B. W. *Helv. Chim. Acta* **2003**, *86*, 1852-1861.

18. Kimmerlin, T.; Sebesta, R.; Campo, M. A.; Beck, A. K.; Seebach D. *Tetrahedron* **2004**, *60*, 7455-7506.

19. Gessier, F.; Schaeffer, L.; Kimmerlin, T.; Flögel, O.; Seebach, D. *Helv. Chim. Acta* **2005**, *88*, 2235-2250.

Appendix
Chemical Abstracts Nomenclature; (Registry Number)

Benzyl hydroxymethyl carbamate: Carbamic acid, (hydroxymethyl)-,
    phenylmethyl ester; (31037-42-0)
Benzyl carbamate: Carbamic acid, phenylmethyl ester; (621-84-1)
Aqueous formaldehyde solution; (50-00-0)
Potassium carbonate; (584-08-7)
p-Toluenesulfonic acid: Benzenesulfonic acid, 4-methyl-; (104-15-4)

Dieter Seebach received his Ph.D. from the Technische Hochschule Karlsruhe, Germany, with Rudolf Criegee. After a postdoctoral stay in Elias J. Corey's group and a lectureship at Harvard University, he returned to the University of Karlsruhe and became a full Professor of Organic Chemistry at the Justus-Liebig-Universität in Giessen, Germany, in 1971. From 1977-2003 he was Professor of Chemistry at the Eidgenössische Technische Hochschule (ETH) in Zürich. Since 2003 he is officially retired; as Professor Emeritus he continues doing research with postdoctoral coworkers. His research activities include: development of new synthetic methods, natural-product synthesis, structure determination, chiral dendrimers, the biopolymer PHB, β-peptides. The results have been described in over 800 research papers.

Hartmut H. Meyer graduated in 1970 from the Technische Hochschule Hannover (Germany) under the supervision of W. Theilacker and F. Klages. After postdoctoral work with Professor D. Seebach at the University of Giessen in 1972, he returned to the University of Hannover and started work on syntheses of enantiopure natural products. In 1987 he became Privatdozent with a Habilitation on studies of the benzidine rearrangement. Since 1996 he has been a Professor at the Leibniz Universität Hannover. His current research interests are synthetic methodology, chemo-enzymatic syntheses and biochemical studies, in cooperation with other research groups.

Albert Karl Beck was born in 1947 in Karlsruhe (Germany), completing a chemistry technician's apprenticeship at the University of Karlsruhe in 1966, and joined the Seebach research group in 1968. Between 1969 and 1972 he continued his education, obtaining official certification as a chemical technician at the Fachschule für Chemotechnik in Karlsruhe. In 1971 he joined the Institute for Organic Chemistry at the University of Giessen, and in 1974 he engaged in a six-month research visit to the California Institute of Technology in Pasadena (USA). He has been an active part of the Laboratory for Organic Chemistry at the ETH in Zürich since 1977. During his association with the Seebach research group he has participated in essentially all of the group's research themes, as evidenced by his coauthorship of more than 80 publications.

Radovan Šebesta was born in 1975 in Myjava, Slovakia. He completed his undergraduate studies at the Comenius University in Bratislava where he also obtained Ph.D. with Prof. Sališová in 2002. Then he worked with Prof. Seebach at the ETH Zürich as a postdoctoral coworker (synthesis of β-aminoacids and peptides) and with Prof. Feringa in Groningen (asymmetric catalysis using phosphoramidites). In 2005 he moved back to Comenius University in Bratislava where he is currently Associate Professor at the Department of Organic Chemistry, working on asymmetric catalysis using ferrocene ligands.

Tyler W. Wilson was born in 1976 in Sacramento, California. After graduating high school, he headed to Portland, Oregon where he worked in carpentry and studied soft-glass blowing at the Pacific Northwest College of Arts. In 1999, he began undergraduate studies at Boise State University in chemistry where he studied the synthesis of conducting polymers. After obtaining a B.S. in chemistry in 2004 he moved to the University of Illinois at Urbana-Champaign and started his graduate education in the laboratories of Scott E. Denmark. His current research is focused on the asymmetric construction of quarternary carbons through the Lewis base catalyzed addition reactions of silyl ketene imines.

(R)-2-(BENZYLOXYCARBONYLAMINO-METHYL)-3-PHENYLPROPANOIC ACID (Z-β²hPHE-OH): PREPARATION OF A β²-AMINO ACID WITH THE AUXILIARY DIOZ

Submitted by J. Constanze D. Müller-Hartwieg,[1] Luigi La Vecchia,[1] Hartmut Meyer,[2] Albert K. Beck,[3] and Dieter Seebach.[3]
Checked by Scott E. Denmark, Nathan Duncan-Gould and Andrew Hoover.

1. Procedure

A. *(4S)-4-Isopropyl-5,5-diphenyl-3-(3-phenylpropionyl)oxazolidin-2-one (1)*. In a 1-L, three-necked, round-bottomed flask equipped with an overhead mechanical stirrer (fitted with an 8-cm Teflon paddle), a Teflon-coated thermocouple, and a 150-mL pressure-equalizing addition funnel fitted with a nitrogen inlet, a solution of (4S)-4-isopropyl-5,5-diphenyloxazolidin-2-one (DIOZ) (Note 1) (22.5 g, 80 mmol) in tetrahydrofuran (THF) (266 mL) (Note 2) is cooled to -40 °C using an ethanol/dry ice bath. *n*-Butyllithium (1.6 M in hexane, 52.75 mL, 84.8

mmol, 1.06 equiv) (Note 3) is added to the well-stirred suspension (Note 4). The resulting reddish solution is allowed to warm to -15 °C (ca 15 min) and is then cooled to -70 °C. A solution of 3-phenylpropanoyl chloride (14.84 g, 87.97 mmol, 1.1 equiv) (Note 5) in THF (42.5 mL) is added via the addition funnel to the solution while the temperature is kept between -70 and -60 °C. Addition is completed after 20 min and stirring is continued for 30 min further (Note 6). The reaction mixture is allowed to warm to room temperature, quenched with half-saturated aq. NH$_4$Cl solution (210 mL) and then is transferred to a 1000–mL separatory funnel where it is extracted twice with EtOAc (160 mL, 25 mL). The combined organic phases are washed consecutively with water (105 mL), sat. aq. NaHCO$_3$ solution (21 mL), and half-saturated brine (105 mL). After being dried over MgSO$_4$ (8 g, 30 min), the organic phase is filtered through an 8-cm medium-porosity glass filter funnel, then the filtrate is concentrated by rotary evaporation (23 °C, 15 mmHg) and then is dried at 0.2 mmHg for 8 h at room temperature. The crude product (35 g) is dissolved in refluxing *tert*-butyl methyl ether (TBME) (105 mL) and hexane (215 mL) is added slowly over 20 min. At the end of the addition, precipitation occurs. The mixture is then cooled in an ice bath for 2 h and is filtered through an 8-cm medium-porosity glass filter funnel to obtain the pure product, which is dried (23 °C at 0.08 mmHg) for 4 h to obtain 29.6 g (89%) of **1** as colorless needles (Note 7).

B.   *(2R,4S)   [2-Benzyl-3-(4-isopropyl-2-oxo-5,5-diphenyl-3-oxazolidinyl)-3-oxopropyl]carbamic acid benzyl ester (2).* A 1-L, three-necked, round-bottomed flask, fitted with a 150-mL, pressure-equalizing, addition funnel fitted with a nitrogen inlet, Teflon-coated thermocouple, a rubber septum, and a magnetic stirring bar (Note 8), is charged under a nitrogen atmosphere with **1** (28.9 g, 70.0 mmol) and dichloromethane (280 mL) (Note 9). The clear, colorless solution is cooled to -50 °C (ethanol/dry ice bath). A solution of TiCl$_4$ in dichloromethane (1 M, 73.5 mL, 73.5 mmol, 1.05 equiv) (Note 10) is added dropwise *via* cannula keeping the temperature between -50 and -40 °C (Note 11). Then a 1 M solution of triethylamine in dichloromethane is added *via* syringe (*ca.* 3 mL) until the dark red color of the enolate persists (Note 12). More triethylamine (10.19 mL, 7.4 g, 73 mmol, 1.05 equiv) in dichloromethane (70 mL) is slowly added over 10 min keeping the temperature between -40 and -35 °C. After stirring the reaction mixture for 20 min further at this temperature, a solution of benzyl isopropoxymethyl carbamate (16.4 g, 73.5 mmol, 1.05 eq) (Note 13) in dichloromethane (80 mL) is added over 10 min *via* the addition

296

funnel, followed by one more equivalent of the 1 M TiCl₄ solution in dichloromethane (73.5 mL, 73.5 mmol, 1.05 equiv) *via* cannula keeping the temperature at -40 °C (Note 10). After 10 min, the cooling bath is replaced by an ice bath and the reaction mixture stirred for 3 h at 0-2 °C (Note 6). The yellow reaction mixture is quenched with a solution of sat. aq. NH₄Cl solution (300 mL) and water (100 mL), then is stirred for 5 min and is transferred to a 1-L separatory funnel. The aqueous phase is separated and extracted with dichloromethane (50 mL). The combined organic layers are washed with half-saturated brine, dried over MgSO₄ (15 g, 30 min), filtered through an 8-cm medium-porosity glass filter funnel, concentrated by rotary evaporation (23 °C, 15 mmHg) and dried *in vacuo* (23 °C, 0.08 mmHg) for 5 h to afford the crude product as a pale-yellow foam (42.7 g) (Note 14). The residue is dissolved, with stirring, in 250 mL of refluxing *tert*-butyl methyl ether (TBME). During the addition of hexane (150 mL) some seed crystals were added (Note 15, 16). The mixture is allowed to cool for 30 min and then is heated back to reflux. Another portion of hexane (100 mL) is added and the mixture is allowed to cool to room temperature overnight. The resulting crystals are collected by suction filtration through an 8-cm medium-porosity glass filter funnel, then are washed with two 100-mL portions of a mixture of ice-cold hexane/TBME, 2:1. After a second recrystallization from hexane/TBME, 1:1 (250 mL each), the isolated crystals are dried *in vacuo* (23 °C, 0.15 mmHg) for 8 h at room temperature to provide 21.05 g (52%) of the product **2** as colorless, diastereomerically pure microcrystals (Notes 17, 18).

C. *(R)-2-(Benzyloxycarbonylaminomethyl)-3-phenylpropanoic acid (3)*. A 1-L, two-necked, round-bottomed flask fitted with an overhead stirrer (fitted with an 8-cm Teflon paddle), and a Teflon-coated thermocouple is charged with **2** (20.0 g, 34.9 mmol), THF (204 mL) and water (102 mL). Hydrogen peroxide (30%, 14.3 mL, 139.6 mmol, 4.0 equiv) (Note 19) is then added and the resulting mixture cooled to 0-5 °C in an ice bath. Lithium hydroxide monohydrate (2.34 g, 55.8 mmol, 1.6 equiv) (Note 20) is added and stirring is continued at 0-5 °C for 2 h (Notes 6, 21). The suspension is allowed to warm to room temperature, then is slowly diluted with aq. Na₂SO₃ solution (20 g in 160 mL, 158 mmol) (Note 22, 23) followed by 1 M aq. NaOH solution (102 mL) and stirring is continued at room temperature for 10 min. The THF is removed on a rotary evaporator (40 °C, 15 mmHg, 30 min) and TBME (635 mL) is added to the resulting suspension. After the mixture is stirred for 10 min, the precipitate (recovered DIOZ auxiliary) is filtered off by suction through an 8-cm medium-porosity glass filter funnel and is successively washed with 1 M aq. NaOH solution (101 mL), water (2 x 83 mL) and TBME (2 x 83 mL) (Note 24). The filtrate and all aqueous washes are combined and acidified to pH 2-3 with 0.52 M aq. citric acid solution (Note 25) and then are transferred to a 4-L separatory funnel and are extracted with TBME (2 x 255 mL). The combined TBME phases are washed with brine (255 mL), dried over MgSO$_4$ (19 g, 10 min), filtered through an 8-cm medium-porosity glass filter funnel and concentrated by rotary evaporation (45 °C, 15 mmHg). The resulting oil slowly solidifies upon standing (Note 26) to afford 10.70 g (98%) of **3** with an enantiomer ratio of ≥ 99.9:0.1 (Notes 27, 28, 29).

2. Notes

1.   This auxiliary (DIOZ) was prepared following the *Organic Syntheses* procedure, ref. 4. Use of the corresponding *R*-auxiliary leads to the *S*- β^2 amino acid derivative *ent*-**3**.

2.   THF was purchased from Fisher Scientific and was dried by percolation through 4Å molecular sieves immediately prior to use.

3.   *n*-Butyllithium in hexane was purchased from Fluka and was titrated by the method of Gilman (1.62 M).

4.   The addition is slightly exothermic, therefore slow addition is recommended.

5.   3-Phenylpropanoyl chloride (98%) was purchased from ACROS and was used as received.

6.   The progress of the reactions can be monitored by TLC analysis (visualization with UV, I$_2$ and KMnO$_4$). Step A: (silica gel, hexane/EtOAc, 4:1); DIOZ: R$_f$ = 0.15, **1**: R$_f$ = 0.75. Step B: (silica gel, hexane/EtOAc, 4:1); **1**: R$_f$ = 0.75, **2**: R$_f$ = 0.3. Step C: (silica gel, hexane/EtOAc, 3:2); **2**: R$_f$ = 0.85, **3**: R$_f$ = 0.11.

7.   The product exhibits the following physicochemical properties: mp 98-99 °C; $[\alpha]_D^{23}$ = –176.2 (c 1, CHCl$_3$); ^{1}H NMR (400 MHz, DMSO) δ: 0.6, 0.85 (2 d, J = 7.1, 6.87 Hz, 6 H, 2 x CH$_3$), 2.03 (dp, J =7.1, 2.5 Hz, 1 H, CH), 2.75 (t, J = 7.9 Hz, 2 H, CH$_2$-phenyl), 2.96, 3.14 (2 m, 2 H, CH$_2$-CO), 5.59 (d, J = 2.6 Hz, 1 H, CHN), 7.05-7.65 (m, 15 H, aryl-H). ^{13}C NMR (125 MHz, DMSO) δ: 171.3, 152.6, 143.1, 140.2, 138.2, 128.8, 128.38, 128.30, 128.2, 128.1, 127.7, 125.9, 125.3, 124.9, 88.7, 64.4, 36.2, 29.8, 29.6, 21.2, 15.4; IR (KBr) cm^{-1}: 2961 (m), 2938 (m), 1781 (s), 1700 (s); MS (EI) *m/z*

(relative intensity, %): 413 (41), 263 (16), 222 (43), 207 (41), 183 (46), 165 (28), 133 (26), 117 (58), 105 (100). Anal. Calc. for $C_{27}H_{27}NO_3$: C 78.42, H 6.58, N 3.39. Found: C, 78.53; H, 6.69; N, 3.54.

8. A 5-cm Rugby-ball-shaped stir bar was used.

9. Dichloromethane purchased from Fisher Scientific and was dried by percolation through 4 Å molecular sieves immediately prior to use.

10. Titanium tetrachloride (1 M in CH_2Cl_2) was purchased from Fluka and was used as received.

11. Although the submitters added the solution *via* syringe, on this scale the checkers found it more convenient to transfer the titanium tetrachloride solution *via* cannula and a 100-mL graduated cylinder (flame dried, with septa and a nitrogen needle inlet.

12. Triethylamine (Reagent Grade) was purchased from Aldrich Chemical Co. and was freshly distilled from CaH_2 prior to use.

13. For the preparation of this electrophile see the preceding procedure in this volume.

14. The crude product contains the desired (*R*,*S*)-diastereomer, as well as the (*S*,*S*)-epimer in a 9:1 ratio. As a by-product (*ca.* 7%) 2-benzyl-3-(4-isopropyl-2-oxo-5,5-diphenyloxazolidin-3-oxopropyl)carbamic acid isopropyl ester is formed. Both by-products can be removed by recrystallization.

15. Seed crystals can be obtained after addition of hexane to a TBME solution of a small amount of crude product.

16. The checkers found that a hot filtration of the TBME solution through ~1.5 g of Celite (acid-washed) in a cotton-plugged funnel removed a small amount (~5%) of triethylamine hydrochloride and made the recrystallization easier.

17. An additional 12% of product **2** can be obtained by chromatography of the mother liquors (column dimensions: 12 cm x 6 cm (h x d), 200 g of SiO_2, hexane/EtOAc, 85:15 (1 L), 80:20 (500 mL), 75:25 (500 mL), 70:30 (500 mL), 65:35 (500 mL) and subsequent crystallization (TBME/hexane). The mother liquors from this crystallization contain a 1:1 mixture of diastereomers that aids in the assignment of the products (see Note 13). Apart from the product, 19% of starting material can be recovered by this chromatography.

18. The product exhibits the following physicochemical properties: mp 129.5-130.5 °C; $[\alpha]_D^{23}$ −96.2 (c 1, $CHCl_3$); ^1H NMR (500 MHz, DMSO) δ: 0.6, 0.85 (2 d, *J* = 7.0 Hz, 6 H, 2 x CH_3), 2.03 (m, 1 H, CH-*i*-Pr), 2.42,

2.59 (2 x dd, 2 H, J = 14.1, 6.5 Hz, CH-CH_2-phenyl), 3.23 (m, 2 H, CH$_2$), 4.11 (m, 1 H, CH), 4.98 (dd, 2 H, J = 21.7, 12.7 Hz, O-CH$_2$-phenyl), 5.54 (d, 1 H, J = 2 Hz, CHN), 6.65-7.63 (m, 21 H, aryl-H + NH). ^{13}C NMR (125 MHz, CDCl$_3$) δ: 173.0, 126.8, 153.1, 143.8, 138.84, 138.81, 138.29, 137.8, 137.8, 129.5, 129.2, 129.1, 129.0, 128.97, 128.7, 128.45, 128.40, 128.3, 126.7, 126.0, 125.4, 89.2, 65.9, 65.6, 44.7, 42.0, 35.1, 30.2, 21.8, 16.0. IR (KBr) cm^{-1}: 3366 (s), 3064 (m), 3029 (m), 2870 (m), 1775 (s), 1716 (s), 1704 (s), 1539 (s) MS (EI) m/z (relative intensity,%): 576 (4), 412 (45), 222 (16), 167 (15), 131 (28), 91 (100). Anal. Calc. for C$_{36}$H$_{36}$N$_2$O$_5$: C 74.98, H 6.29, N 4.86. Found: C 75.02, H 6.32, N 4.94.

19. Hydrogen peroxide (Trace Select ≥ 30%) was purchased from Fluka and was used as received.

20. Lithium hydroxide (Purum p.a. ≥ 99%) was purchased from Fluka and was used as received.

21. This procedure was used as described in ref. 6 and 7.

22. The temperature was maintained at 18-24 °C by re-immersion in the ice bath.

23. Sodium sulfite (Reagent Grade, 98%) was purchased from Sigma-Aldrich Co. and was used as received.

24. The recovered DIOZ-auxiliary (9.34 g, 95%, >95% purity by ^{1}H NMR) can be reused.

25. Citric acid (99%) was purchased from the Aldrich Chemical Co. and was used as received.

26. The checkers found that rotary evaporation (45 °C, 15 mmHg) followed by high vacuum (0.2 mmHg, 12 h) was sufficient to obtain a solid. Suspending this solid in 150 mL of TBME followed by filtration and further rinsing with TBME (2 x 20 mL) generated analytically pure material after removing the solvent (45 °C, 15 mmHg, 3 h; then 0.2 mmHg, 23 °C, 30 h).

27. The enantiomer ratio er was determined by HPLC: t_R (S)-**3** = 23.33 min (<0.1%), (R)-**3** = 27.69min (>99.9%) (Chiralpak AD-H; hexane/EtOH/MeOH, 92:4:4 + 0.1% TFA; 1 mL/min, 10 μg in 10 μL of MeOH, UV 210 nm.

28. If sodium hydroxide is used instead of LiOH/H$_2$O$_2$ (as described in ref. 5) partial racemization was observed, and the er dropped to 90:10.

29. The product exhibits the following physicochemical properties: mp 73.0-75.0 °C; [α]$_D^{23}$ = -4.1 (c 1, EtOH); ^{1}H NMR (500 MHz, DMSO) δ: 2.77 (m, 3 H, CH-CH$_2$-phenyl), 3.15, 3.22, (2 m, 2 H, CH$_2$), 5.02 (s, 2 H, O-CH$_2$-phenyl), 7.15-7.46 (m, 11 H, aryl-H + NH), 12.3 (bs, 1 H, -CO$_2$H); ^{13}C

NMR (125 MHz, DMSO) δ: 174.8, 156.3, 139.2, 137.3, 128.9, 128.4, 128.3, 127.84, 127.75 126.2, 65.3, 47.2, 42.2, 35.2; IR (KBr) cm^{-1}: 3346 (br, s), 3031 (br, s), 2662 (m), 1700 (s); MS *m/z* (relative intensity, %): 313 (2.9), 149 (38), 131 (23), 117 (29), 108 (29), 107 (27), 103 (11), 92 (20), 91 (100), 79 (31), 78 (10), 77 (27), 74 (10), 65 (23). Anal. Calc. for $C_{18}H_{19}NO_4$: C 68.99, H 6.11, N 4.47. Found: C 68.93, H 6.75, N 4.62.

Safety and Waste Disposal Information

All hazardous materials should be handled and disposed of in accordance with "Prudent Practices in the Laboratory"; National Academy Press; Washington, DC, 1995.

3. Discussion

The procedure described herein is an application of the Evans oxazolidinone-auxiliary method to the overall enantioselective derivatization of an achiral carboxylic acid with electrophiles (4 → 5).[8] Instead of one of the classical auxiliaries the valine-derived 4-isopropyl-5,5-diphenyl-oxazolidinone (DIOZ) is used, which has many practical advantages outlined in previous publications of our group[6,9] and in an *Organic Syntheses* procedure for its preparation.[4] The transformation of the present procedure is a carbamidomethylation (*Mannich* reaction) of 3-phenylpropionic acid (R = CH$_2$, R^E = CH$_2$NHCO$_2$Bn in 4, 5). The actual electrophile (Bn–O–CO–$^+$NH=CH$_2$), generated *in situ* by the action of the Lewis acid TiCl$_4$ on benzyl isopropoxymethyl carbamate (Bn–O–CO–NH–CH$_2$–O–CHMe$_2$), reacts with the Ti-enolate of the N-(3-phenylpropionyl)oxazolidinone 1 to give β-amino-acid derivative 2 with a diastereoselectivity of 9:1. Starting material 1, the minor stereoisomer *epi*-2 and other by-products can be readily removed by crystallization when the benzyl *isopropoxy*methyl carbamate is employed, rather than the *methoxy*methyl carbamate, which was originally used[10] for this type of reaction (Note 13).

The product, described herein, for preparing β^2-amino-acid derivatives is of general importance: it has been applied to the following building blocks for β-peptide synthesis, with Fmoc protecting group PG and, where applicable, with appropriate acid-labile protection in the side chain R: β^2hAla, β^2hPhe, β^2hTyr, β^2hVal, β^2hLeu, β^2hIle, β^2hMet, β^2hArg, β^2hPro, β^2hLys.[5,6,11,12,13] Actually, all β^2-amino acids with proteinogenic side chains have been obtained with the help of the auxiliary DIOZ.[13] For numerous other methods of preparing certain β^2-amino acids, including derivatives of β^2hPhe, such as compound **3**, we refer the reader to the literature.[13]

α-amino acid   β³-amino acid   β²-amino acid
(homoamino acids)

1. Novartis Institutes for BioMedical Research (NIBR), Preparations Laboratories, CH-4002 Basel, Switzerland
2. Fachbereich Chemie, Institut für Organische Chemie, Leibniz Universität Hannover, Schneiderberg 1B, D-30167 Hannover, Germany.
3. Departement für Chemie und Angewandte Biowissenschaften, Laboratorium für Organische Chemie, ETH-Hönggerberg HCI, Wolfgang-Pauli-Str. 10, CH-8093 Zürich, Switzerland
4. Brenner, M.; La Vecchia, L.; Leutert, T.; Seebach, D. *Org. Synth.* **2003**, *80*, 57-65.
5. Seebach, D.; Schaeffer, L.; Gessier, F.; Bindschädler, P.; Jäger, C.; Josien, D.; Kopp, S.; Lelais, G.; Mahajan, Y.; Micuch, P.: Sebesta, R.; Schweizer, B. W. *Helv. Chim. Acta* **2003**, *86*, 1852-1861.
6. Hintermann, T.; Seebach, D. *Helv. Chim. Acta* **1998**, *81*, 2093-2126.
7. Lelais, G.; Campo, M. A.; Kopp, S.; Seebach, D. *Helv. Chim. Acta* **2004**, *87*, 1545-1560.
8. Evans, D. A.; Bartroli, J.; Shih, T. L.; *J. Am. Chem. Soc.* **1981**, *103*, 2127-2129.
9. Gaul, C.; Schweizer, B. W.; Seiler, P.; Seebach, D. *Helv. Chim. Acta* **2002**, *85*, 1546-1566.
10. Barnett, C. J.; Wilson, T. M.; Evans, D. A.; Somers, T. C.; *Tetrahedron Lett.* **1997**, *38*, 735-738.

302

11. Kimmerlin, T.; Sebesta, R.; Campo, M. A.; Beck, A: K.; Seebach, D. *Tetrahedron* **2004**, *60*, 7455-7506.
12. Gessier, F.; Schaeffer, L.; Kimmerlin, T.; Flögel, O.; Seebach, D. *Helv. Chim. Acta* **2005**, *88*, 2235-2250.
13. Lelais, G.; Seebach, D. *Biopolymers* (Peptide Science) **2004**,*76*, 206-243; Seebach, D.; Beck, A. K.; Bierbaum, D. J. *Chem. Biodiv.* **2004**, *1*, 1111-1239.

Appendix
Chemical Abstracts nomenclature (Registry Number)

(4*S*)-4-Isopropyl-5,5-diphenyloxazolidin-2-one (DIOZ): 2-Oxazolidinone, 4-(1-methylethyl)-5,5-diphenyl-, (4*S*)-; (184346-45-0)

n-Butyllithium: Butyllithium; (109-72-8)

3-Phenylpropanoyl chloride: Benzenepropanoyl chloride; (645-45-4)

(4*S*)-4-Isopropyl-5,5-diphenyl-3-(3-phenyl-propionyl)oxazolidin-2-one: 2-Oxazolidinone, 4-(1-methylethyl)-3-(1-oxo-3-phenylpropyl)-5,5-diphenyl-, (4*S*)-; (213887-81-1)

Titanium tetrachloride; (7550-45-0)

Triethylamine: Ethanamine, *N,N*-diethyl-; (121-44-8)

(2*R*,4*S*) [2-Benzyl-3-(4-isopropyl-2-oxo-5,5-diphenyl-3-oxazolidinyl)-3-oxopropyl]carbamic acid benzyl ester: Carbamic acid, [(2*R*)-3-[(4*S*)-4-(1-methylethyl)-2-oxo-5,5-diphenyl-3-oxazolidinyl]-3-oxo-2-(phenylmethyl)propyl]-, phenylmethyl ester; (218800-56-7)

Hydrogen peroxide; (7722-84-1)

Lithium hydroxide monohydrate; (1310-66-3)

(*R*)-2-(Benzyloxycarbonylaminomethyl)-3-phenylpropanoic acid: Benzenepropanoic acid, α-[[[(phenylmethoxy)carbonyl]amino]methyl]-, (α*R*)-; (132696-47-0)

Dieter Seebach received his Ph.D. from the Technische Hochschule Karlsruhe, Germany, with Rudolf Criegee. After a postdoctoral stay in Elias J. Corey's group and a lectureship at Harvard University, he returned to the University of Karlsruhe and became a full Professor of Organic Chemistry at the Justus-Liebig-Universität in Giessen, Germany, in 1971. From 1977-2003 he was Professor of Chemistry at the Eidgenössische Technische Hochschule (ETH) in Zürich. Since 2003 he is officially retired; as Professor Emeritus he continues doing research with postdoctoral coworkers. His research activities include: development of new synthetic methods, natural-product synthesis, structure determination, chiral dendrimers, the biopolymer PHB, β-peptides. The results have been described in over 800 research papers.

J. Constanze D. Müller-Hartwieg studied chemistry at the University of Konstanz, Germany. After her Ph.D. at the Max-Planck-Institute of Biochemistry in Martinsried/Munich, Germany, in the field of bioorganic chemistry and peptide synthesis, she joined Novartis in Basel, Switzerland as a Postdoc working in the Combinatorial Chemistry Unit. Since 2002 she has been working as a Lab Head in the Preparation Laboratories for Novartis Institutes for BioMedical Research, Basel. She is responsible for the scale up of organic syntheses, in order to produce larger amounts of building blocks, intermediates and potential drug candidates for a wide range of disease areas.

Luigi La Vecchia did his Ph.D. 1989 at the University of Saarbrücken, Germany. Thereafter, he joined Sandoz as Lab Head of the Research Kilo lab. Between 1993 -1994 he did a sabbatical at the Sandoz Research Institute/Chemical Development in East Hanover, NJ. In 1996 he took over a position as Lab Head Process R&D in Chemical Development in Basel, Switzerland. Since 1998 he has been assuming various positions within the Novartis Institute for Biomedical Research, Basel, Switzerland. Currently, he is Director of the Preparations Laboratories. He was involved in projects such as RAD001 (Certican), QAB149 (Indacaterol) and NIM811.

Hartmut H. Meyer graduated in 1970 from the Technische Hochschule Hannover (Germany) under the supervision of W. Theilacker and F. Klages. After postdoctoral work with Professor D. Seebach at the University of Giessen in 1972, he returned to the University of Hannover and started work on syntheses of enantiopure natural products. In 1987 he became Privatdozent with a Habilitation on studies of the benzidine rearrangement. Since 1996 he has been a Professor at the Leibniz Universität Hannover. His current research interests are synthetic methodology, chemo-enzymatic syntheses and biochemical studies, in cooperation with other research groups.

Albert Karl Beck was born in 1947 in Karlsruhe (Germany), completing a chemistry technician's apprenticeship at the University of Karlsruhe in 1966, and joined the Seebach research group in 1968. Between 1969 and 1972 he continued his education, obtaining official certification as a chemical technician at the Fachschule für Chemotechnik in Karlsruhe. In 1971 he joined the Institute for Organic Chemistry at the University of Giessen, and in 1974 he engaged in a six-month research visit to the California Institute of Technology in Pasadena (USA). He has been an active part of the Laboratory for Organic Chemistry at the ETH in Zürich since 1977. During his association with the Seebach research group he has participated in essentially all of the groups research themes, as evidenced by his coauthorship of more than 80 publications.

Nathan W. Duncan-Gould received his bachelor's degrees in Molecular Biology, Biochemistry and Chemistry from the College of Charleston. During this time he performed research with Professor Justin K. Wyatt on the total synthesis of Cytosporone E. In 2004 he began his graduate studies in Organic Chemistry at the University of Illinois, unde the mentorship of Professor Scott E. Denmark. The focus of his research is the investigation of phase transfer catalysis (PTC) by quantitative structure activity relationships (QSAR).

Andrew J. Hoover was born in New London, Connecticut, in 1987. He enrolled at the University of Illinois in 2005, and beginning in 2006 he conducted research on deoxyribozyme catalysis with Professor Scott Silverman. As an Amgen scholar during the summer of 2007, he worked with Professor Richmond Sarpong at UC Berkeley on efficient syntheses of cyclopentenones, and in the fall of 2007 he joined the research group of Scott Denmark, where he currently is investigating the use of Lewis bases as catalysts for enantioselective functionalization of alkenes. He will graduate in December 2008 with a chemistry BS degree, and will enroll in graduate school in 2009.

ERRATA

SYNTHESIS OF (–)-(*S*,*S*)- BIS(4-ISOPROPYLOXAZOLINE)

David A. Evans, Keith A. Woerpel, Bernd Nosse, Andreas Schall,
Yogesh Shinde, Eva Jezek, Mohammad Mahbubul Haque, R. B. Chhor,
and Oliver Reiser

The amount of 2,2-dimethylpropanedioyl dichloride used in the procedure
was incorrectly recordedto be 025 mol. The correct volume used should be
3.3 mL, which translates to 0.025 mol. A corrected version of this procedure
is available on the *Organic Syntheses* website (orgsyn.org).

AUTHOR INDEX

Javed, M. I., **85**, 189
Johnson, J. S., **85**, 278

Kitching, M. O., **85**, 72
Kocienski, P. J., **85**, 45
Knauber, T., **85**, 196
Krause, H., **85**, 34

Langle, S., **85**, 231
Lautens, M., **85**, 172
La Vecchia, L., **85**, 295
Ley, S. V., **85**, 72
Linder, C., **85**, 196
Longbottom, D. A., **85**, 72
Lu, C. -D., **85**, 158

Mani, N. S., **85**, 179
Mans, D. J., **85**, 238, 248
Marin, J., **85**, 147
Matsunaga, S., **85**, 118
Maw, G., **85**, 219
McAllister, G. D., **85**, 15
McDermott, R. E., **85**, 138
McNaughton, B. R., **85**, 27
Meyer, H., **85**, 287, 295
Miller, B. L., **85**, 27
Montchamp, J.-L., **85**, 96
Moore, D. A., **85**, 10
Morra, N. A., **85**, 53
Mosa, F., **85**, 219
Movassaghi, M., **85**, 88
Mudryk, B., **85**, 64
Müller-Hartwieg, J. C. D., **85**, 295

Zakarian, A., **85**, 158
Zhang, A., **85**, 248
Zhang, H., **85**, 147
Zimmermann, B., **85**, 196

Benzyl carbamate: Carbamic acid, phenylmethyl ester; (621-84-1) **85**, 287

Benzyl chloromethyl ether: Benzene, [(chloromethoxy)methyl]-; (3587-60-8) **85**, 45

1-Benzyl-3-(4-chloro-phenyl)-5-*p*-tolyl-1*H*-pyrazole (908329-95-3) **85**, 179

Benzylhydrazine dihydrochloride; (20570-96-1) **85**, 179

Benzyl hydroxymethyl carbamate: Carbamic acid, (hydroxymethyl)-, phenylmethyl ester; (31037-42-0) **85**, 287

(2*R*,4*S*) [2-Benzyl-3-(4-isopropyl-2-oxo-5,5-diphenyl-3-oxazolidinyl)-3-oxopropyl]carbamic acid benzyl ester (218800-56-7) **85**, 295

Benzyl isopropoxymethyl carbamate **85**, 287

1-Benzyloxymethoxy-1-hexyne: Benzene, [[(1-hexyn-1-yloxy)methoxy]methyl]-; (162552-11-6) **85**, 45

Benzyloxymethoxy-2,2,2-trifluoromethyl ether: Benzene, [[(2,2,2-trifluoroethoxy)methoxy]methyl]-: (153959-88-7) **85**, 45

(*R*)-(+)-1,1'-Bi(2-naphthol); (18531-94-7) **85**, 238

(*R*)-BINOL; (18531-94-7) **85**, 238

(*S*)-BINOL: [1,1'-Binaphthalene]-2,2'-diol, (1*S*)-: (18531-99-2) **85**, 118

(*R*)-(1,1'-Binaphthalene-2,2'-dioxy)chlorophosphine: (*R*)-Binol-P-Cl; (155613-52-8) **85**, 238

[1,1'-Binaphthalene]-2,2'-diol, (1*S*)-: (18531-99-2) **85**, 118

(*R*)-2,2-Binaphthoyl-(*S,S*)-di(1-phenylethyl)aminoylphosphine (415918-91-1) **85**, 238

Bis[1,2:5,6-η-(1,5-cyclooctadiene)]nickel: [bis(1,5-cyclooctadiene)nickel (0)]; (1295-35-8) **85**, 248

Bis(diphenylphosphino)methane; (2071-20-7) **85**, 196

Bis-(Hydroxymethyl)-cyclopropane; (2345-75-7) **85**, 15

(-)-Bis[(*S*)-1-phenylethyl]amine (56210-72-1) **85**, 238

(-)-Bis[(*S*)-1-phenylethyl]amine hydrochloride (40648-92-8) **85**, 238

3,5-Bis(trifluoromethyl)bromobenzene; (328-70-1) **85**, 248

Bromination, **85**, 53

Bromine; (7726-95-6) **85**, 231

2-Bromopropene; (557-93-7) **85**, 1, 172

N-Bromosuccinimide: 1-Bromo-2,5-pyrrolidinedione; (128-08-5) **85**, 53, 267

(-)-2-*tert*-Butyl-(4*S*)-benzyl-(1,3)-oxazoline: 4,5-Dihydrooxazole, (4*S*)-benzyl, 2-*tert*-butyl; (75866-75-0) **85**, 267

4-Bromotoluene; (106-38-7) **85**, 196

tert-Butyl acetoacetate; (1694-31-1) **85**, 278

tert-Butyl bromoacetate; (5292-43-3) **85**, 10

tert-Butyl (1*R*)-2-cyano-1-phenylethylcarbamate (126568-44-3) **85**, 219

tert-Butyl diazoacetate; (35059-50-8) **85**, 278

tert-Butyl (1*S*)-2-hydroxy-1-phenylethylcarbamate (117049-14-6) **85**, 219

tert-Butyllithium: Lithium, (1,1-dimethylethyl)-; (5944-19-4) **85**, 1, 209

n-Butyllithium: Butyllithium; (109-72-8) **85**, 1, 45, 53, 158, 238, 248, 295

2-*tert*-Butoxycarbonylamino-4-(2,2-dimethyl-4,6-dioxo-[1,3]dioxan-5-yl)-4-oxo-butyric
    acid *tert*-butyl ester; (10950-77-9) **85**, 147

(2*S*)-2-[(*tert*-Butoxycarbonyl)amino]-2-phenylethyl methanesulfonate (110143-62-9) **85**,
    219

(*R*)-2-(Benzyloxycarbonylaminomethyl)-3-phenylpropanoic acid: Benzenepropanoic
    acid, α-[[[(phenylmethoxy)carbonyl]amino]methyl]-, (a*R*)-; (132696-47-0) **85**,
    295

N-α-*tert*-Butoxycarbonyl-L-aspartic acid α-*tert*-butyl ester (Boc-L-Asp-O*t*-Bu); (34582-
    32-6) **85**, 147

1-*tert*-Butoxycarbonyl-2,3-dihydropyrrole: 1*H*-Pyrrole-1-carboxylic acid, 2,3-dihydro-,
    1,1-dimethylethyl ester; (73286-71-2) **85**, 64

tert-Butyl *tert*-butyldimethylsilylglyoxylate **85**, 278

tert-Butyldimethylsilyl trifluoromethanesulfonate; (69739-34-0) **85**, 278

N-(*tert*-Butyloxycarbonyl)pyrrolidin-2-one; (85909-08-6) **85**, 64

Carbamic acid, (hydroxymethyl)-, phenylmethyl ester; (31037-42-0) **85**, 287

(Carbethoxymethylene)triphenylphosphorane; (1099-45-2) **85**, 15

Cbz-L-proline: 1,2-Pyrrolidinedicarboxylic acid, 1-(phenylmethyl) ester, (2*S*)-; (1148-11-
    4) **85**, 72

4-Chlorobenzaldehyde; (104-88-1) **85**, 179

2-Chloropyridine; (109-09-1) **85**, 88

Cinnamyl alcohol: 3-Phenyl-2-propen-1-ol; (104-54-1) **85**, 96

Cinnamyl-*H*-phosphinic acid: [(2*E*)-3-phenyl-2-propenyl]-Phosphinic acid; (911128-46-
    6) **85**, 96

314

Condensation, **85**, 27, 34, 179, 248, 267

Copper(I) bromide; (7787-70-4) **85**, 196

Copper chloride: Cuprous chloride; (7758-89-6) **85**, 209

Copper Cyanide; (544-92-3) **85**, 131

Coupling **85**, 158, 196

Cuprous chloride; (7758-89-6) **85**, 209

(*S*)-2-Cyano-pyrrolidine-1-carboxylic acid benzyl ester: (63808-36-6) **85**, 72

Cyanuric chloride: 2,4,6-Trichloro-1,3,5-triazine; (108-77-0) **85**, 72

Cyclen: 1,4,7,10-Tetraazacyclododecane; (294-90-6) **85**, 10

Cycloaddition, **85**, 72, 131, 138, 179

Cycloheptane-1,3-dione (1194-18-9) **85**, 138

Cyclohexene oxide; (286-20-4) **85**, 106

Cyclopropanation **85**, 172

Cyclopropanecarboxylic acid, 2-bromo-2-methyl-, ethyl ester; (89892-99-9) **85**, 172

Cyclopropanecarboxylic acid, 2-methylene-, ethyl ester; (18941-94-1) **85**, 172

Dehydration, **85**, 34, 72

Di(*μ*-bromo)bis(*η*-allyl)nickel(II): [allylnickel bromide dimer]; (12012-90-7) **85**, 248

(*E*)-2,3-Dibromobut-2-enoic acid: (2-Butenoic acid, 2,3-dibromo-, (2*E*)- (9); (24557-17-3) **85**, 231

2,5-Dibromo-1,1-dimethyl-3,4-diphenylsilole: Silacyclopenta-2,4-diene, 2,5-dibromo-1,1-dimethyl-3,4-diphenyl-; (686290-22-2) **85**, 53

Di-*tert*-butyl dicarbonate: Dicarbonic acid, C,C'-bis(1,1-dimethylethyl) ester; (24424-99-5) **85**, 72, 219

Di(*tert*-butyl) (2*S*)-4,6-dioxo-1,2-piperidinedicarboxylate; (653589-10-7) **85**, 147

Di(*tert*-butyl) (2*S*,4*S*)-4-hydroxy-6-oxo-1,2-piperidinedicarboxylate; (653589-16-3) **85**, 147

Dichloroacetyl chloride (79-36-7) **85**, 138

Dichlorodimethylsilane; (75-78-5) **85**, 53

Dicyclohexylmethylamine: Cyclohexanamine, *N*-cyclohexyl-*N*-methyl-; (7560-83-0) **85**, 118

Diene formation, **85**, 1

2-[(Diethylamino)methyl]benzene thiolato-copper(I) **85**, 209

(±)-Diethyl (*E,E,E*)-cyclopropane-1,2-acrylate, **85**, 15

Diethyl *trans*-1,2-cyclopropanedicarboxylate; (3999-55-1) **85**, 15

Diethyl(2-[(trimethylsilanyl)sulfanyl]benzyl)amine **85**, 209

N,N-Diisopropylethylamine: *N*-Ethyl-*N*-(1-methylethyl)-2-propanamine; (7087-68-5) **85**, 64, 158, 278

DMAP: *N, N*-Dimethyl-4-Pyridinamine: (1122-58-3) **85**, 64

N-(3-Dimethylaminopropyl)-*N'*-ethylcarbodiimide hydrochoride (EDC·HCl); (25952-53-8) **85**, 147

9,9-Dimethyl-4,5-bis(diphenylphosphino)xanthene: Xantphos; (161265-03-8) **85**, 96

Dimethyl-bis-phenylethynyl silane: Benzene, 1,1'-[(dimethylsilylene)di-2,1-ethynediyl]bis-; (2170-08-3) **85**, 53

(2*R*,3*R*)-2,3-Dimethylbutane-1,4-diol: (2*R*,3*R*) 2,3-Dimethyl-1,4-butanediol; (127253-15-0) **85**, 158

2,2-Dimethyl-1,3-dioxane-4,6-dione (Meldrum's acid); (2033-24-1) **85**, 147

N, N-Dimethyl-4-Pyridinamine: (1122-58-3) **85**, 64

2*R*,3*R*)-2,3-Dimethylsuccinic acid; (5866-39-7) **85**, 158

Dimethyl sulfoxide: Methyl sulfoxide; Methane, sulfinybis-; (67-68-5) **85**, 189

1,3-Diphenylacetone *p*-tosylhydrazone: Benzenesulfonic acid, 4-methyl-, [2-phenyl-1-(phenylmethyl)ethylidene]hydrazide; (19816-88-7) **85**, 45

Diphenyldiazomethane (883-40-9) **85**, 189

(*S*)-(−)-1,3-Diphenyl-2-propyn-1-ol: Benzenemethanol, α-(2-phenylethynyl)-, (αS)-; (132350-96-0) **85**, 118

Elimination **85**, 45, 172

1,2-Epoxydodecane: Oxirane, decyl-; (2855-19-8) **85**, 1

Ethyl diazoacetate: Acetic acid, 2-diazo-, ethyl ester; (623-73-4) **85**, 172

Ethylene; (74-85-1) **85**, 248

2-Ethyl-3-Quinolinecarboxylic acid (888069-31-6) **85**, 27

2-Ethyl-3-Quinolinecarboxylic acid hydrochloride (888014-11-7) **85**, 27

2-Ethyl-3-Quinolinecarboxylic acid, methyl ester (119449-61-5) **85**, 27

Formaldehyde solution; (50-00-0) **85**, 287

Halogenation, **85**, 53, 231

Heterocycle, **85**, 10, 27, 34, 53, 64, 72, 88

Heterocyclic carbene, **85**, 34

Hydrogen peroxide; (7722-84-1) **85**, 158, 295

3-Hydroxybutan-2-one; (513-86-0) **85**, 34

Hypophosphorous acid: Phosphinic acid; (6303-21-5) **85**, 96

Indium bromide: (13465-09-3) **85**, 118

Iodine; (7553-56-2) **85**, 219, 248

(4*S*)-4-Isopropyl-5,5-diphenyloxazolidin-2-one (DIOZ): 2-Oxazolidinone, 4-(1-
   methylethyl)-5,5-diphenyl-, (4*S*)-; (184346-45-0) **85**, 295

(4*S*)-4-Isopropyl-5,5-diphenyl-3-(3-phenyl-propionyl)oxazolidin-2-one: 2-
   Oxazolidinone, 4-(1-methylethyl)-3-(1-oxo-3-phenylpropyl)-5,5-diphenyl-, (4*S*)-;
   (213887-81-1) **85**, 295

(4*S*)-Isopropyl-2-oxazolidinone: (4*S*)-4-(1-Methylethyl)-2-oxazolidinone; (17016-83-0)
   85, 158

(4*S*)-Isopropyl-3-propionyl-2-oxazolidinone: (4*S*)-4-(1-Methylethyl)-3-(1-oxopropyl)-2-
   oxazolidinone; (77877-19-1) **85**, 158

Lithium; (7439-93-2) **85**, 53

Lithium aluminum hydride; (16853-85-3) **85**, 158

Lithium hydroxide monohydrate; (1310-66-3) **85**, 295

Lithium triethylborohydride; (22560-16-3) **85**, 64

(*R*)-(-)-Mandelic acid; (611-71-2) **85**, 106

(*S*)-(+)-Mandelic acid; (17199-29-0) **85**, 106

(*R*)-Mandelic acid salt of (1*S*,2*S*)-*trans*-2-(*N*-benzyl)amino-1-cyclohexanol; (882409-00-9) **85**, 106

(*S*)-Mandelic acid salt of (1*R*,2*R*)-*trans*-2-(*N*-benzyl)amino-1-cyclohexanol; (882409-01-0) **85**, 106

Manganese(IV) oxide; (1313-13-9)

Mesitylamine: Benzenamine, 2,4,6-trimethyl-; (88-05-1) **85**, 34

3-(Mesitylamino)butan-2-one: 2-Butanone, 3-[(2,4,6-trimethylphenyl)-amino]-; (898552-96-0) **85**, 34

Mesitylene (108-67-8) **85**, 196

N-Mesityl-*N*-(3-oxobutan-2-yl)formamide: Formamide, N-(1-methyl-2-oxopropyl)-N-(2,4,6-trimethylphenyl)-; (898553-01-0) **85**, 34

Metallation, **85**, 1, 45, 209

Methanesulfonyl chloride; (124-63-0) **85**, 219

2-Methyl-3-butyn-2-ol; (115-19-5) **85**, 118

(*E*)-3,4-Methylenedioxy-β-nitrostyrene; (22568-48-5) **85**, 179

Methylhydrazine; (60-34-4) **85**, 179

4-Methyl-2'-nitrobiphenyl; (70680-21-6) **85**, 196

trans-p-Methyl-β-nitrostyrene: Benzene, 1-methyl-4-[(1*E*)-2-nitroethenyl]-; (5153-68-4) **85**, 179

(*R*)-3-Methyl-3-phenylpentene: [(1*R*)-1-ethyl-1-methyl-2-propenyl]-benzene]; (768392-48-9) **85**, 248

(*S*)-(−)-4-Methyl-1-phenyl-2-pentyn-1,4-diol: (321855-44-1) **85**, 118

Methyl propionylacetate: Pentanoic acid, 3-oxo-, methyl ester; (30414-53-0) **85**, 27

1-Methyl-2-pyrrolidone; (872-50-4) **85**, 196

1-Methyl-2-pyrrolidinone (872-50-4) **85**, 238

Methyltriphenylphosphonium bromide; (1779-49-3) **85**, 248

Naphthalene; (91-20-3) **85**, 53

2-Nitrobenzaldehyde; (552-89-6) **85**, 27

2-Nitrobenzoic acid; (552-16-9) **85**, 196

Oxalyl chloride: HIGHLY TOXIC; Ethanedioyl dichloride; (79-37-8) **85**, 189

Oxidation, **85**, 15, 189, 267, 278

Oxone® monopersulfate; (37222-66-5) **85**, 278

Palladium(II) acetate: Pd(OAc)₂; (3375-31-3) **85**, 96

Palladium acetylacetonate; (140024-61-4) **85**, 196

1,10-Phenanthroline; (66-71-7) **85**, 196

(1,10-Phenanthroline)bis(triphenylphosphine)copper(I) nitrate; (33989-10-5) **85**, 196

Phenylacetylene: Ethynylbenzene; (536-74-3) **85**, 53, 118, 131

(*S*)-Phenylalaninol: (3182-95-4) **85**, 267

N-Phenylbenzenecarboxamide (benzanilide); (93-98-1) **85**, 88

2-Phenyl-1-butene; (2039-93-2) **85**, 248

(*S*)-Phenylglycine: Benzeneacetic acid, α-amino-, (α*S*)-; (2935-35-5) **85**, 219

3-Phenylpropanoyl chloride: Benzenepropanoyl chloride; (645-45-4) **85**, 295

N-[1-Phenyl-3-(trimethylsilyl)-2-propyn-1-ylidene]-benzeneamine; (77123-64-9) **85**, 88

Phosphorus trichloride (7719-12-2) **85**, 238

Potassium carbonate; (584-08-7) **85**, 287

Potassium hydroxide; (1310-58-3) **85**, 196

Propanol, 2-amino-, 3-phenyl, (*S*); (3182-95-4) **85**, 267

2-Propenoic acid, 3,3'-(1,2-cyclopropanediyl)bis-, diethyl ester, [1α(E),2β(E)]-; (58273-88-4) **85**, 15

Propionyl chloride: Propanoyl chloride; (79-03-8) **85**, 158

1*H*-Pyrazole, 5-(1,3-benzodioxol-5-yl)-3-(4-chlorophenyl)-1-methyl-; (908329-89-5) **85**, 179

1-Pyrrolidinecarboxylic acid, 2-(aminocarbonyl)-, phenylmethyl ester, (2*S*)-; (34079-31-7) **85**, 72

1,2-Pyrrolidinedicarboxylic acid, 1-(phenylmethyl) ester, (2*S*)-; (1148-11-4) **85**, 72

(*S*)-5-Pyrrolidin-2-yl-1*H*-tetrazole: 2*H*-Tetrazole, 5-[(2*S*)-2-pyrrolidinyl]-; (33878-70-5) **85**, 72

Quinoline; (91-22-5) **85**, 196

p-Toluenesulfonic acid: Benzenesulfonic acid, 4-methyl-; (104-15-4) **85**, 287

Triethylamine; (121-44-8) **85**, 131, 189, 219, 295

Triethylamine hydrochloride: Ethanamine, *N,N*-diethyl-, hydrochloride (1:1); (554-68-7) **85**, 72

Triflic anhydride; (358-23-6) **85**, 88

Trifluoroacetic anhydride; (407-25-0) **85**, 64

2,2,2-Trifluoroethanol (75-89-8) **85**, 45

Trifluoromethanesulfonic acid anhydride; (358-23-6) **85**, 88

Trimethylacetaldehyde: Pivaldehyde: Propanal, 2,2-dimethyl; (630-19-3) **85**, 267

1-(Trimethylsilyl)acetylene; (1066-54-2) **85**, 88

Trimethylsilyl chloride: Silane, chlorotrimethyl-; (75-77-4) **85**, 209

1-Trimethylsilyloxybicyclo[3.2.0]heptan-6-one (125302-44-5) **85**, 138

1-Trimethylsilyloxycyclopentene (19980-43-9) **85**, 138

1-Trimethylsilyloxy-7,7-dichlorobicyclo[3.2.0]heptan-6-one (66324-01-4) **85**, 138

Triphenylphosphine oxide; (791-28-6) **85**, 248

Zinc (II) chloride; (7646-85-7) **85**, 27; **85**, 53